GRAPHS OF TRIGONOMETRIC FUNCTIONS

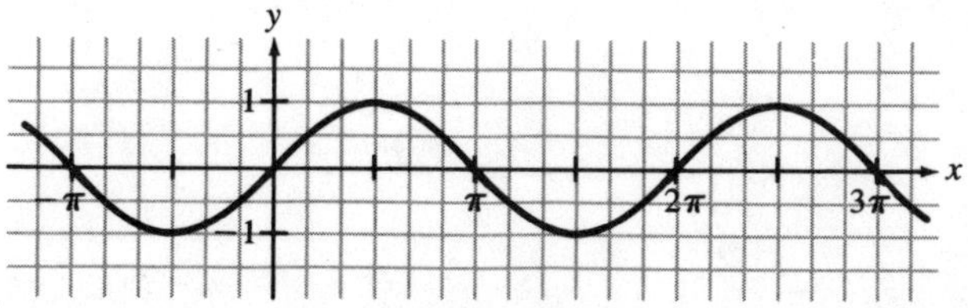

$y = \sin x$

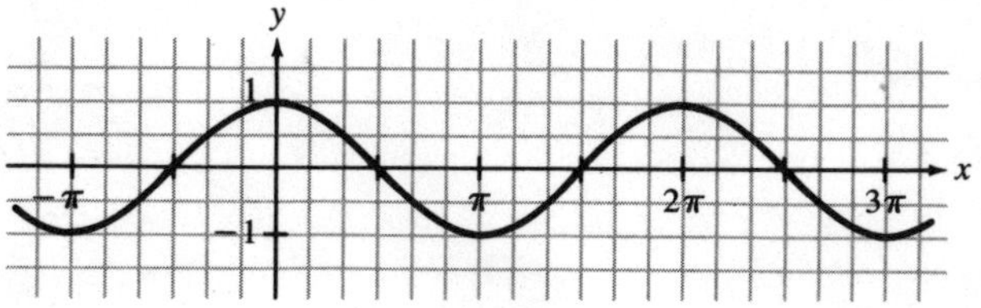

$y = \cos x$

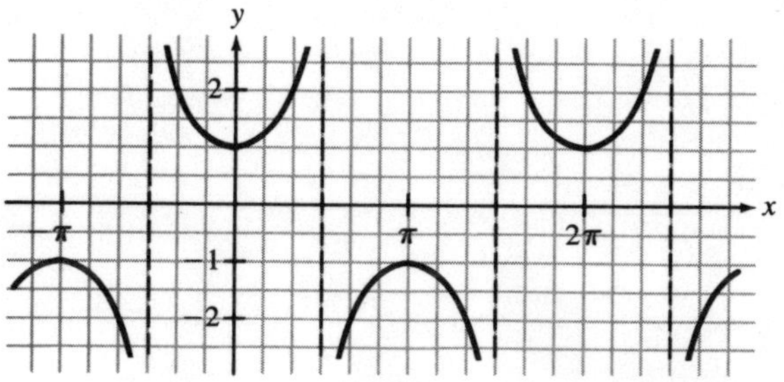

$y = \sec x$

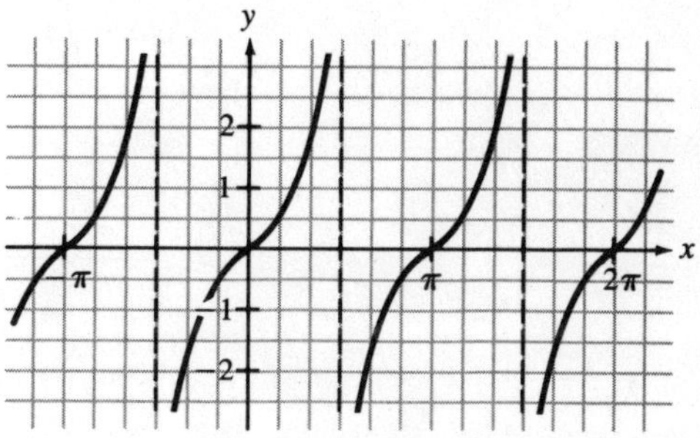

$y = \tan x$

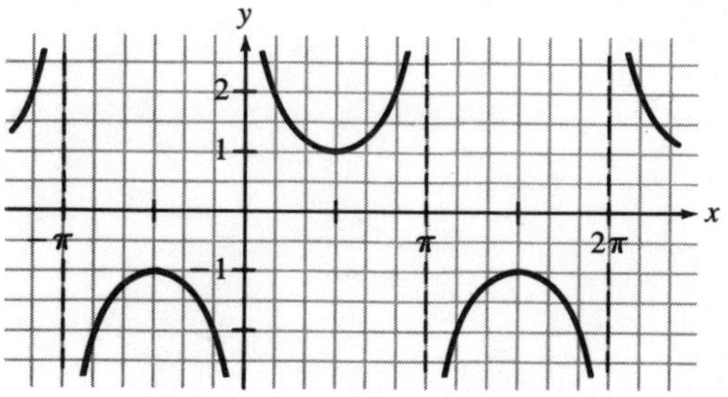

$y = \csc x$

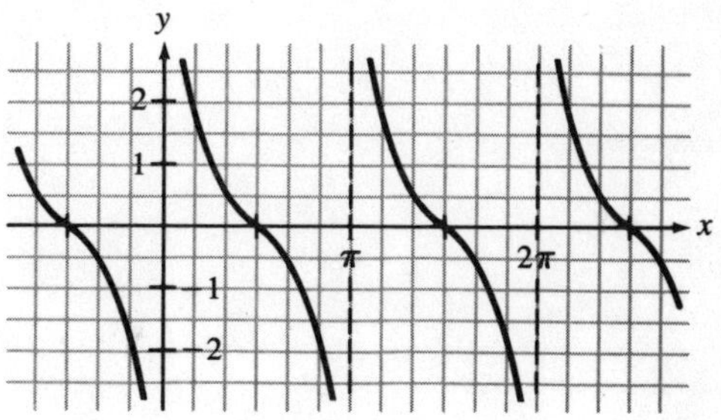

$y = \cot x$

TRIGONOMETRY

Chris Vancil

University of Kentucky

WEST PUBLISHING COMPANY

St. Paul • New York • Los Angeles • San Francisco

DESIGN AND PRODUCTION: Schneider & Co.
ART STUDIO: ANCO/Boston
COMPOSITION: Polyglot Pte Ltd
COVER ART: Delor Erickson/Studio West

50 W. Kellogg Boulevard
P.O. Box 64526
St. Paul, MN 55164-1003

Printed in the United States of America

Library of Congress Cataloging-in-Publication Data

Vancil, Chris N.
Trigonometry.

Includes index.
1. Trigonometry. I. Title.
QA531.V33 1988 516.2′4 87-27938
ISBN 0-314-61443-5

CONTENTS

PREFACE

This book is a presentation of the fundamentals of plane trigonometry. It is designed to provide a solid foundation for students who will continue into calculus as well as for students who will apply trigonometry in a technical area.

The prerequisite is two years of high school algebra or intermediate algebra. A course in geometry is desirable but not essential. Geometric concepts are presented as needed.

The principal objective has been to write a text that students find understandable and instructors find easy to use. I have class tested preliminary versions of this text over the last five years. My experience indicates that the organization plays an important role in the students' comprehension and the instructors' satisfaction. Consequently, much attention has been devoted to developing topics in a natural, logical progression.

Text Organization

- Chapter 1 contains a review of angles and triangles plus some college algebra topics. Depending on the mathematical background of the students, some or all of this material can be skipped. Chapter 1 material can also be postponed until needed. Section 1.1 is needed for Chapter 2. Section 1.2 must be covered before beginning Chapter 3. Sections 1.3 and 1.4 are needed to begin Section 4.4, and Section 1.5 is required for Chapter 5.
- Chapter 2 is traditional right-triangle trigonometry. However, only the three principal trigonometric ratios are initially introduced. Since the reciprocal ratios are not actually needed for solving right triangles, their introduction is postponed. I find that students benefit from this uncluttered beginning.
- Chapter 3 extends the concept of angle measurement and the definitions of the trigonometric ratios. It also introduces trigonometric equations. This early introduction enables me to include trigonometric equations in Chapters 4, 5, and 6, thus providing extensive coverage of a topic many students find difficult.
- Chapter 4 covers radian measure and its applications. I find that this late presentation serves two purposes. First, it provides a better start in the course because students are familiar with degree measure. Second, it provides an opportunity to review the basics of trigonometry in a different setting. Section 4.3 is a review of the Chapter 3 material with the only difference being the use of radians instead of degrees.
- Chapters 5–8 are traditional topics, and because radians are introduced late, real numbers and radians are stressed in these chapters.

Text Features

- The presentation is precise, yet the writing style is informal.
- Important results are set off.
- Figures are included wherever possible to help visualize a concept or problem.
- Applications from a variety of fields are included throughout the text.
- Each section contains exercises (with answers) integrated into the narrative. These problems are similar to preceding examples or illustrations and allow students to test their understanding before proceeding. They may also be used by instructors to provide additional classroom examples.
- Each section is followed by a set of carefully written problems that provide practice in the skills, concepts, and applications of the subject. For the most part, the problems occur in matching pairs, with an even-numbered problem very similar to the problem immediately preceding it.
- Each chapter concludes with a review consisting of two parts. The first part is a summary of the important terms or formulas of the chapter. Page references are included. The second part is a problem set, beginning with true-false questions, which spans the entire chapter.
- Excluding proofs, answers to all odd-numbered problems of the section problem sets and to all problems of the chapter review problem sets are included in the appendix.

Ancillaries

- Proofs and answers not in the text appear in the Instructor's Manual, which is available to instructors upon request.
- A Student Solutions Manual is also available. It provides detailed solutions to approximately half of the odd-numbered problems in the section problem sets and to all of the odd-numbered problems in the chapter reviews. In addition, it includes two tests, with answers, which cover each chapter.

Comments about Calculators

With the availability of inexpensive, electronic calculators, this text is written with the assumption that each student has a scientific calculator. There are no specially marked problems designed for calculators because the calculator is thoroughly integrated into the course. However, many of the problems require exact answers and must be worked without electronic aid.

Trigonometry tables and their instructions appear in the appendix but are not used anywhere else in the text. They are included only for students who might be temporarily without a calculator.

Finally, there are no detailed instructions about how to use the calculator. Such comments are best left to the instruction manual that accompanies it.

Acknowledgments

A special thanks goes to my good friend Tom Moss. Tom taught from the material, served as a consultant, worked all the problems, and wrote the Student Solutions Manual. I am very grateful for his contributions.

I wish to thank the following individuals who reviewed the manuscript and offered helpful suggestions:

Neal Brand, North Texas State University
Paul Britt, Louisiana State University
Robert A. Chaffer, Central Michigan University
Patricia Clark, Indiana State University
Romae J. Cormier, Northern Illinois University
Duane E. Deal, Ball State University
Katherine Franklin, Los Angeles Pierce College
David B. Johnson, Diablo Valley College
Herbert Kasube, Bradley University
Chris Kolaczewski, University of Akron
Charles Luning, Sam Houston State University
Richard Marshall, Eastern Michigan University
Barbara Meininger, University of Oregon
Vincent Polito, North Harris County College
Thomas Radin, San Joaquin Delta College
Ivan D. Stones, Kearney State College
Monty J. Strauss, Texas Tech University
Robert B. Thompson, Lane Community College
Arnold Wendt, Western Illinois University

I also wish to thank the staff at West Publishing Company, and in particular, the following people: Jerry Westby, the acquisitions editor; Barbara Fuller, the production editor; and Fran Deziel, the promotion manager. Further, I am indebted to Debbie Schneider of Schneider & Co., production coordinator, whose advice and expertise were extremely valuable.

Finally, I wish to thank Brian Moses, who typed the manuscript, my wife Cathy, who provided much-needed encouragement and release time from domestic chores, and the numerous students who used the preliminary versions. Their comments, suggestions, and support are sincerely appreciated.

C.V.

1

FOUNDATIONS

CHAPTER OUTLINE

This chapter deals with some introductory ideas that are needed for a successful study of trigonometry. It begins with a review of some of the significant properties of angles and triangles. Section 1.2 follows with the development of a coordinate system. After a discussion of equations, Section 1.4 covers one of the most important topics in mathematics—the concept of function. This idea is treated in further detail in the concluding section.

1.1 ANGLES AND TRIANGLES

An **angle** consists of two rays that originate from a common point. In Figure 1.1 we see the angle formed by the two rays $\overrightarrow{AB}$ and $\overrightarrow{AC}$. Each ray is called a **side**, and the common point is called the **vertex**. For the angle in Figure 1.1, the sides are $\overrightarrow{AB}$ and $\overrightarrow{AC}$, and the vertex is A.

Figure 1.1

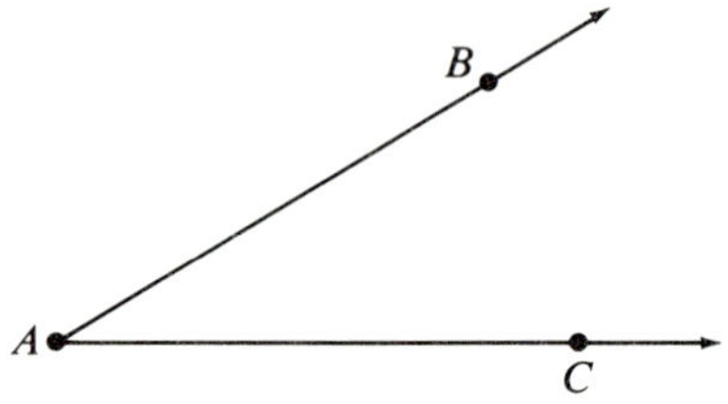

The notation for angle is $\angle$. In specifying an angle, we generally use three points: the vertex and a point on each side (different from the vertex). This notation is written with the vertex given as the middle point. Thus the angle in Figure 1.1 is labeled $\angle BAC$ or $\angle CAB$. When the vertex belongs to one angle only (and hence there is no possibility for confusion), we sometimes refer to an angle by its vertex only; for example, we can write $\angle A$ for the angle in Figure 1.1.

A **straight angle** is one in which the two rays lie along the same line and point in *opposite* directions. A **zero angle** is one in which the two rays lie along the same line but point in the *same* direction. The two rays are thus coincident. See Figure 1.2.

Figure 1.2

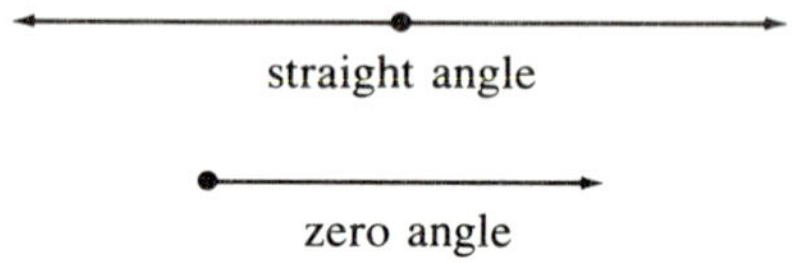

Angles are commonly measured in units called degrees. A **degree** is the measure of an angle that, when its vertex is placed at the center of a circle, intercepts an arc whose length is $\frac{1}{360}$ of the circumference of the circle. Thus the measure of a straight angle is 180°, and the measure of a zero angle is 0°. Other angles may be measured with a protractor, as illustrated in Figure 1.3. We see that

$$m\,\angle BAC = 40^\circ,$$

$$m\,\angle BAD = 70^\circ,$$

$$m\,\angle BAE = 130^\circ.$$

(The notation for measure is m.)

The protractor could be repositioned to measure $\angle CAD$, $\angle DAE$, and

Figure 1.3

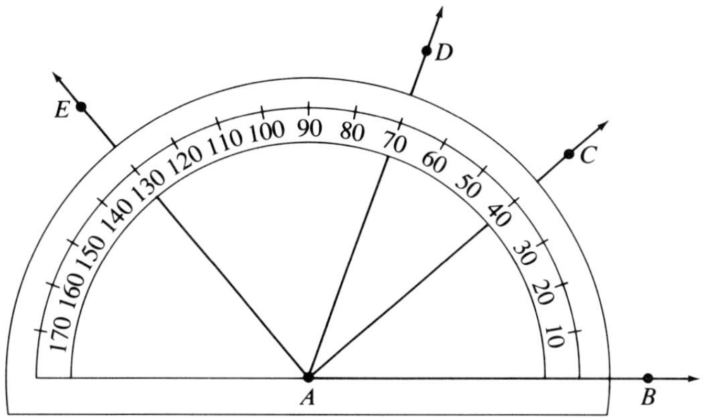

$\angle CAE$, but subtracting works as well. We have

$$m\,\angle CAD = 70^\circ - 40^\circ = 30^\circ,$$
$$m\,\angle DAE = 130^\circ - 70^\circ = 60^\circ,$$
$$m\,\angle CAE = 130^\circ - 40^\circ = 90^\circ.$$

Each degree is divided into minutes; a **minute** is $\frac{1}{60}$ of a degree. Each minute is subdivided into seconds; a **second** is $\frac{1}{60}$ of a minute. Using $'$ to denote minutes and $''$ to denote seconds, we have

$$1' = \left(\frac{1}{60}\right)^\circ \quad \text{or} \quad 60' = 1^\circ,$$
$$1'' = \left(\frac{1}{60}\right)' \quad \text{or} \quad 60'' = 1'.$$

However, when measurements more accurate than to the nearest degree are needed, we will use decimal notation. For example, we will say 62.42° instead of $62^\circ 25' 12''$.

An angle whose measure is 90° is called a **right angle**. An angle whose measure is between 0° and 90° is said to be **acute**. An angle is called **obtuse** if its measure is between 90° and 180°.

A **triangle** consists of three line segments connecting three noncollinear points. Each of the line segments is called a **side**, and each of the points is called a **vertex**.

An important property concerning the angles of a triangle is given in the following statement.

The sum of the measures of the three angles of a triangle is 180°.

So if we label a triangle as in Figure 1.4, this property says that

$$m\,\angle A + m\,\angle B + m\,\angle C = 180^\circ.$$

Figure 1.4

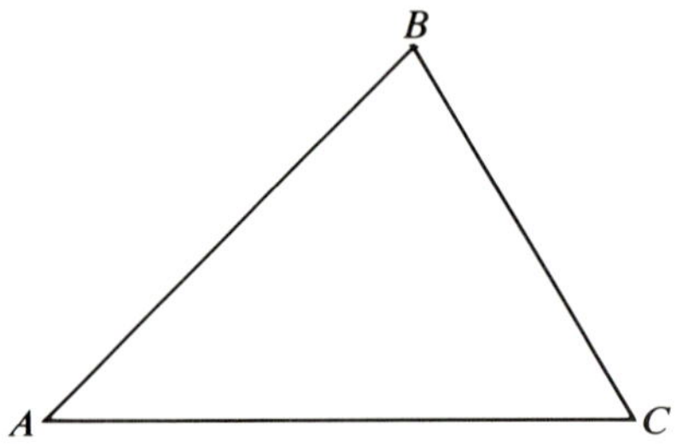

Two triangles are said to be **similar** provided that there exists a correspondence between vertices such that the measures of corresponding angles are equal and the lengths of corresponding sides are proportional. Thus we say that triangle ABC is similar to triangle DEF provided that

$$m\,\angle A = m\,\angle D, \quad m\,\angle B = m\,\angle E, \quad m\,\angle C = m\,\angle F,$$

and

$$\frac{AB}{DE} = \frac{AC}{DF} = \frac{BC}{EF}.$$

(It should be pointed out that the notation AB refers to the *distance* from A to B while $\overline{AB}$ refers to the *segment*.) This definition means that two triangles are similar when they have exactly the same shape (but not necessarily the same size), as can be seen in Figure 1.5.

Figure 1.5
Similar triangles

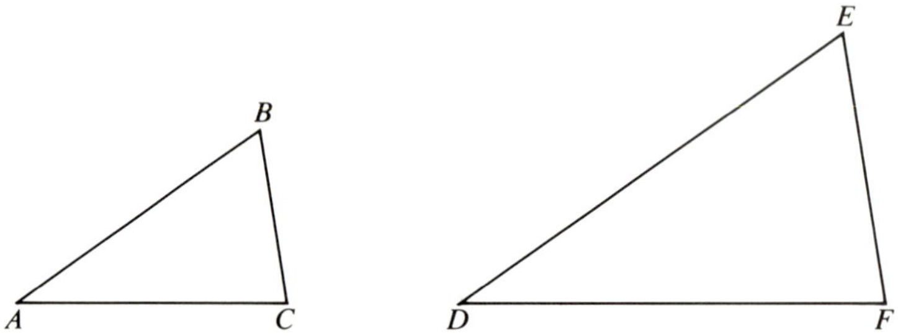

EXAMPLE 1 Suppose $\triangle ABC$ is similar to $\triangle DEF$. (The notation for triangle is $\triangle$.) Given $AB = 6$, $BC = 9$, and $DE = 10$, find EF.

SOLUTION We have shown the situation in Figure 1.6, where we have let $x = EF$.

Figure 1.6

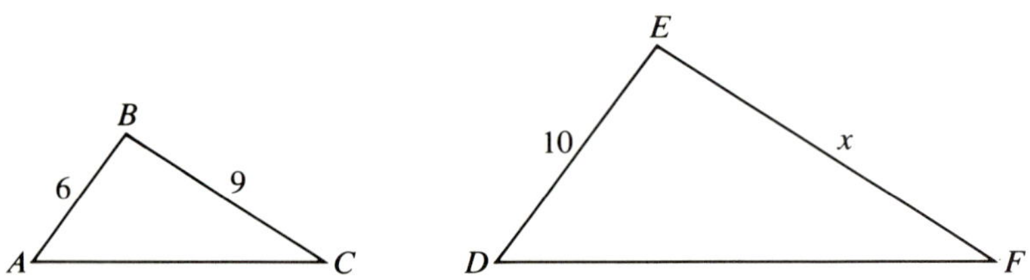

Because corresponding sides are proportional,

$$\frac{AB}{DE} = \frac{BC}{EF}.$$

Thus

$$\frac{6}{10} = \frac{9}{x}$$

$$6x = 90$$

$$x = \frac{90}{6} = 15.$$

▲

Another point worth noting is that for similar triangles, the ratio between any two sides of one triangle is equal to the ratio between corresponding sides of the other. For example, if $\triangle ABC$ is similar to $\triangle DEF$, then

$$\frac{AB}{BC} = \frac{DE}{EF}.$$

Since

$$\frac{AB}{DE} = \frac{BC}{EF},$$

$$(AB)(EF) = (BC)(DE).$$

Upon dividing by $(BC)(EF)$, we obtain

$$\frac{AB}{BC} = \frac{DE}{EF}.$$

In a similar fashion,

$$\frac{AB}{AC} = \frac{DE}{DF} \quad \text{and} \quad \frac{BC}{AC} = \frac{EF}{DF}.$$

Thus in Example 1 we could have written

$$\frac{6}{9} = \frac{10}{x},$$

which leads to the same result: $x = 15$.

The lengths of the sides of the triangles in Example 1 are taken to be exact. We may therefore provide an exact answer. However, when the triangles represent physical situations, the lengths are considered approximate because they are determined with mechanical devices. Thus each applied problem will specify a degree of accuracy or level of accuracy for the answer that is consistent with the rules for operations with approximations. For a discussion of approximations and computations with approximations, see Appendixes A.1 and A.2.

▲ *Exercise* Suppose $\triangle ABC$ is similar to $\triangle DEF$. Given $AB = 5$, $BC = 12$, and $DE = 3$, find EF.

Answer $7\frac{1}{5}$

To determine that two triangles are similar, we may use any of the following **Similarity Theorems**.

SSS (*side-side-side*): If there exists a correspondence between the vertices of two triangles such that corresponding sides are proportional, then the triangles are similar.

AA (*angle-angle*): If there exists a correspondence between the vertices of two triangles such that the measures of two angles of one triangle are equal to the measures of the corresponding angles of the other, then the triangles are similar.

SAS (*side-angle-side*): If there exists a correspondence between the vertices of two triangles such that two sides of one triangle are proportional to the corresponding sides of the other, and the measures of the included angles are equal, then the triangles are similar.

Note that the angle-angle theorem requires only *two* pairs of corresponding angles to have equal measures. However, it follows that the third pair must also have equal measures, since the sum of the measures of the three angles of a triangle is 180°.

We see that although the definition of similar triangles requires two conditions, equal angles and proportional sides, the first two Similarity Theorems indicate that each condition implies the other.

Note: If we say that two different angles are equal, we mean that their *measures* are equal. Technically such angles are *congruent*, not equal. Furthermore, when we use α (alpha) to represent the measure of an angle, we can refer to the angle as "angle α" if there is no possibility for confusion.

EXAMPLE 2 Find the length (to two significant digits) of a flagpole that casts a 12-foot shadow if a vertical yardstick casts a 2.0-foot shadow at the same time. (Assume that both the flagpole and the yardstick are perpendicular to level ground.)

SOLUTION We have illustrated the situation in Figure 1.7, where we have let

$$x = \text{length of flagpole (in feet)}.$$

Angles A and D are equal because the sun's rays are parallel. Since angles C and F are right angles, the two triangles are similar by the AA Similarity

Figure 1.7

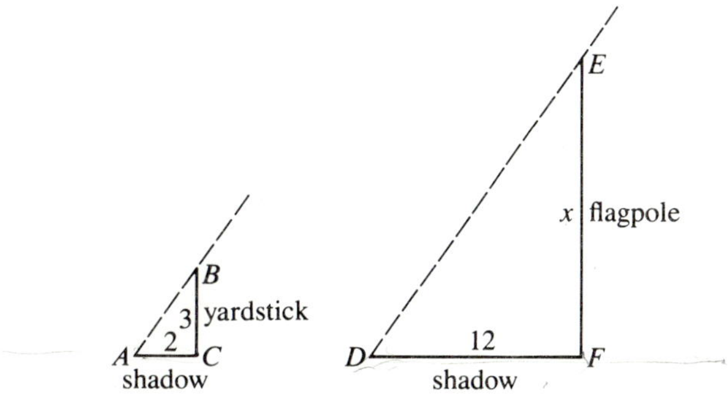

Theorem. Therefore,

$$\frac{3}{2} = \frac{x}{12}$$

$$2x = 36$$

$$x = 18.$$

The flagpole is 18 feet long.

▲

▲ ***Exercise*** Find the length (to two significant digits) of a flagpole that casts a 10-foot shadow if a vertical yardstick casts a 2.0-foot shadow at the same time. (Assume that both the flagpole and the yardstick are perpendicular to level ground.)

Answer 15 ft

Another important property concerns *right* triangles. A **right triangle** is a triangle that has a right angle. The side opposite the right angle is called the **hypotenuse**, and the other two sides are called **legs**. See Figure 1.8. Note the symbol at C to indicate that $\angle C$ is the right angle.

Figure 1.8
Right triangle

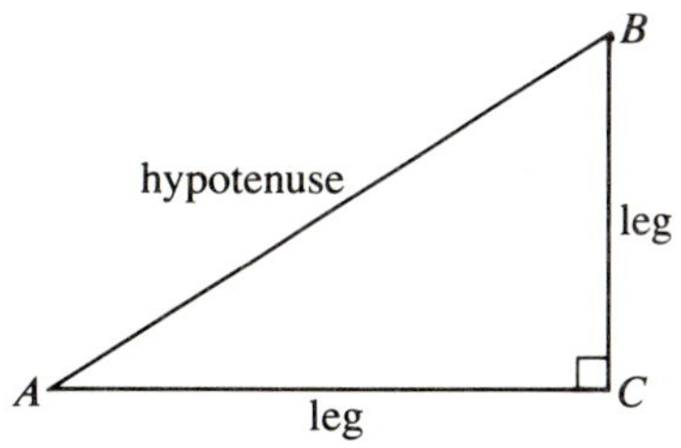

An important property of right triangles, called the **Pythagorean Theorem**, follows.

In a right triangle the sum of the squares of the lengths of the two legs is equal to the square of the length of the hypotenuse.

Figure 1.9

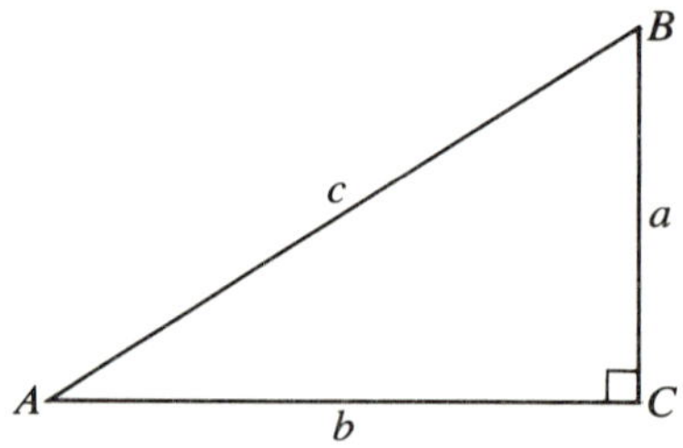

Thus, if we label a right triangle as in Figure 1.9, the Pythagorean Theorem says that $a^2 + b^2 = c^2$.

EXAMPLE 3 Find a for the triangle in Figure 1.10.

Figure 1.10

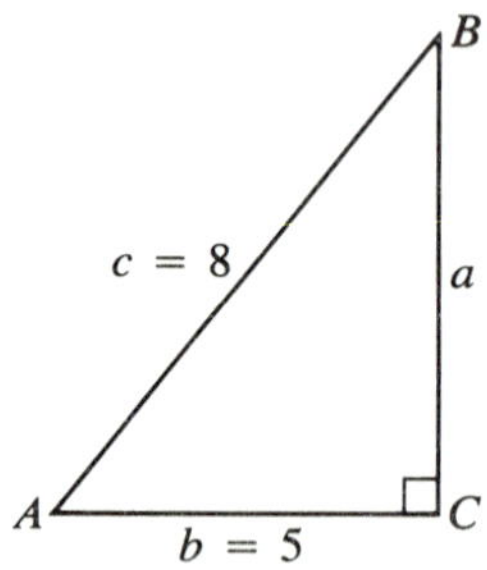

SOLUTION Because we have a right triangle, we know that

$$a^2 + b^2 = c^2.$$

Thus

$$\begin{aligned} a^2 + 25 &= 64 \\ a^2 &= 64 - 25 = 39 \\ a &= \sqrt{39} \quad (\approx 6.245). \end{aligned}$$

▲

▲ ***Exercise*** Find b for the triangle in Figure 1.11.

Figure 1.11

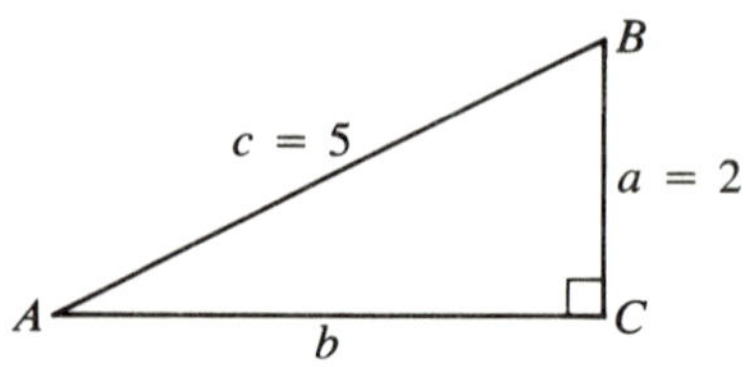

Answer $\sqrt{21}$ (≈ 4.583)

EXAMPLE 4 Suppose a satellite is orbiting the earth at an altitude of 110 miles. How far (to three significant digits) is the satellite from the horizon? (Assume that the earth's radius is 3960 miles.)

SOLUTION We have the situation depicted in Figure 1.12, where

$$x = \text{distance from satellite to horizon (in miles)}.$$

Figure 1.12

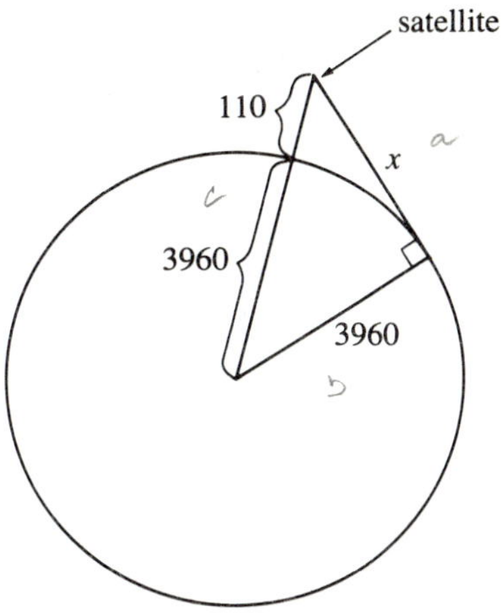

The line from the satellite to the horizon is tangent to the circle and thus perpendicular to the radius. So we have a right triangle. By the Pythagorean Theorem,

$$x^2 + 3960^2 = (3960 + 110)^2$$
$$x^2 + 3960^2 = 4070^2$$
$$x^2 = 4070^2 - 3960^2 = 883{,}300$$
$$x = \sqrt{883{,}300} \approx 940.$$

Hence, the satellite is approximately 940 miles from the horizon.

▲

▲ ***Exercise*** Suppose a satellite is orbiting the earth at an altitude of 155 miles. How far (to three significant digits) is the satellite from the horizon? (Assume that the earth's radius is 3960 miles).

Answer 1120 mi

A special triangle that appears frequently is the 45°-45°-90° triangle. All such right triangles are similar to the one in Figure 1.13, where, given legs of length 1, the Pythagorean Theorem tells us that the hypotenuse has length $\sqrt{2}$.

Figure 1.13

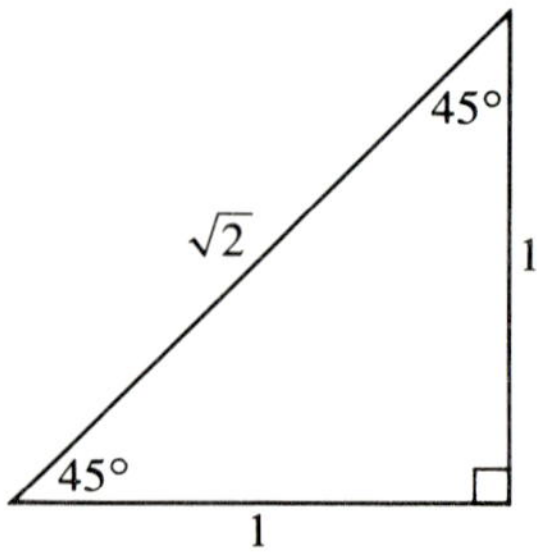

Another frequently occurring triangle is the 30°-60°-90° triangle. We'll examine it by considering a triangle having all sides of length 2. The altitude from one vertex to the opposite side bisects both the angle at that vertex and the opposite side. See Figure 1.14.

Figure 1.14

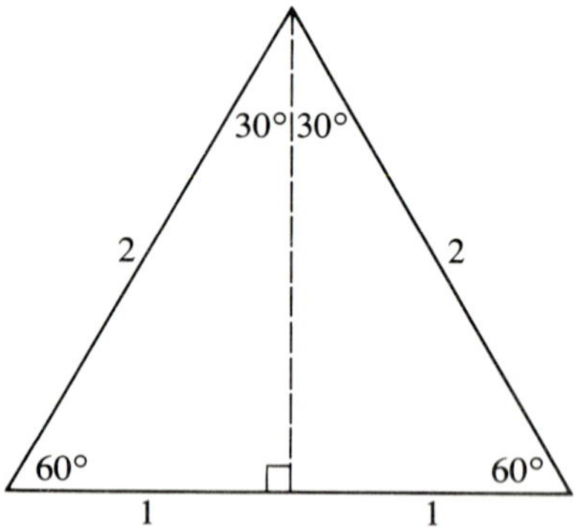

By the Pythagorean Theorem, the length of this altitude is $\sqrt{2^2 - 1^2} = \sqrt{3}$. Hence all 30°-60°-90° triangles are similar to the one in Figure 1.15.

Figure 1.15

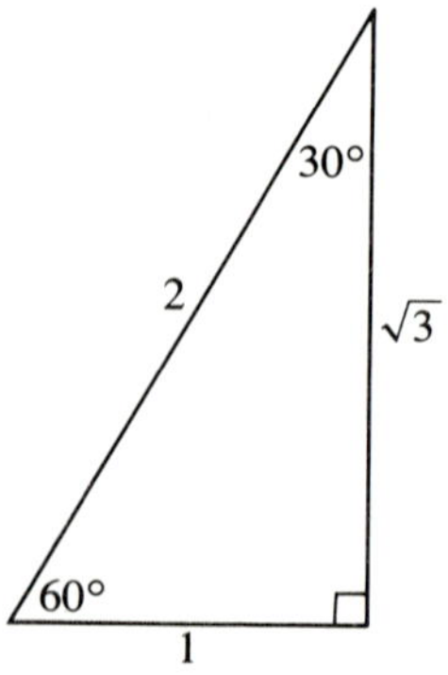

EXAMPLE 5 What are the exact lengths of the legs of a 30°-60°-90° triangle if the hypotenuse has length 6?

SOLUTION Consider the similar triangles in Figure 1.16.

Figure 1.16

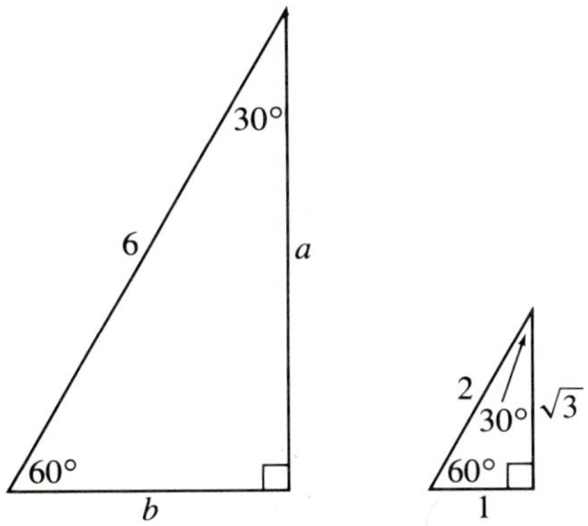

We have

$$\frac{b}{6} = \frac{1}{2}$$

$$b = \frac{1}{2} \cdot 6 = 3.$$

Furthermore,

$$\frac{a}{3} = \frac{\sqrt{3}}{1}$$

$$a = 3\sqrt{3}.$$

Thus the lengths of the legs are 3 and $3\sqrt{3}$.

▲

▲ ***Exercise*** What are the exact lengths of the legs of a 30°-60°-90° triangle if the hypotenuse has length 4?

Answer 2 and $2\sqrt{3}$

Problem Set 1.1

In Problems 1–4, $\triangle ABC$ is similar to $\triangle DEF$.

1. Given $AB = 5$, $BC = 6$, and $DE = 12$, find EF.

2. Given $AB = 10$, $AC = 6$, and $DE = 3$, find DF.

3. Given $AC = 7$, $BC = 5$, and $EF = 6$, find DF.

4. Given $AC = 10$, $AB = 12$, and $DE = 15$, find DF.

5. Find the length (to two significant digits) of a radio tower that casts a 40-foot shadow if a vertical yardstick casts a 30-inch shadow at the same time. (Assume that both the tower and the yardstick are perpendicular to level ground.)

6. Find the height (to two significant digits) of a building that casts a 50-foot shadow if a 6.0-foot vertical pole casts a 2.0-foot shadow at the same time. (Assume that both the building and the pole are perpendicular to level ground.)

7. A man is walking away from a streetlight. When he is 25 feet away from a point directly beneath the light, his shadow is 10 feet long. If the man is 6.0 feet tall, how high (to two significant digits) is the light?

8. A woman is standing near a streetlight that is 20 feet high. If the woman is 5.0 feet tall and is casting a 4.0-foot shadow, how far (to two significant digits) is she from a point directly beneath the light?

In Problems 9–12, *a*, *b*, and *c* refer to the following triangle.

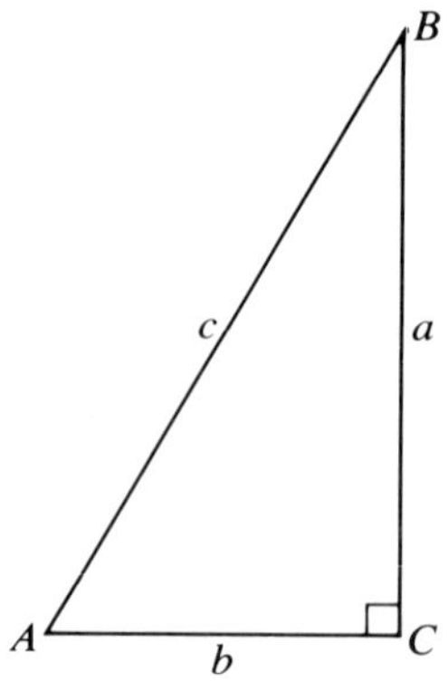

9. Given $a = 2$ and $b = 3$, find c.

10. Given $a = 5$ and $b = 4$, find c.

11. Given $a = 2$ and $c = 6$, find b.

12. Given $b = 5$ and $c = 7$, find a.

13. A guy wire is fastened to a telephone pole at a point 20 feet above the ground. If the wire is anchored 8.0 feet from the base of the pole, how long (to two significant digits) is the wire?

14. A bookcase that is 30 inches wide, 18 inches deep, and 7 feet 9 inches high is carried into a room and placed on its front. It will be raised into position by pushing the top upward, rotating the case about the lower front edge. If the ceiling is 8.0 feet high, will there be enough clearance?

15. In the following triangle, $AB = 12$ and $BC = 8$. Find CD.

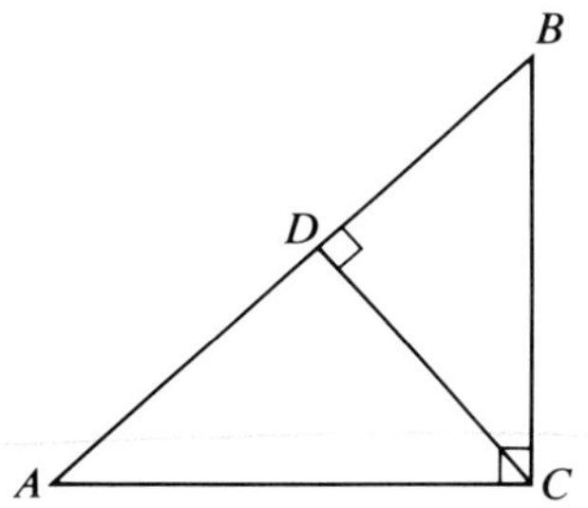

16. Suppose that a stretch of highway is rising 15 feet for every 100 horizontal feet. How much vertical distance (to the nearest 10 feet) is gained along this stretch as a car travels 1 mile?

In Problems 17–20, *a*, *b*, and *c* refer to the following special triangle.

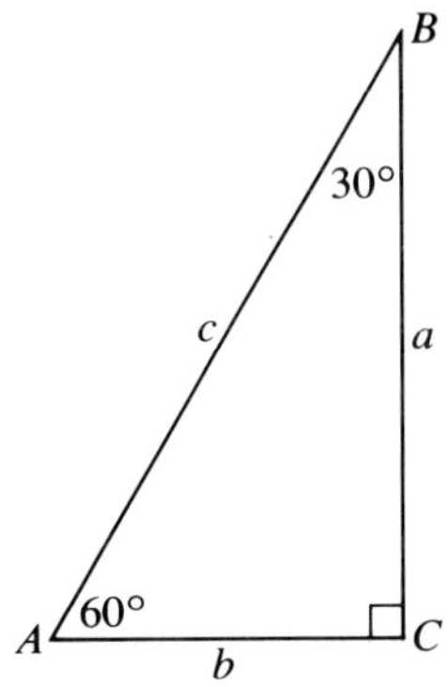

17. Given $c = 12$, find a and b, exactly.

18. Given $c = \sqrt{5}$, find a and b, exactly.

19. Given $b = 3$, find a and c, exactly.

20. Given $a = 2$, find b and c, exactly.

21. What is the exact length of each leg of a 45°-45°-90° triangle if the hypotenuse has length 6?

22. What is the exact length of the hypotenuse of a 45°-45°-90° triangle if each leg has length 4?

23. Consider the situation depicted on the following diagram of a pool table. P' is the point beyond the cushion such that the cushion is the perpendicular bisector of $\overline{PP'}$, Q is a side pocket, and R is the intersection

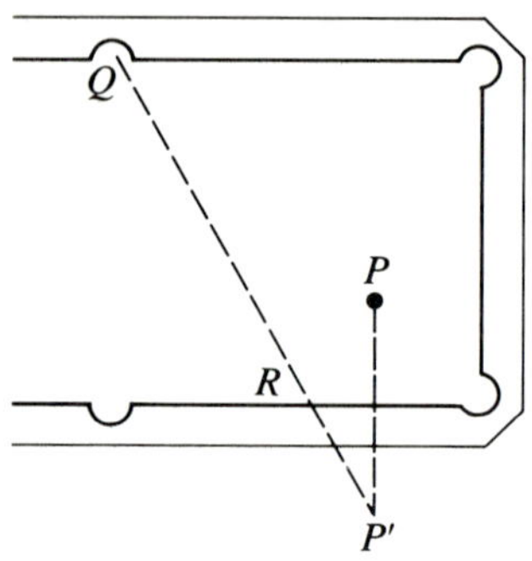

of $\overline{P'Q}$ with the side cushion. Show that if the cue ball strikes a ball at P, propelling it into the cushion at R (with sufficient speed and no English), then it will go into the side pocket at Q (assuming no interference).

1.2 THE CARTESIAN COORDINATE SYSTEM

The symbolism (a, b) is used to designate an **ordered pair**, a pair of numbers a and b in which the order is important. Thus (1, 2), (5, −3), and (2, 1) are examples of ordered pairs. Because the order is significant, (1, 2) is different from (2, 1).

When $a < b$, the notation (a, b) also denotes the open interval from a to b. However, within the context of a particular problem, it will be apparent whether the symbolism refers to an open interval or to an ordered pair.

Let us consider two perpendicular number lines intersecting at their origins. Such a system, as in Figure 1.17, is called a **Cartesian coordinate system**. By convention the number lines are vertical and horizontal with the positive directions up and to the right, respectively. It is useful to have the unit lengths of each number line the same, but this is not always practical or convenient.

Figure 1.17

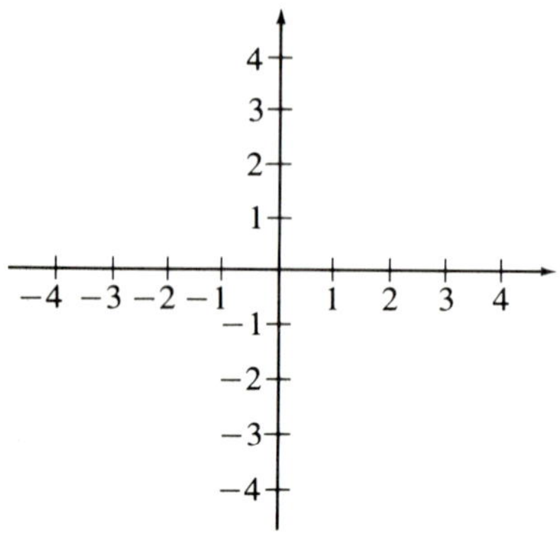

In this system each number line is called an **axis**. The horizontal line is called the ***x*-axis**, and the vertical line is called the ***y*-axis**. The point of intersection is called the **origin**. The plane determined by such a system is called the **Cartesian** or **coordinate plane** or the ***xy*-plane**. The two axes divide the plane into four distinct regions called **quadrants**, numbered I through IV, as indicated in Figure 1.18. The points on the axes are not part of any quadrant.

Figure 1.18

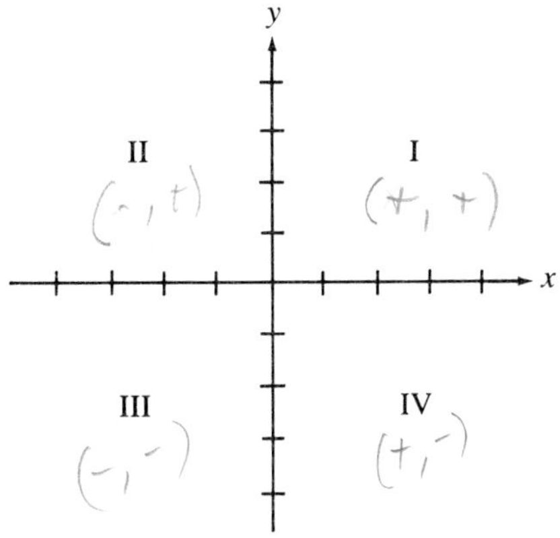

In this system each point in the plane is associated with an ordered pair, and each ordered pair corresponds to a point in the plane. Given a point P, the ordered pair associated with it can be found as follows. Construct the line passing through P perpendicular to the x-axis. The x-axis coordinate of the point of intersection of the x-axis with the constructed line is the first entry of the ordered pair that corresponds to P. Construct the line passing through P perpendicular to the y-axis. The y-axis coordinate of the point of intersection of the y-axis with the constructed line is the second entry of the ordered pair corresponding to P. In Figure 1.19 we see that the point P is associated with (a, b).

Figure 1.19

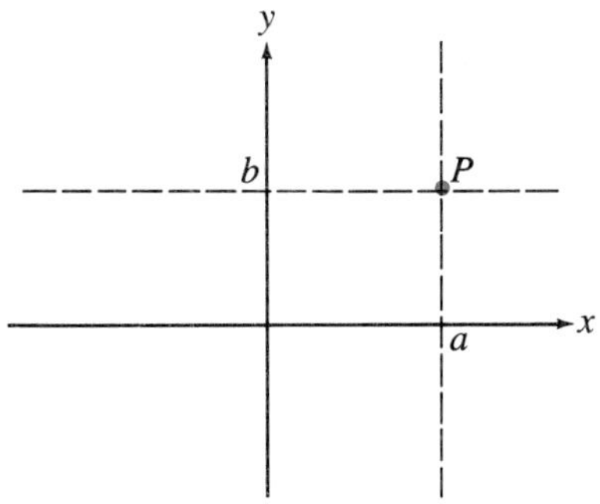

Note that every point on the x-axis is associated with an ordered pair of the form $(a, 0)$ for some a, and every point on the y-axis is associated with an ordered pair of the form $(0, b)$ for some b. The ordered pair corresponding to

Figure 1.20

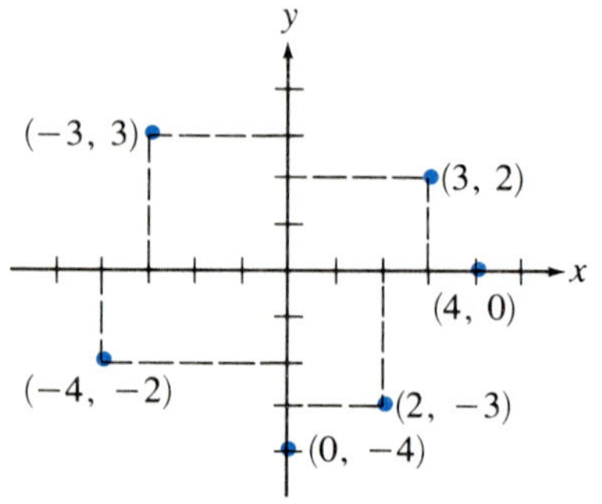

the origin is (0, 0). Some points and associated ordered pairs are indicated in Figure 1.20.

Whenever the axis coordinates do not appear, as in Figure 1.20, we shall assume that the distance between consecutive hash marks is 1 unit. If the axes are not labeled, we shall assume that the horizontal axis is the x-axis and that the vertical axis is the y-axis.

Because of the one-to-one relationship that exists between points in the plane and ordered pairs, we shall frequently interchange the terms. When we refer to the point $P(a, b)$, we mean that the point is labeled P, and its corresponding ordered pair is (a, b). Furthermore, we shall call the first entry of an ordered pair the ***x*-coordinate** and the second entry the ***y*-coordinate**.

In the Cartesian coordinate system the distance between any two points in the plane may be calculated by using the coordinates. If $P_1(x_1, y_1)$ and $P_2(x_2, y_2)$ determine a nonhorizontal and nonvertical line, then we may construct a right triangle, as shown in Figure 1.21.

Figure 1.21

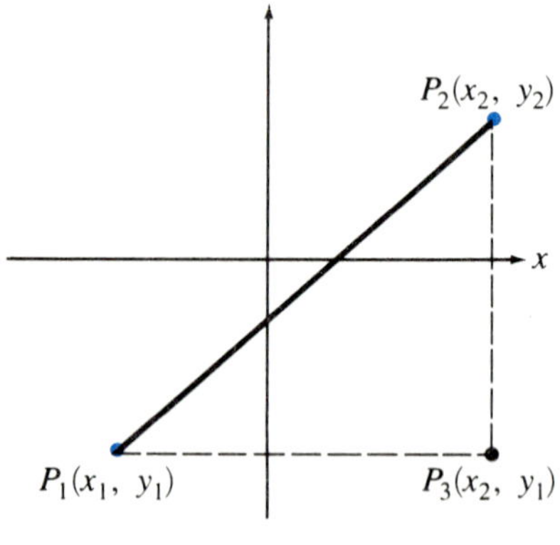

The point P_3 is the intersection of two perpendicular lines; the line passing through P_1 that is perpendicular to the y-axis and the line passing through P_2 that is perpendicular to the x-axis. Thus the y-coordinate of P_3 must be the same as the y-coordinate of P_1, and the x-coordinate of P_3 must be the same as the x-coordinate of P_2.

Because P_1 and P_3 lie along the same horizontal line,

$$P_1P_3 = |x_2 - x_1|,$$

and since P_2 and P_3 lie along the same vertical line,

$$P_2P_3 = |y_2 - y_1|.$$

By the Pythagorean Theorem, we have

$$(P_1P_2)^2 = (P_1P_3)^2 + (P_2P_3)^2.$$

Substituting, we get

$$\begin{aligned}(P_1P_2)^2 &= |x_2 - x_1|^2 + |y_2 - y_1|^2\\ &= (x_2 - x_1)^2 + (y_2 - y_1)^2.\end{aligned}$$

Therefore,

$$P_1P_2 = \sqrt{(x_2 - x_1)^2 + (y_2 - y_1)^2}, \qquad \textbf{(1.1)}$$

since distance is nonnegative.

If P_1 and P_2 determine a horizontal line, then $y_1 = y_2$, and the distance between P_1 and P_2 is $|x_2 - x_1|$. Thus

$$P_1P_2 = |x_2 - x_1| = \sqrt{(x_2 - x_1)^2} = \sqrt{(x_2 - x_1)^2 + (y_2 - y_1)^2}.$$

A similar analysis may be made if P_1 and P_2 determine a vertical line.

Equation (1.1) is called the **distance formula**. Summarizing, it says the following:

The distance between the points $P_1(x_1, y_1)$ and $P_2(x_2, y_2)$, denoted by P_1P_2, is

$$P_1P_2 = \sqrt{(x_2 - x_1)^2 + (y_2 - y_1)^2}.$$

EXAMPLE 1 Determine the distance between $P_1(2, -3)$ and $P_2(-1, 3)$.

SOLUTION Using the distance formula, we have

$$\begin{aligned}P_1P_2 &= \sqrt{(-1 - 2)^2 + (3 - (-3))^2}\\ &= \sqrt{(-3)^2 + 6^2} = \sqrt{45} = 3\sqrt{5} \quad (\approx 6.708).\end{aligned}$$

▲

Because $(x_2 - x_1)^2 = (x_1 - x_2)^2$ and $(y_2 - y_1)^2 = (y_1 - y_2)^2$, the order in which the coordinates are subtracted is of no consequence. Thus in Example 1 we could have written

$$P_1P_2 = \sqrt{(2 - (-1))^2 + (-3 - 3)^2}$$

and obtained the same result.

▲ ***Exercise*** Determine the distance between $P_1(-3, 4)$ and $P_2(1, -2)$.

Answer $2\sqrt{13}$ (≈ 7.211)

Problem Set 1.2

1. Let $P(x, y)$ be a point in the plane. In which quadrant is P if we have the following?

(a) $x > 0$ (b) $y < 0$

(c) $x > 0$ and $y < 0$ (d) $x > 0$ or $y < 0$

(e) $xy > 0$ (f) $\dfrac{x}{y} < 0$

2. (a) List the ordered pair that corresponds to each of the points in the plane in the following figure.

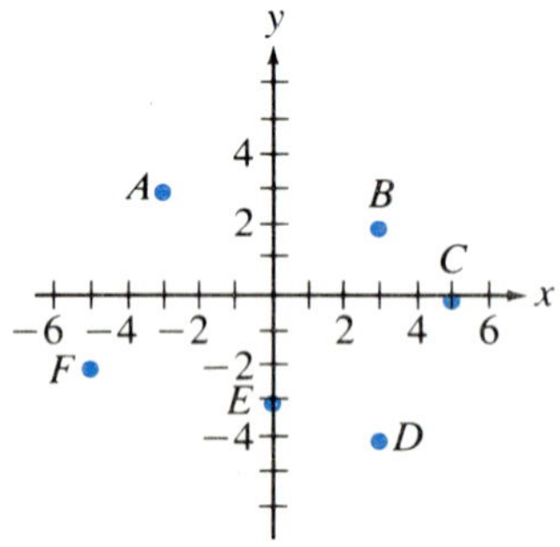

(b) Using a Cartesian coordinate system, plot the following points:

$$A(-2, 4),\quad B(0, 2),\quad C\left(1, \frac{5}{2}\right),\quad D(2, -4),\quad E(-3, 0),\quad F\left(\frac{-3}{2}, -1\right).$$

In Problems 3–16, determine the distance between the given points.

3. $(1, 2), (4, 6)$ **4.** $(2, 5), (7, 17)$

5. $(-2, -3), (10, -8)$ **6.** $(-2, 3), (1, -1)$

7. $(-7, -5), (-6, -2)$ **8.** $(3, 6), (-2, 2)$

9. $(1, -1), (3, 3)$ **10.** $(-5, 2), (1, 6)$

11. $(3\sqrt{5}, -1), (\sqrt{5}, 3)$ **12.** $(0, -2), (\sqrt{5}, 3)$

13. $(2\sqrt{6}, \sqrt{5}), (-\sqrt{6}, 3\sqrt{5})$ **14.** $(0, -\sqrt{3}), (-\sqrt{13}, \sqrt{3})$

15. $\left(\frac{3}{5}, -1\right), \left(\frac{7}{5}, \frac{1}{2}\right)$ **16.** $\left(\frac{-1}{4}, \frac{1}{6}\right), \left(\frac{3}{2}, \frac{2}{3}\right)$

17. Find all points on the x-axis 5 units from $(-2, 3)$.

18. Find all points on the y-axis 3 units from $(-1, -2)$.

19. Find the point on the x-axis equidistant from $(-3, -5)$ and $(2, 7)$.

20. Find the point on the y-axis equidistant from $(2, 3)$ and $(-3, 7)$.

21. Find the distance between (a, b) and $(a, -b)$.

22. Find the distance between (a, b) and $(-a, b)$.

1.3 EQUATIONS IN TWO VARIABLES AND THEIR GRAPHS

The relationship between two quantities is frequently expressed in the form of an equation. In this section we examine equations relating two variables: x and y. Examples include

$$2x - y = x + y - 3, \quad y = x^2, \quad \text{and} \quad x^2 + y^2 = 9.$$

A **solution** of an equation in x and y is an ordered pair (a, b) that makes the equation true when x is replaced by a and y is replaced by b. Hence $(2, 1)$ is a solution of $2x + y = 5$, but $(1, 2)$ is not.

Given an equation in x and y, we shall use its solutions and the Cartesian coordinate system to give us a "picture." The **graph of an equation** in x and y is the set of all points in the xy-plane that are solutions of the equation. Most equations have infinitely many solutions; hence the graph cannot be constructed by plotting them all. We will plot a few solutions and connect them with a curve that follows the general trend of the plotted points.

EXAMPLE 1 Graph $y = x + 1$.

SOLUTION We begin by constructing a table that lists some of the infinitely many solutions of this equation. We will choose some values for x and use the equation to find the corresponding value for y. For example, when $x = -3$, $y = -3 + 1 = -2$. When $x = -2$, $y = -2 + 1 = -1$, and so on. We get the following table:

x	-3	-2	-1	0	1	2	3
y	-2	-1	0	1	2	3	4

This table is simply another way of representing the ordered pairs. Each column refers to an ordered pair, with the entry in the top row being the x-coordinate and the entry in the bottom row being the y-coordinate. The graph of these ordered pairs is shown in Figure 1.22.

Figure 1.22

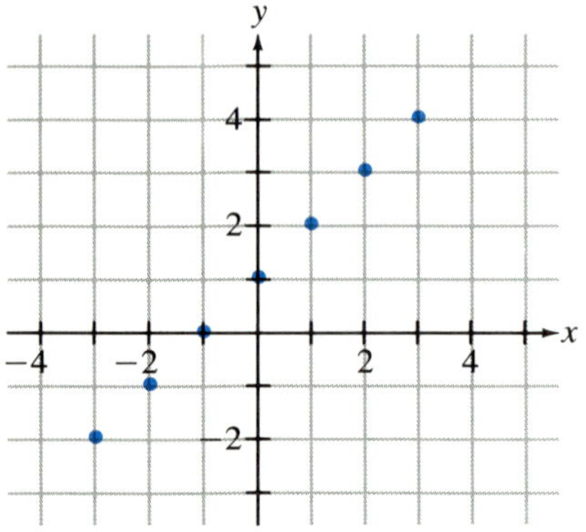

Since the solution set in infinite, and the solutions appear to lie along a straight line, we suspect that the graph of $y = x + 1$ is the line determined by these points. See Figure 1.23.

Figure 1.23
$y = x + 1$

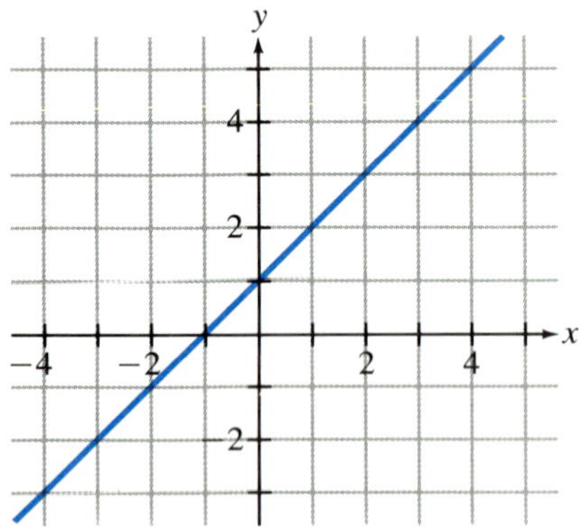

▲

When graphing equations by finding arbitrary solutions, how can we be certain that we have plotted representative points? And how can we be sure that we have connected them properly to give us an accurate sketch? The fact is that we cannot unless we have a general idea of what the graph looks like before we begin. In Sections 5.1 and 5.3, we'll obtain a general idea of the graph by examining the equation. Meanwhile, suggested values for plotting points will be included with the problems.

EXAMPLE 2 Graph $x = y^2$.

SOLUTION We will construct a table that lists some of the solutions by selecting some values for y and finding the corresponding values for x. We get

x	4	1	0	1	4
y	−2	−1	0	1	2

The graph of these ordered pairs is shown in Figure 1.24.

Figure 1.24

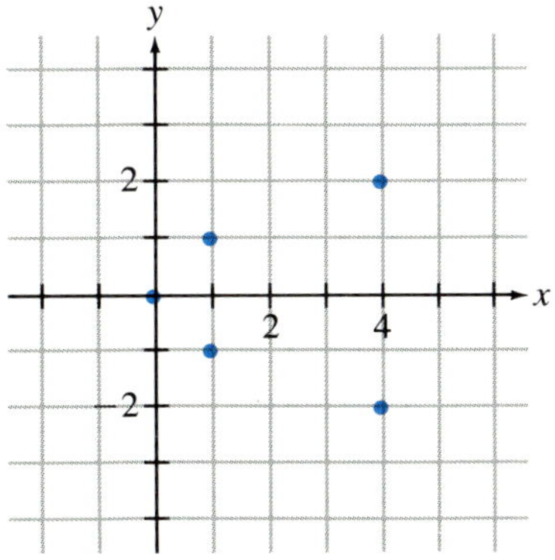

Although these points do not lie along a straight line, we can still make a guess about the graph. See Figure 1.25.

Figure 1.25
$x = y^2$

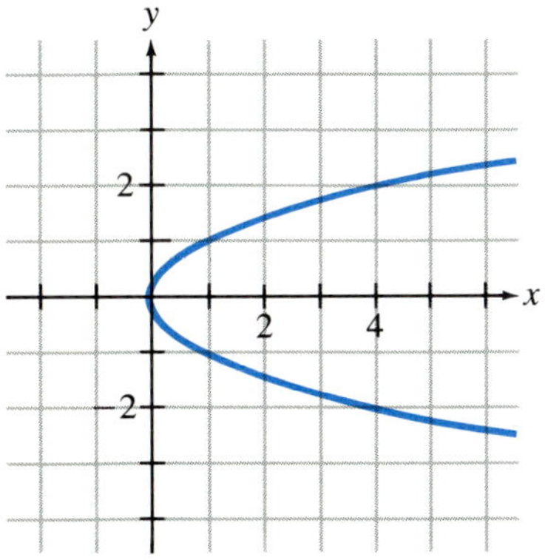

▲

▲ *Exercise* Graph the equation after plotting solutions with the given values.

(a) $y = -x + 1$; x-values: -2, -1, 0, 1, and 2
(b) $x = -y^2$; y-values: -2, -1, 0, 1, and 2 (Recall that $-y^2 = -(y^2)$.)

Answer See Figure 1.26.

Figure 1.26

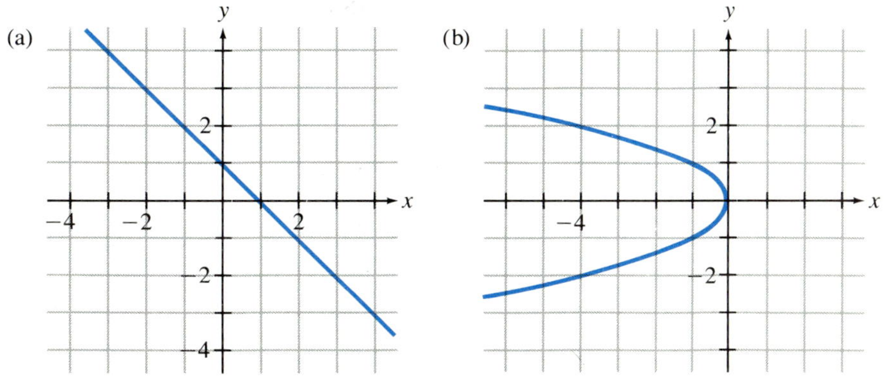

An equation in x and y is a statement of the relationship between two variables. If the relationship is such that for each permissible value of x, there corresponds exactly one value of y that makes the equation true, then we say that **y is a function of x**. Examples include

$$x + y = 3, \quad y = |x|, \quad \text{and} \quad x = y^3.$$

Once we select a value for x, the value for y is uniquely determined. We have but one choice for y to make the equation true. In this regard, x will be referred to as the **independent variable**, and y will be called the **dependent variable**.

On the other hand, we don't have this property in any of the following:

$$x^2 + y^2 = 9, \quad x = |y|, \quad \text{and} \quad x = y^2.$$

In the first, for instance, when $x = 0$, we have $y = 3$ or $y = -3$. The y-value is not uniquely determined. In the second, when $x = 1$, $y = 1$ or $y = -1$. In the third, when $x = 4$, $y = 2$ or $y = -2$. The values for y are not uniquely determined. Thus, for each of these last three equations, y is *not* a function of x.

▲ ***Exercise*** Determine whether y is a function of x.

(a) $x^2 + y^2 = 4$ (b) $|x| = |y|$ (c) $y = x^3$

Answer (a) no (b) no (c) yes

We can determine whether y is a function of x by looking at the graph of an equation. In fact, y is a function of x if and only if no vertical line intersects the graph of the equation in more than one point. Indeed, if there exists a vertical line that intersects the graph in two or more points, there will be two or more y-values that correspond to the same x-value. Such a situation indicates that y is *not* a function of x.

▲ ***Exercise*** Determine whether y is a function of x for the equations graphed in Figure 1.27.

Figure 1.27

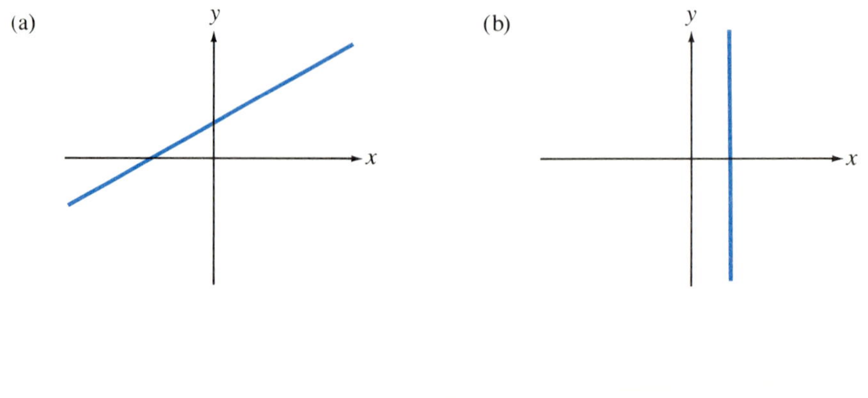

Figure 1.27 *(continued)*

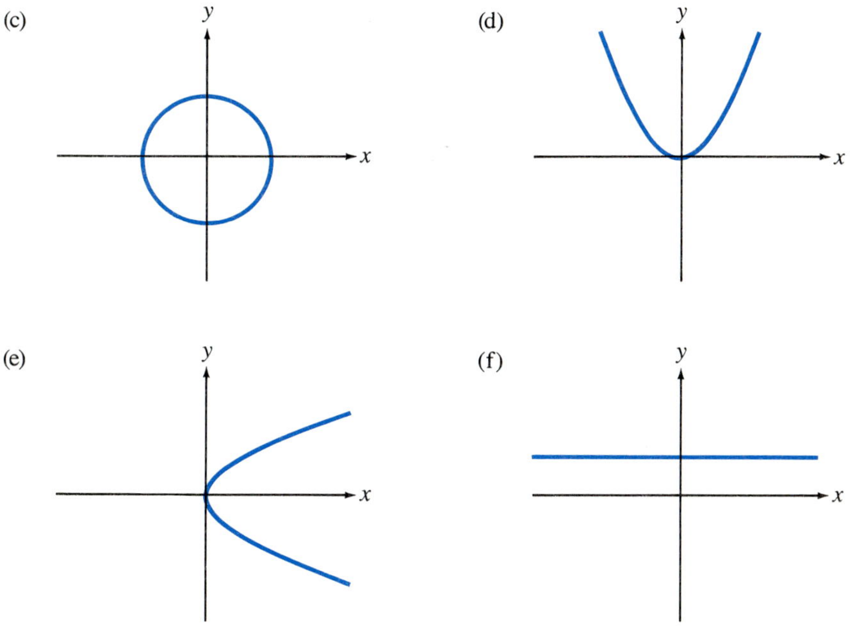

Answer (a) yes (b) no (c) no (d) yes (e) no (f) yes

Problem Set 1.3

In Problems 1–4, list three solutions to the given equation.

1. $y = \sqrt{x + 2}$

2. $y = \sqrt{x - 3}$

3. $x^2 + y^2 = 1$

4. $x = y^3$

In Problems 5–12, sketch the graph of the equation after plotting points with the given values.

5. $y = x + 3$; x-values: $-4, -2, 0, 2, 4$

6. $y = 3 - 2x$; x-values: $-4, -2, 0, 2, 4$

7. $y = -x^2$; x-values: $-2, -1, 0, 1, 2$

8. $y = x^2$; x-values: $-2, -1, 0, 1, 2$

9. $x = |y|$; y-values: $-2, -1, 0, 1, 2$

10. $y = x^3$; x-values: $-2, -1, 0, 1, 2$

11. $x^2 + y^2 = 4$; x-values: $-2, -1, 0, 1, 2$

12. $y = \sqrt{x}$; x-values: 0, 1, 4, 9, 16

In Problems 13–20, determine whether y is a function of x.

13. $y = -x^3$

14. $y^2 = -x$

15. $x^2 + y^2 = 4$

16. $y = 2x^2 + 8x - 1$

17. $x^2 + 4y^2 = 1$

18. $x^2 - 4y^2 = 1$

19. $y = |x|$

20. $x = y^2$

In Problems 21–34, determine whether y is a function of x for the equation graphed.

21.

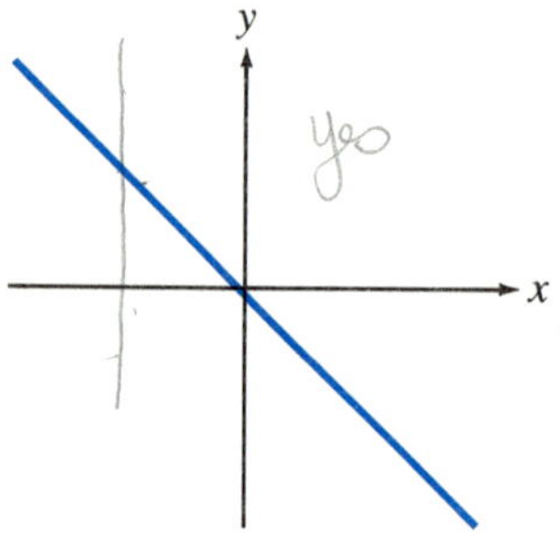

22.

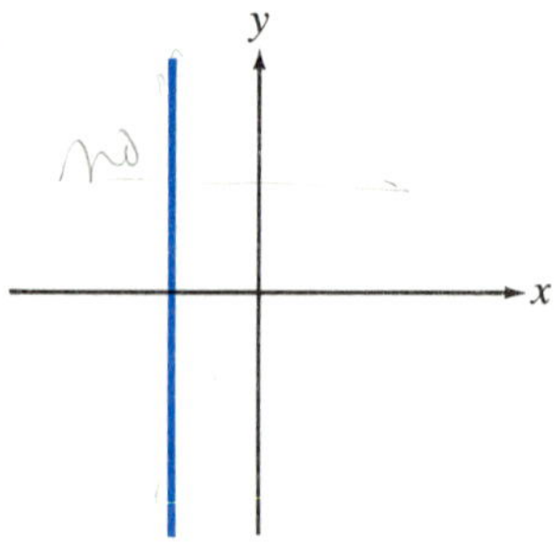

23.

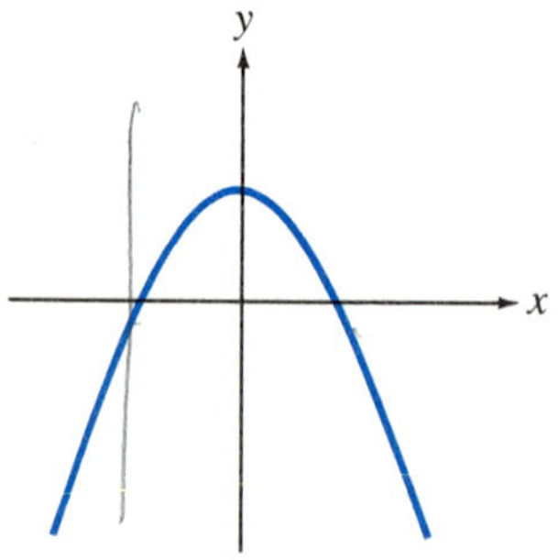

24.

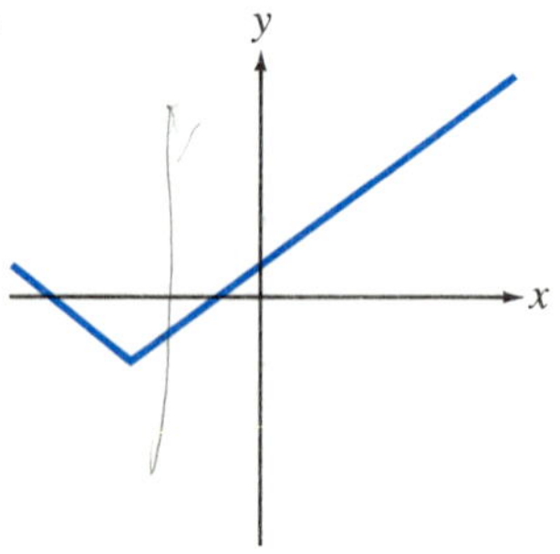

25.

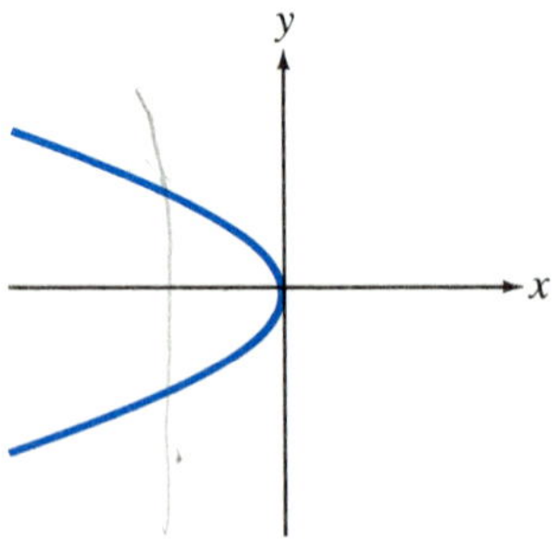

26.

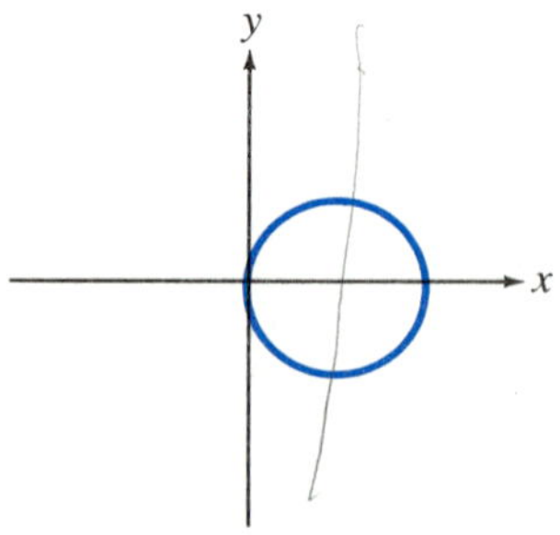

27.

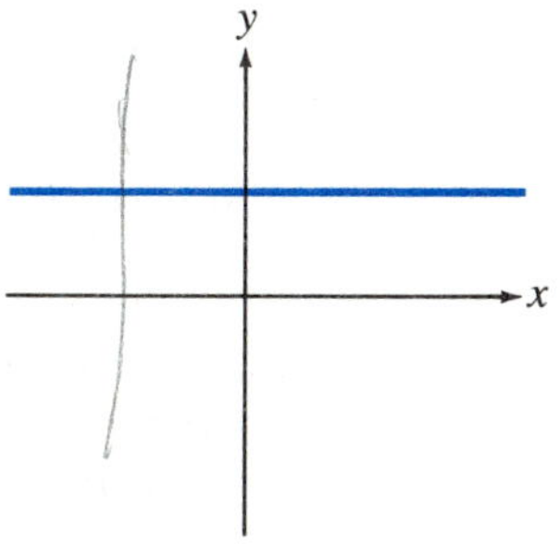

28.

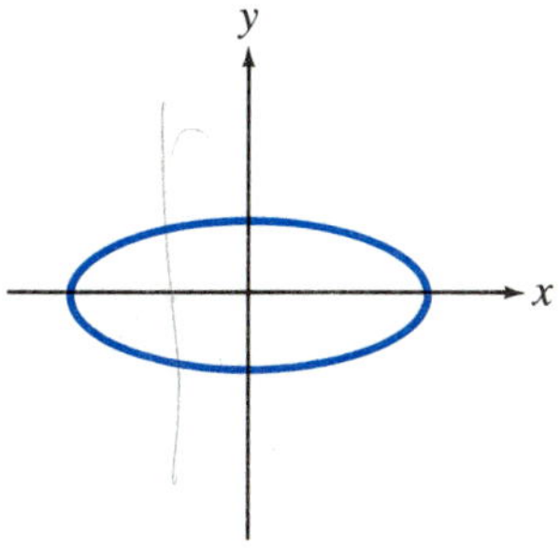

29.

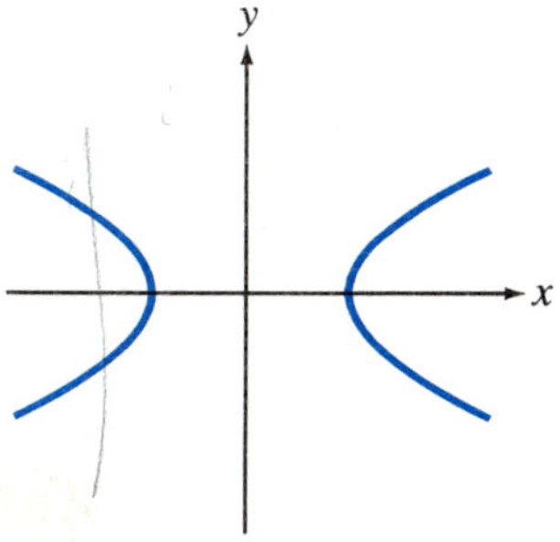

30.

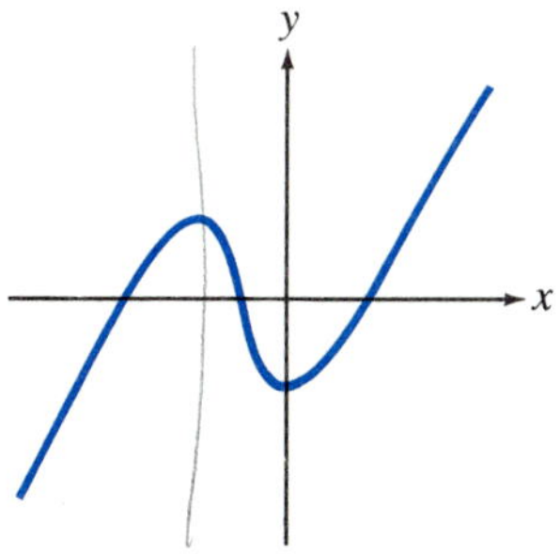

31.

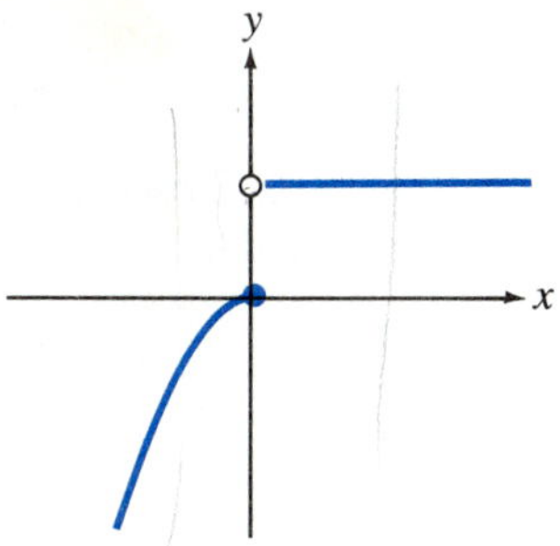

32.

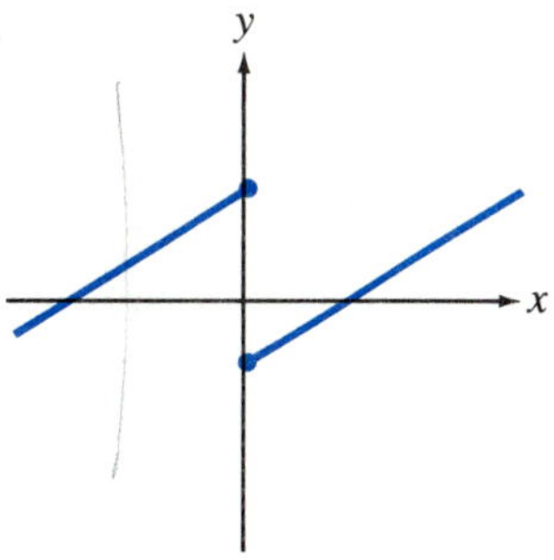

33.

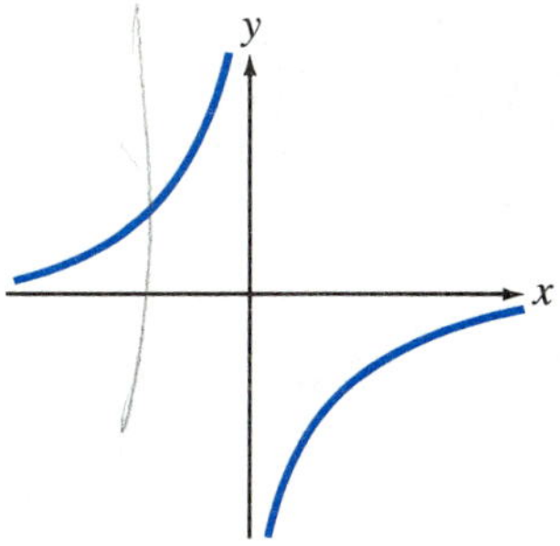

34.

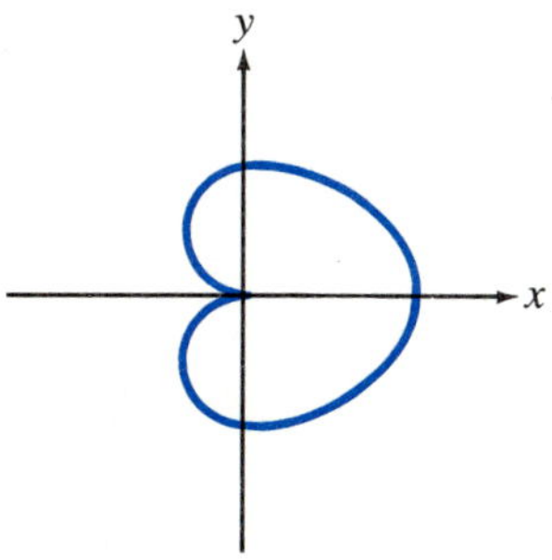

1.4 FUNCTIONS

The discussion at the end of Section 1.3 leads us to a formal definition of the term *function*. Let X and Y be nonempty sets. A **function** from X to Y is a correspondence that associates each element of X with a unique (exactly one) element of Y.

The two correspondences depicted in Figure 1.28 are functions. Each element of X is associated with one and only one element of Y.

Figure 1.28

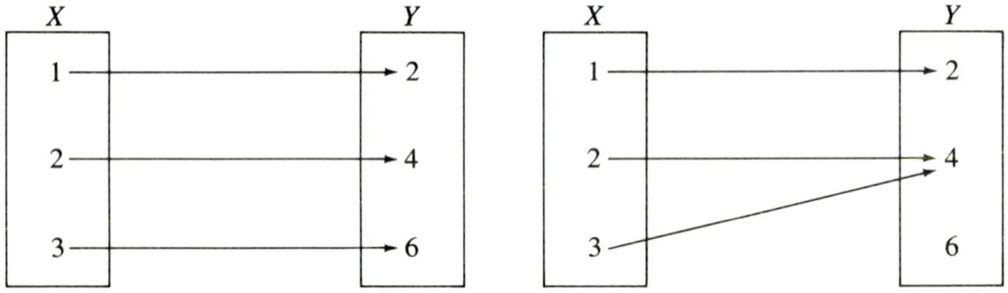

(i) (ii)

However, the correspondence shown in Figure 1.29 is not a function. There is an element of X (in particular, 2) that is associated with two different elements of Y.

Figure 1.29

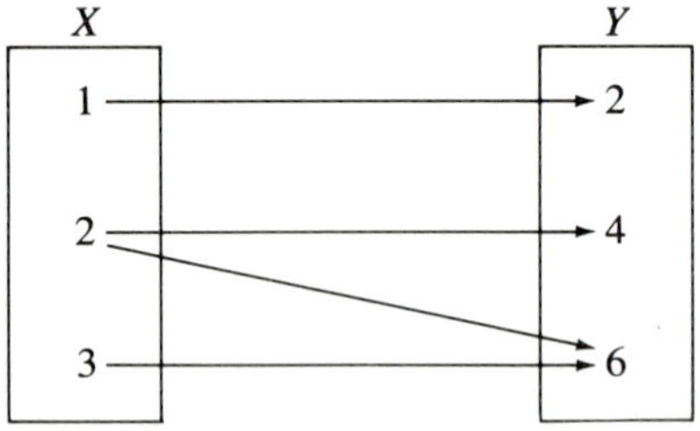

Let f be a function from X to Y. If an element a of X is associated with an element b of Y, then we say that b is the **value of f at a**. We shall denote this value by $f(a)$, read "f at a," and call a, the element in parentheses, the **argument**. So if f denotes the function in Figure 1.28(ii), then

$$f(1) = 2, \quad f(2) = 4, \quad \text{and} \quad f(3) = 4.$$

The set X is called the **domain** of the function, and the set of values of f, $\{f(a) \mid a \in X\}$, is called the **range** of f. Referring to the function of Figure 1.28(ii) again, we see that the domain is $\{1, 2, 3\}$, and the range is $\{2, 4\}$.

Figure 1.30

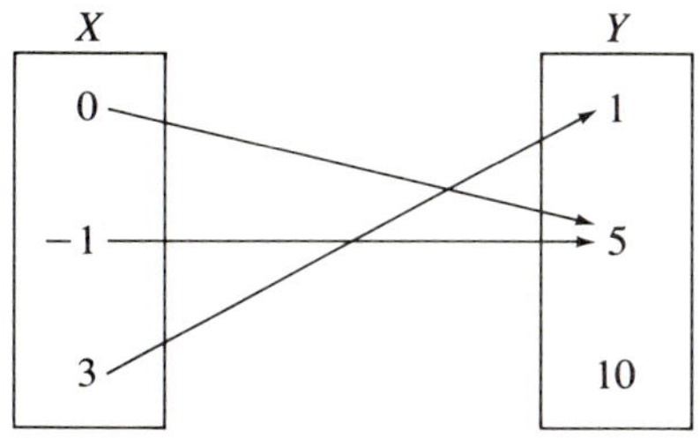

▲ *Exercise* Let f be the function defined in Figure 1.30. Determine each of the following:

(a) $f(0)$ (b) $f(3)$ (c) $f(1)$ (d) domain (e) range

Answer (a) $f(0) = 5$ (b) $f(3) = 1$ (c) f isn't defined at 1.
(d) $\{0, -1, 3\}$ (e) $\{1, 5\}$

Functions are often defined by writing a **formula of definition**. For example, suppose that we want to associate each real number with its square. Let x be a variable with universe (replacement set) R (the set of all real numbers). Hence x is to be associated with x^2. This correspondence is a function. Let's call it f. Thus the value of f at x is x^2. In mathematical terms, we say that

$$f(x) = x^2. \tag{1.2}$$

Equation (1.2) is the formula of definition for f. Because the variable x represents each real number, statement (1.2) says that the value of f at each real number is its square. Thus the function is defined.

To find $f(-2)$, the value of the function at -2, we use the formula of definition and substitute the argument -2 for the variable x. We have

$$f(-2) = (-2)^2 = 4.$$

When a function is defined by a formula, the domain of the function is the universe of the variable. And although the variable chosen is frequently x, it does not have to be. For the previous function our formula of definition could have been

$$f(t) = t^2 \quad \text{or} \quad f(u) = u^2.$$

Each formula describes the same correspondence.

EXAMPLE 1 Let x be a variable with universe R, and let f be defined by $f(x) = 2x + 1$. Determine each of the following:

(a) $f(3)$ (b) $f(-2)$ (c) $f(a)$ (d) $f(x + h)$ (e) $f(x) + f(h)$
(f) $f(x) + h$ (g) $f(f(0))$

SOLUTION (a) To calculate $f(3)$, we simply use the formula of definition and replace x by 3. Thus

$$f(3) = 2 \cdot 3 + 1 = 7.$$

(b) Similarly, $f(-2) = 2(-2) + 1 = -3$
(c) Similarly, $f(a) = 2a + 1$
(d) To find $f(x + h)$ we do the same. That is, for each x in the formula we substitute the argument $x + h$. Hence

$$f(x + h) = 2(x + h) + 1 = 2x + 2h + 1.$$

(e) $f(x) + f(h) = (2x + 1) + (2h + 1) = 2x + 2h + 2$
(f) $f(x) + h = 2x + 1 + h$
(g) We may work from the inside to the outside or vice versa. Working from the inside to the outside, we get

$$f(f(0)) = f(2 \cdot 0 + 1) = f(1) = 2 \cdot 1 + 1 = 3.$$

The same result is obtained by working from the outside to the inside. We have

$$f(f(0)) = 2 \cdot f(0) + 1 = 2(2 \cdot 0 + 1) + 1 = 2 + 1 = 3.$$

▲

Note that for the function of Example 1, $f(x + h) \neq f(x) + f(h)$ and $f(x + h) \neq f(x) + h$.

▲ *Exercise* Let x be a variable with universe R, and let f be defined by $f(x) = 3x - 2$. Determine each of the following:

(a) $f(-3)$ (b) $f(1)$ (c) $f(a)$ (d) $f(x + h)$ (e) $f(x) + f(h)$
(f) $f(x) + h$ (g) $f(f(0))$

Answer (a) -11 (b) 1 (c) $3a - 2$ (d) $3x + 3h - 2$ (e) $3x + 3h - 4$
(f) $3x - 2 + h$ (g) -8

When the domain of a function is not specified explicitly, it is assumed to be the set of all real numbers whose functional values are defined and real. Thus if $h(x) = 1/x$, the domain of h is $R - \{0\}$ since $1/x$ is not defined if x is zero. And if $g(x) = \sqrt{x}$, then the domain of g is $[0, \infty)$ because $\sqrt{x}$ is not real if x is negative. (Refer to Appendix A.3 for a discussion of intervals and their notations.)

▲ *Exercise* Determine the domain of the following functions:

(a) $f(x) = \dfrac{1}{x + 2}$ (b) $g(x) = \sqrt{x - 1}$

Answer (a) $R - \{-2\}$ (b) $[1, \infty)$

In some cases the domain must be specified. For example, suppose that we want to associate each *positive* number with its square. We could write

$$g(x) = x^2, \qquad x > 0,$$

where the specification $x > 0$ tells us that x represents positive numbers only. That is, the domain of g is $(0, \infty)$. So $g(-2)$, for instance, doesn't exist because g isn't defined at -2. The formula may be applied only to those elements in the domain. And although a domain may be any set, all the functions considered in this text have domains that are intervals or unions of intervals.

Consider an equation in x and y such that y is a function of x. We say that the equation *defines* a function f because $y = f(x)$ for some function f. The formula of definition for f may be found by solving the equation for y in terms of x. For example, let f be the function defined by

$$2x + 3y = 5.$$

If we want to write the formula of definition for f, we solve the equation for y in terms of x. Here we get

$$y = \frac{-2}{3}x + \frac{5}{3}.$$

The right-hand side of the equation will be the right-hand side of our formula if we use the variable x. Thus

$$f(x) = \frac{-2}{3}x + \frac{5}{3}.$$

▲ ***Exercise*** For each of the following, y is a function of x. Let f be the function defined and determine $f(x)$.

(a) $3x + y = 2$ (b) $x^2 - 2y + 3x = 5$ (c) $\dfrac{y-2}{3} = x^2$

Answer (a) $f(x) = 2 - 3x$ (b) $f(x) = \dfrac{x^2 + 3x - 5}{2}$ (c) $f(x) = 3x^2 + 2$

The **graph of a function** f whose domain and range are subsets of R is the set of all points $(x, f(x))$ in the xy-plane such that x is in the domain of f. With x restricted to this domain, this is exactly the same as the graph of the equation $y = f(x)$, in which y is a function of x. Thus no vertical line intersects the graph of a function in more than one point.

EXAMPLE 2 Graph $f(x) = x^2$.

SOLUTION We will construct a table that lists some of the points by selecting some values for x and computing $f(x)$. We get

x	-3	-2	-1	0	1	2	3
$f(x)$	9	4	1	0	1	4	9

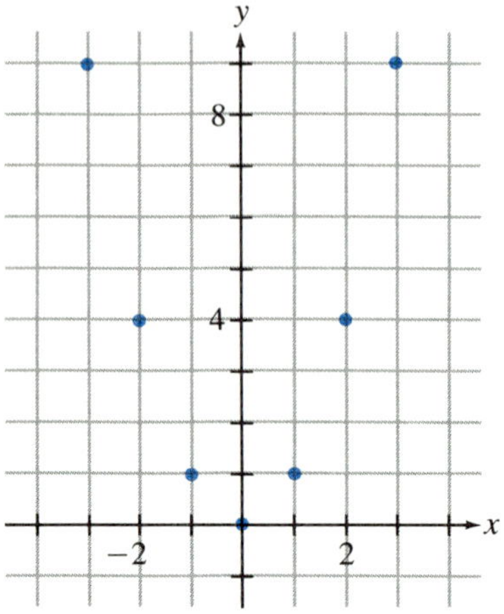

Figure 1.31

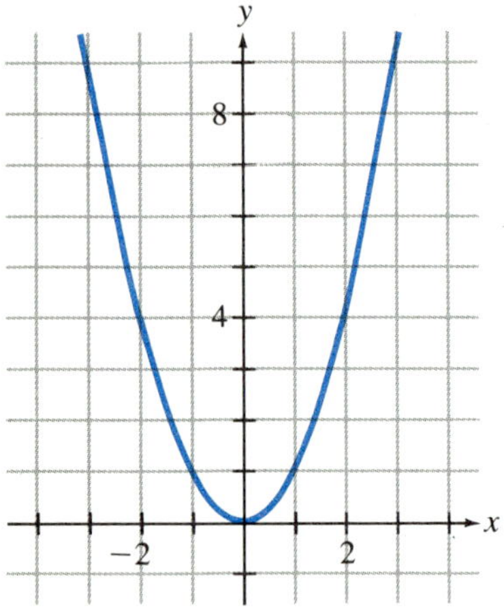

Figure 1.32
$f(x) = x^2$

The graph of these ordered pairs is shown in Figure 1.31.

Connecting these with a curve that follows the general trend of the plotted points gives us the graph. See Figure 1.32.

▲

▲ *Exercise* Graph $f(x) = x^3$ after plotting points with x-values of $-2, -1, 0,$ 1, and 2.

Answer See Figure 1.33.

Figure 1.33

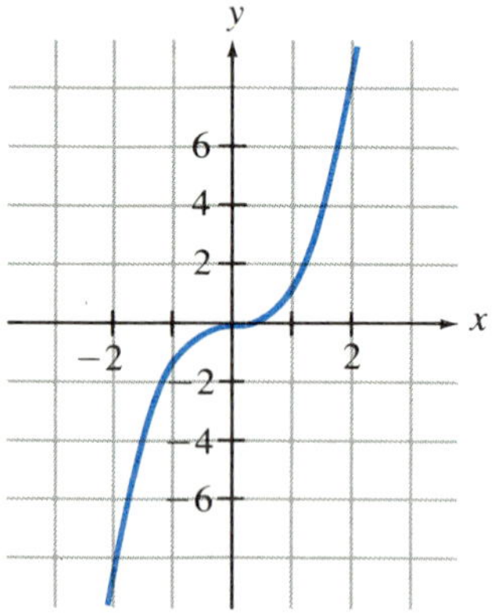

Problem Set 1.4

1. Let f be the function defined in the following figure. Determine each of the following:

(a) $f(-1)$ (b) $f(-2)$ (c) $f(2)$

(d) domain (e) range

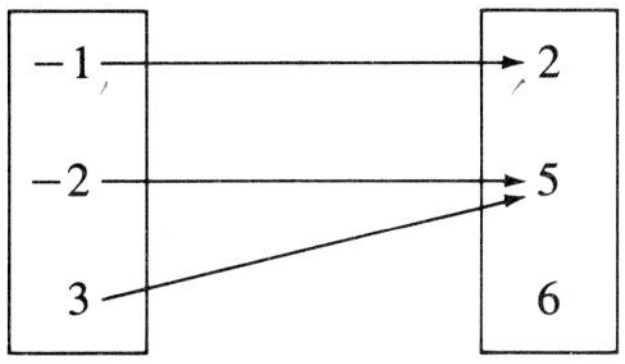

2. Let f be the function defined in the following figure. Determine each of the following:

(a) $f(-3)$ (b) $f(2)$ (c) $f(-1)$
(d) domain (e) range

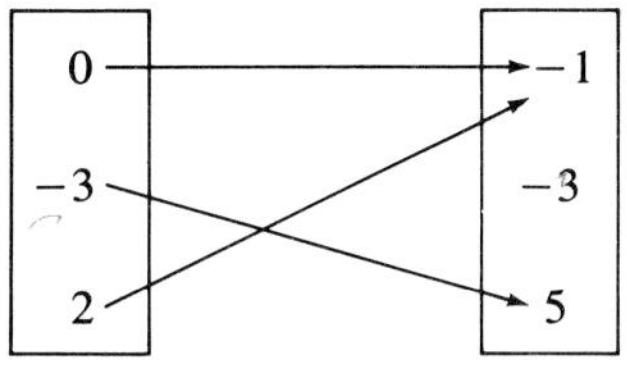

In Problems 3–10, determine the domain of the function f.

3. $f(x) = |x|$

4. $f(x) = -x^3$

5. $f(x) = \dfrac{1}{x^2}$

6. $f(x) = \dfrac{2}{x-3}$

7. $f(x) = -\sqrt{x}$

8. $f(x) = \sqrt{-x}$

9. $f(x) = \sqrt[3]{2x+3}$

10. $f(x) = \sqrt{2x+3}$

11. Let $f(x) = 2x^2 - x + 3$. Determine each of the following:

(a) $f(0)$ (b) $f(a)$ (c) $f(\sqrt{2})$
(d) $f(1)$ (e) $f(f(1))$ (f) $\dfrac{f(t+h) - f(t)}{h}$

12. Let $g(x) = \dfrac{1}{x^2}$. Determine each of the following:

(a) $g(a)$ (b) $\dfrac{1}{g(a)}$ (c) $g\left(\dfrac{1}{a}\right)$
(d) $g(\sqrt{a})$ (e) $\sqrt{g(a)}$
(f) $g(2)$ (g) $g(g(2))$

13. Let $f(x) = x^2 + 2x - 1$. Determine each of the following:

(a) $\dfrac{f(x+h) - f(x)}{h}$ (b) $f(x^2)$ (c) $f(f(x))$

14. Let $f(x) = 5$. Determine each of the following:

(a) $f(0)$ (b) $f(-3)$ (c) $f(5)$
(d) $f(x^2)$ (e) $f(f(x))$

In Problems 15–18, let f be the function defined by the given equation and find $f(x)$.

15. $2x - y + 5 = 0$

16. $x - 3y + 2 = 0$

17. $2x^2 + x - 2y = 7$

18. $\dfrac{y-1}{2} = x^2$

In Problems 19–22, sketch the graph of the function after plotting points with the given x-values.

19. $f(x) = -2x + 1;\ x = -2, -1, 0, 1, 2$

20. $f(x) = 3x - 2;\ x = -2, -1, 0, 1, 2$

21. $f(x) = -x^3;\ x = -2, -1, 0, 1, 2$

22. $f(x) = -x^4;\ x = -2, -1, 0, 1, 2$

23. Let f be the function graphed in the following figure.

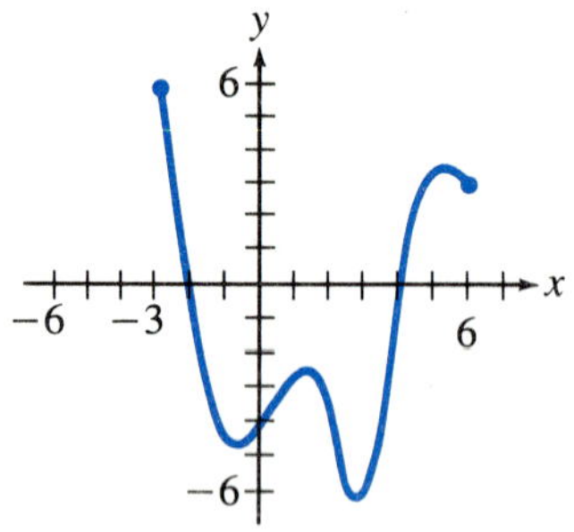

(a) What is the domain of f?

(b) What is the range of f?

(c) $f(0) = ?$

(d) $f(5) = ?$

(e) If $f(a) = 0$, then $a = ?$

24. Let $f(x) = (x - 1, x^2)$. Determine each of the following:

(a) $f(0)$ (b) $f(2)$ (c) $f(-3)$

1.5 SPECIAL FUNCTIONS AND INVERSES

A function f with the property that

$$f(-x) = f(x)$$

for *all* x in its domain is called an **even function**. For example, $f(x) = x^2$ is even because

$$f(-x) = (-x)^2 = x^2 = f(x).$$

A function f is called **odd** provided that

$$f(-x) = -f(x)$$

for *all* x in its domain. For example, $f(x) = x^3$ is odd because

$$f(-x) = (-x)^3 = -x^3 = -f(x).$$

Even and odd functions may be recognized by their graphs.

Consider an even function. For each point (a, b) on the graph, the point $(-a, b)$ is on the graph. This means that for each point P on the graph (excluding any point on the y-axis), the point Q that has the property that $\overline{PQ}$ is perpendicular to and bisected by the y-axis lies on the graph. Such a graph is said to be **symmetric about the *y*-axis**. See Figure 1.34. Note that if we folded Figure 1.34 along the y-axis, the two parts of the graph would coincide.

Figure 1.34
Symmetry about the y-axis

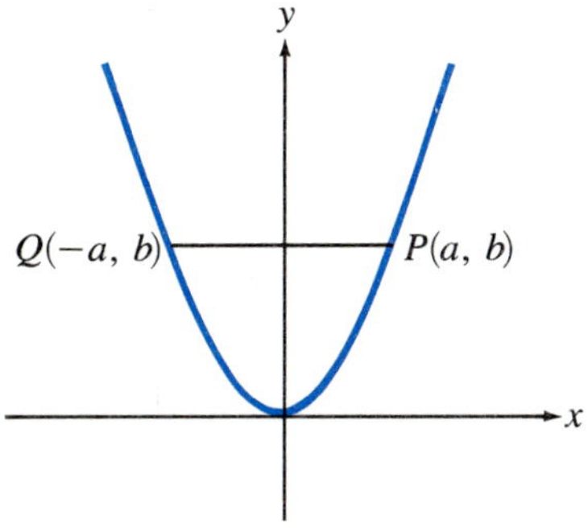

Now consider an odd function. For each point (a, b) on the graph, the point $(-a, -b)$ is on the graph. This means that for each point P on the graph (excluding the origin), the point Q that has the property that $\overline{PQ}$ is bisected by the origin is also on the graph. Such a graph is said to be **symmetric about the origin**. See Figure 1.35. If we folded Figure 1.35 along the y-axis and then along the x-axis (or vice versa), the two parts of the graph would coincide.

Figure 1.35
Symmetry about the origin

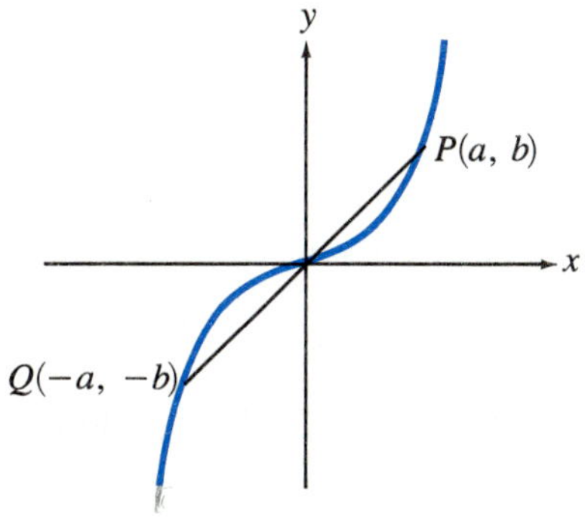

So the relationship between even/odd and symmetry is as follows:

1. A function is even if and only if its graph is symmetric about the y-axis.
2. A function is odd if and only if its graph is symmetric about the origin.

▲ ***Exercise*** Determine whether the function graphed (Figure 1.36(a)–(d)) is even, odd, or neither.

Figure 1.36

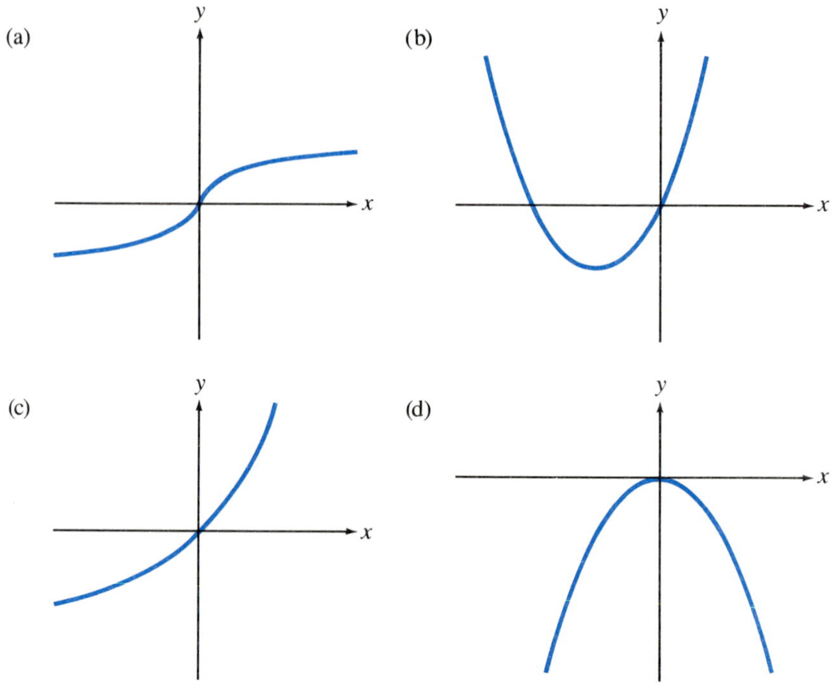

Answer (a) odd (b) neither (c) neither (d) even

A function f is called **one-to-one** provided that

$$f(a) = f(b) \quad \text{implies} \quad a = b$$

for any a and b in the domain.

EXAMPLE 1 Determine whether the function is one-to-one:

(a) $f(x) = 2x - 1$ (b) $g(x) = x^2$

SOLUTION (a) Let a and b be any two elements in the domain of f such that

$$f(a) = f(b).$$

Can we conclude that $a = b$? We have

$$2a - 1 = 2b - 1$$
$$2a = 2b$$
$$a = b.$$

Thus $f(a) = f(b)$ implies that $a = b$, and f is one-to-one.

(b) Let a and b be elements of the domain such that

$$g(a) = g(b).$$

We have

$$a^2 = b^2.$$

So must $a = b$? No. We can have $a = 2$ and $b = -2$ and still have

$$a^2 = b^2$$
$$g(a) = g(b).$$

Thus, $g(a) = g(b)$ does *not* imply that $a = b$, so g is *not* one-to-one.

▲

▲ ***Exercise*** Determine whether the function is one-to-one:

(a) $f(x) = \dfrac{1}{x^2}$ (b) $g(x) = 3x + 2$

Answer (a) not one-to-one (b) one-to-one

A function associates each element of the domain with a unique element of the range. But an element in the range may be paired with more than one element of the domain. However, *for one-to-one functions,* each element of the range corresponds to exactly one element of the domain. Thus the function depicted in Figure 1.37(i) is *not* one-to-one, while the function in Figure 1.37(ii) *is* one-to-one.

Figure 1.37

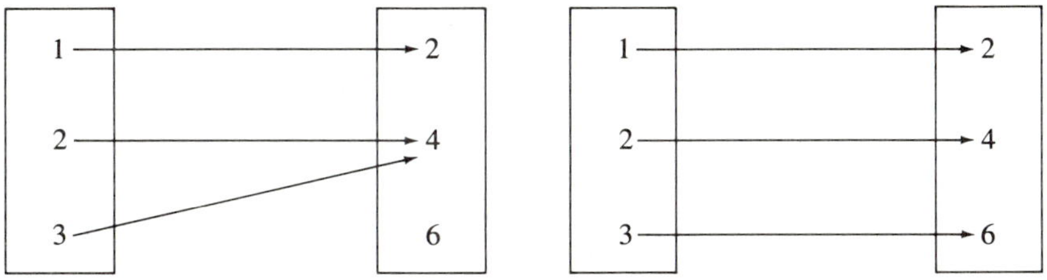

(i) not one-to-one (ii) one-to-one

Examination of the graph of a function will indicate whether it is one-to-one. A function is one-to-one provided that no horizontal line intersects the graph in more than one point.

▲ ***Exercise*** Determine whether the function graphed is one-to-one (Figure 1.38(a)–(d)).

Figure 1.38

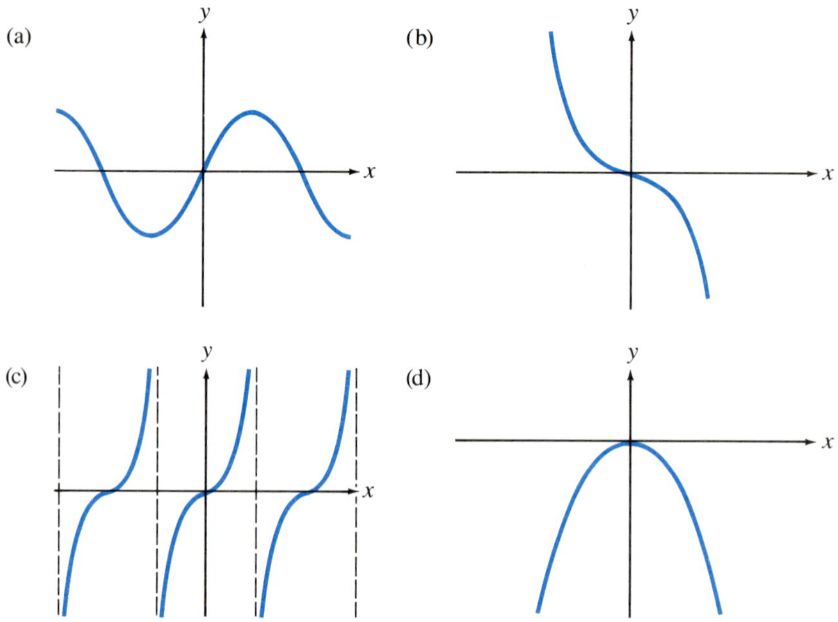

Answer Only (b) is one-to-one.

Given an equation in x and y, its **inverse** is the equation obtained by switching the variables. For example, given

$$x + y = 3, \quad y = |x|, \quad \text{and} \quad x = y^3,$$

the respective inverses are

$$y + x = 3, \quad x = |y|, \quad \text{and} \quad y = x^3.$$

Note that although y is a function of x in each of the original three equations, this is not the case for all of the inverses. Indeed, for the inverse $x = |y|$, y is *not* a function of x.

▲ *Exercise* Find the inverse and specify whether the inverse defines a function:

(a) $y = x^2 + 7$ (b) $y = 2x^3 - 1$

Answer (a) $x = y^2 + 7$; no (b) $x = 2y^3 - 1$; yes

If $y = f(x)$, where f is a one-to-one function, then the inverse $x = f(y)$ also defines a function. Denoted by f^{-1}, this **inverse function** is a function from the range of f to the domain of f. See Figure 1.39.

Figure 1.39

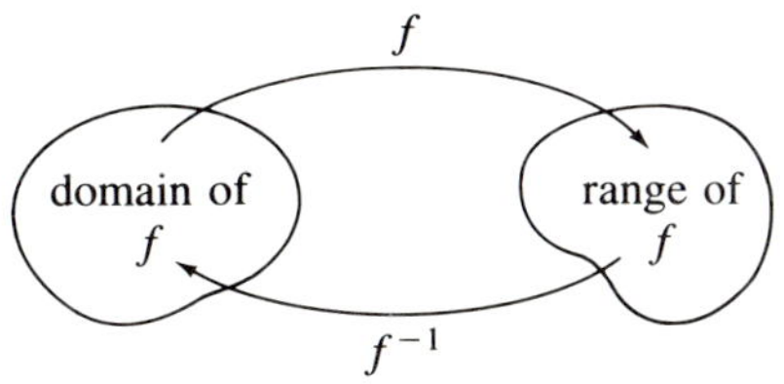

The one-to-one function of Figure 1.37(ii) and its inverse are shown in Figure 1.40.

Figure 1.40

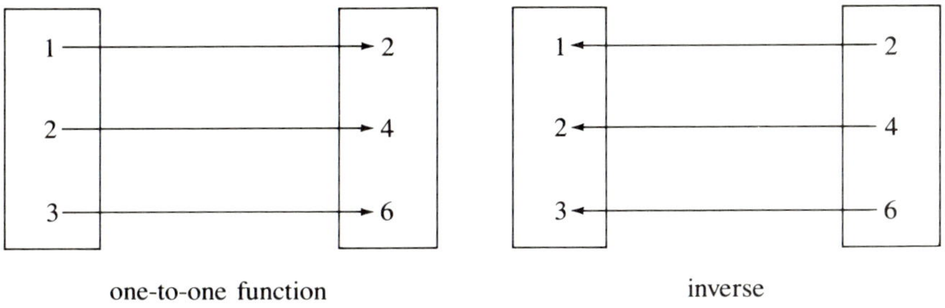

So the inverse function is just the reverse of the given one-to-one function. Furthermore, $f^{-1}(b) = a$ if and only if $f(a) = b$.

There is an interesting relationship between the graph of a function and the graph of its inverse. Let's graph $y = \sqrt{x}$ and its inverse $x = \sqrt{y}$ on the same coordinate system, with the same unit distance on each axis. See Figure 1.41.

Figure 1.41

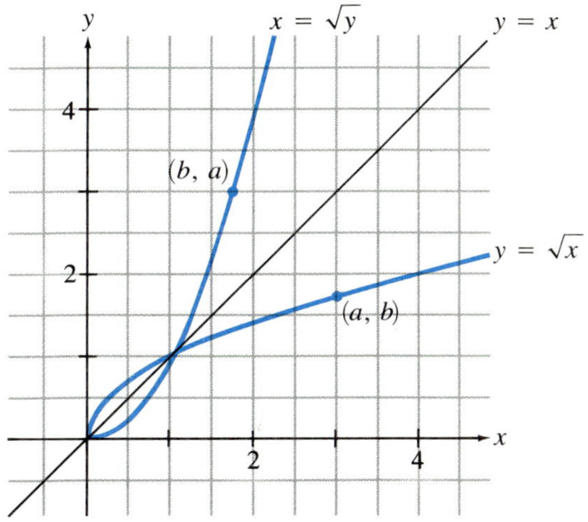

The point (a, b) is on the graph of $y = \sqrt{x}$ if and only if (b, a) is on the graph of $x = \sqrt{y}$. We see that the points (1, 1), (4, 2) and $(2, \sqrt{2})$ are on the graph of $y = \sqrt{x}$, while (1, 1), (2, 4), and $(\sqrt{2}, 2)$ are on the graph of $x = \sqrt{y}$.

Consequently, if Figure 1.41 were folded along the line $y = x$, the graphs would coincide. In this sense, we say that each graph is the reflection of the other through the line $y = x$.

Problem Set 1.5

In Problems 1–4, determine whether the function is even, odd, or neither.

1. $f(x) = -x^4$

2. $f(x) = (x - 1)^2$

3. $f(x) = \sqrt[3]{x}$

4. $f(x) = 2^x$

In Problems 5–8, determine whether the function graphed is even, odd, or neither.

5.

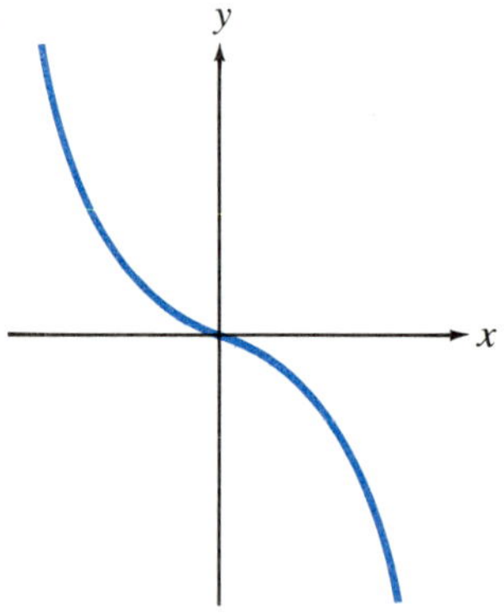

6.

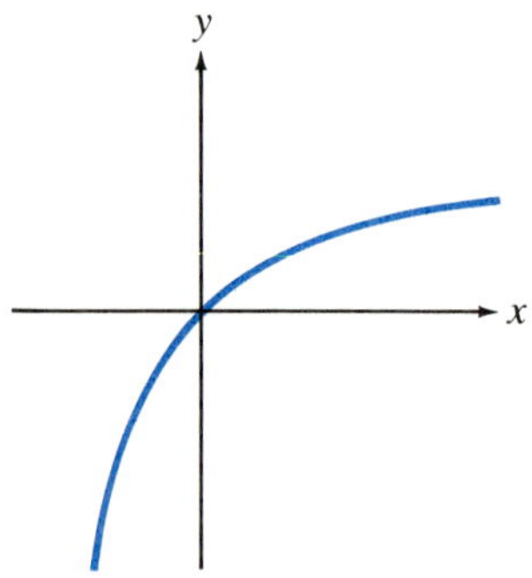

7.

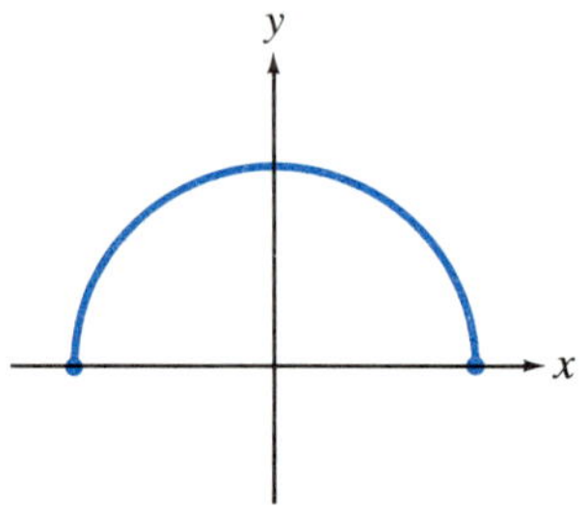

8.

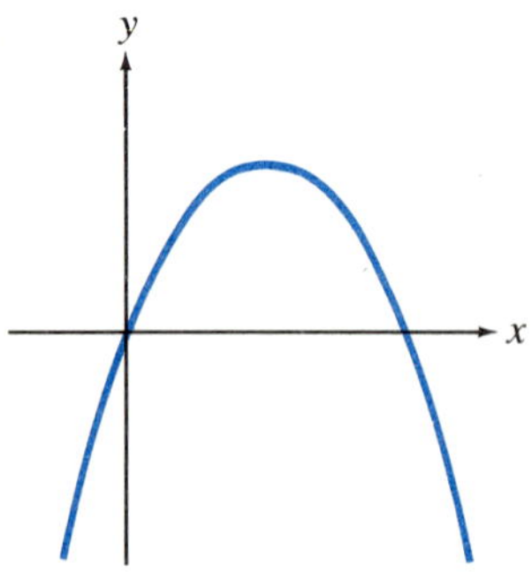

In Problems 9–12, determine whether the function is one-to-one.

9. $f(x) = 5 - 2x$

10. $f(x) = -3$

11. $f(x) = x^4$

12. $f(x) = \dfrac{1}{x}$

In Problems 13–16, determine whether the function graphed is one-to-one.

13.

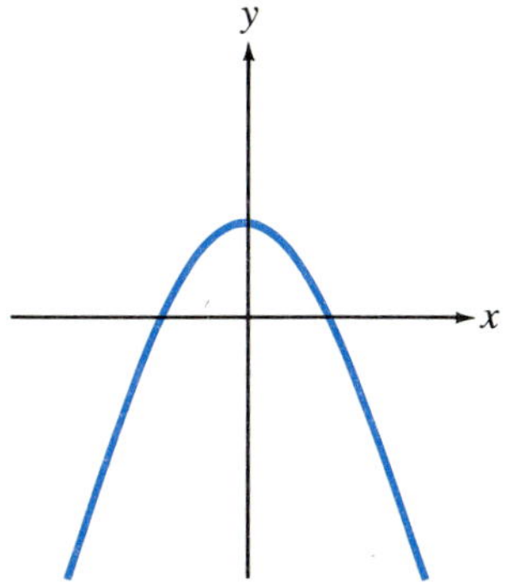

14.

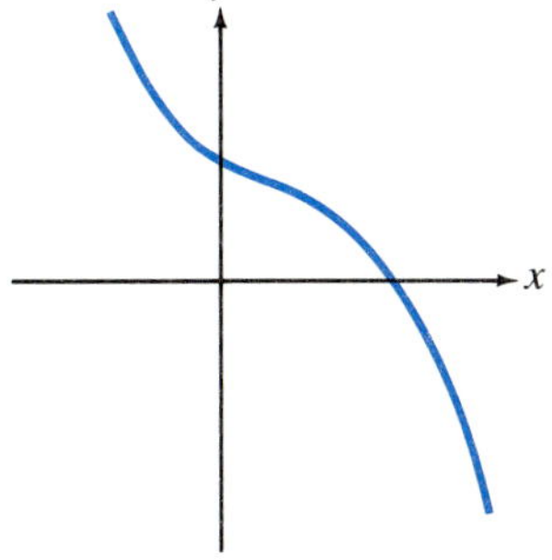

15.

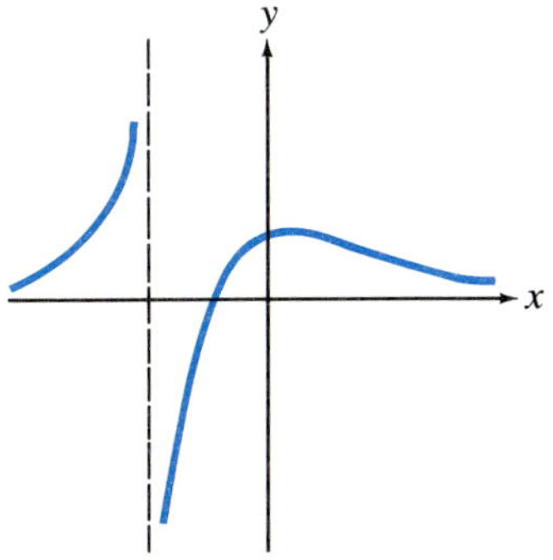

16.

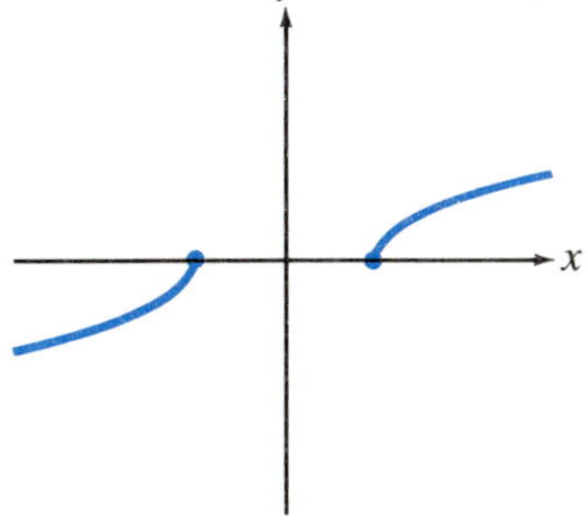

In Problems 17 and 18, restrict the domain of the function so that the restricted function is one-to-one but has the same range.

17. $f(x) = x^2 + 2$

18. $f(x) = (x - 1)^2$

In Problems 19–22, find the inverse and specify whether the inverse defines a function.

19. $y = x^2 - 5$

20. $y = 2x - 3$

21. $y = x^3 - 1$

22. $y = -x^4$

In Problems 23 and 24, determine f^{-1} for the one-to-one function.

23. $f(x) = 3x - 5$

24. $f(x) = x^3 + 1$

25. Let f be the function defined in the figure. Find

(a) $f^{-1}(2)$

(b) $f^{-1}(f(3))$

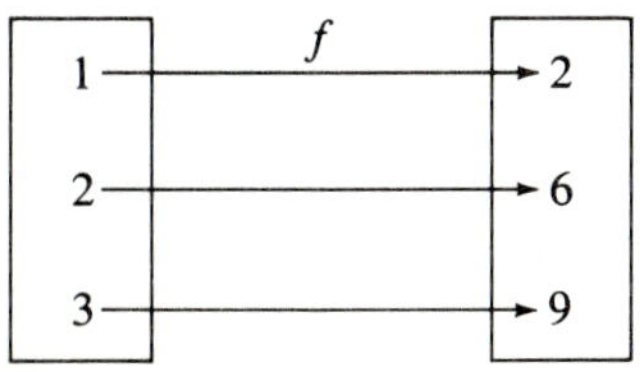

26. Let f be a one-to-one function. Show each of the following:
   (a) $f^{-1}(f(x)) = x$ for all x in the domain of f
   (b) $f(f^{-1}(x)) = x$ for all x in the domain of f^{-1}

In Problems 27 and 28, sketch the inverse of the function graphed.

27.

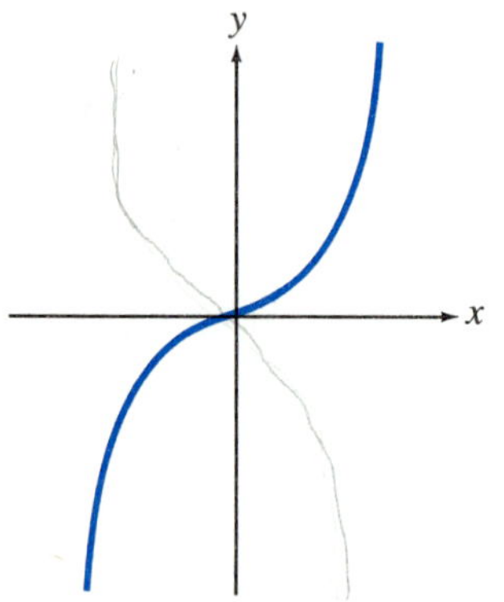

28.

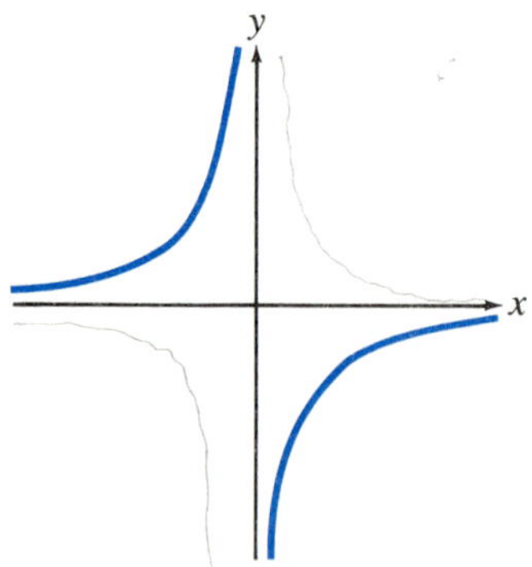

CHAPTER 1 REVIEW

Key Terms and Formulas

acute angle, page 3
angle, page 2
Cartesian coordinate system, page 14
coordinate plane, page 15
degrees, page 2
distance formula, page 17
$P_1P_2 = \sqrt{(x_2 - x_1)^2 + (y_2 - y_1)^2}$
domain, page 26
even function, page 32
formula of definition, page 27
function, page 26
graph of a function, page 29
graph of an equation, page 19
inverse function, page 36
obtuse angle, page 3
odd function, page 33
one-to-one function, page 34
Pythagorean Theorem, page 7
$a^2 + b^2 = c^2$
quadrant, page 15
range, page 26
right angle, page 3
right triangle, page 7
similar triangles, page 4
Similarity Theorems, page 6
solution, page 19
symmetry about the origin, page 33
symmetry about the y-axis, page 33
y is a function of x, page 22

Problem Set

1. Specify "true" or "false."

(a) An angle of measure 120° is called acute.

(b) A measure of 25°30′ is the same as 25.5°.

(c) If a triangle has angles of 25° and 55°, then the third angle must be obtuse.

(d) If there exists a correspondence between the vertices of two triangles such that two sides of one triangle are proportional to the corresponding sides of the other triangle, and the measures of one pair of corresponding angles are equal, then the triangles must be similar.

(e) If $x < 0$, then the point $(x^2, -x)$ is in quadrant IV.

(f) If $P_1(x_1, y_1)$ and $P_2(x_2, y_2)$ determine a vertical line, then $P_1P_2 = |x_2 - x_1|$.

(g) The area of a circle is a function of its radius.

(h) A graph is symmetric about the y-axis provided that for each point (a, b) on the graph the point $(-a, -b)$ is on the graph.

(i) If f is a one-to-one function and $f(1) = 3$, then $f^{-1}(3) = 1$.

(j) If $(2, 5)$ is on the graph of f^{-1}, then $(5, 2)$ is on the graph of f.

(k) $(-2, 2)$ is a solution of $x = |y|$.

(l) If a graph is symmetric about the origin, then it is symmetric about the y-axis.

(m) A function cannot equal its own inverse.

In Problems 2 and 3, $\triangle ABC$ is similar to $\triangle DEF$.

2. Given $AB = 3$, $BC = 7$, and $DE = 5$, find EF.

3. Given $AC = 6$, $BC = 4$, and $EF = 5$, find DF.

4. Find the height (to two significant digits) of a tree that casts a 35-foot shadow if a 5.0-foot vertical pole casts a 2.0-foot shadow at the same time. (Assume that both the tree and the pole are perpendicular to level ground.)

In Problems 5 and 6, *a*, *b*, and *c* refer to the following triangle.

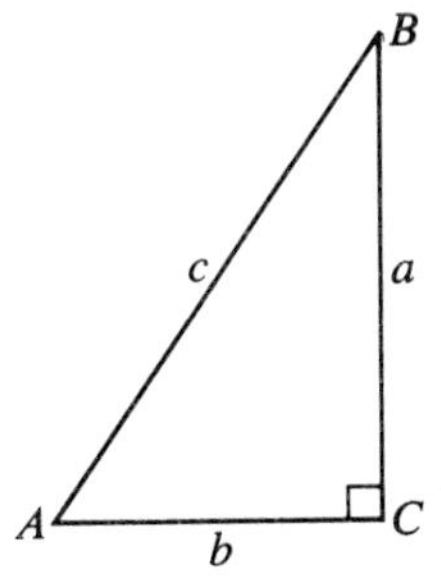

5. Given $a = 4$ and $b = 2$, find c.

6. Given $a = 3$ and $c = \sqrt{15}$, find b.

7. Suppose plane A has flown 5.2 miles north of Lexington and is 8.1 miles from plane B, which is flying east of Lexington. How far (to two significant digits) is plane B from Lexington?

8. In the following triangle, $AC = 5$ and $BC = 10$. Find CD.

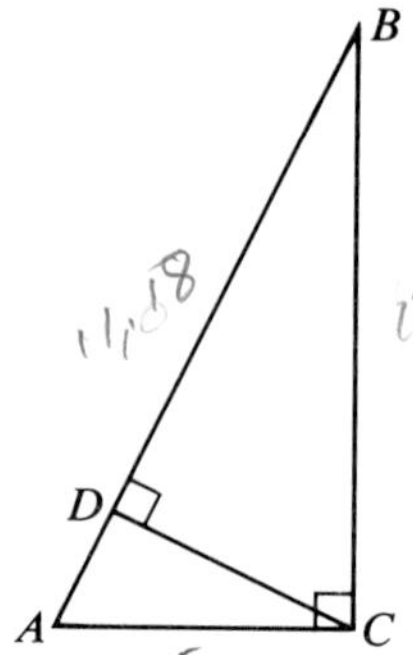

In Problems 9 and 10, *a*, *b*, and *c* refer to the following triangle.

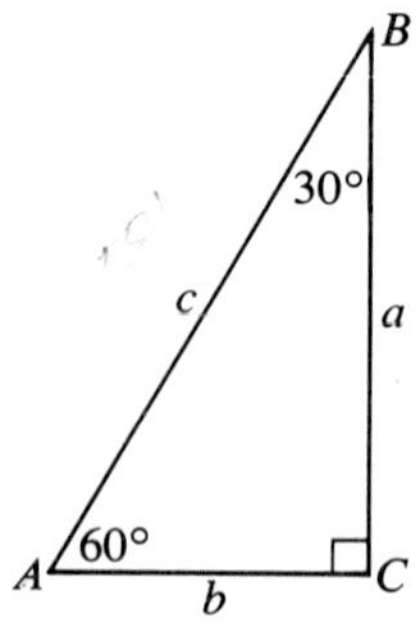

9. Given $c = 14$, find a and b, exactly.

10. Given $a = \sqrt{6}$, find b and c, exactly.

11. What is the exact length of each leg of a 45°-45°-90° triangle if the hypotenuse has length 5?

In Problems 12–14, determine the distance between the given points.

12. $(-2, 1), (4, 5)$

13. $\left(-3, \frac{5}{2}\right), \left(\frac{-4}{3}, 1\right)$

14. $(\sqrt{5}, -2), (3\sqrt{5}, 1)$

In Problems 15 and 16, sketch the graph of the equation after plotting points with the given values.

15. $y = x - 3$, x-values: $-2, 0, 2, 4, 6$

16. $x = 1 - y^2$, y-values: $-2, -1, 0, 1, 2$

In Problems 17–20, determine whether y is a function of x.

17. $x^2 + y^2 = 5$

18. $y = -x^2$

19. $y = |x - 2|$

20. $2x^2 + y^2 = 10$

In Problems 21–24, determine whether y is a function of x for the equation graphed.

21.

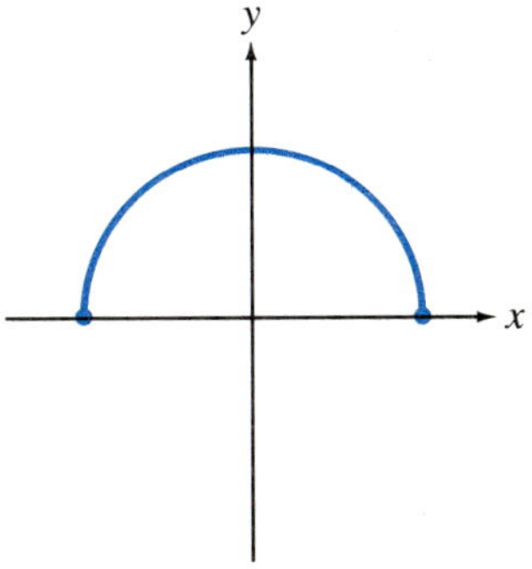

22.

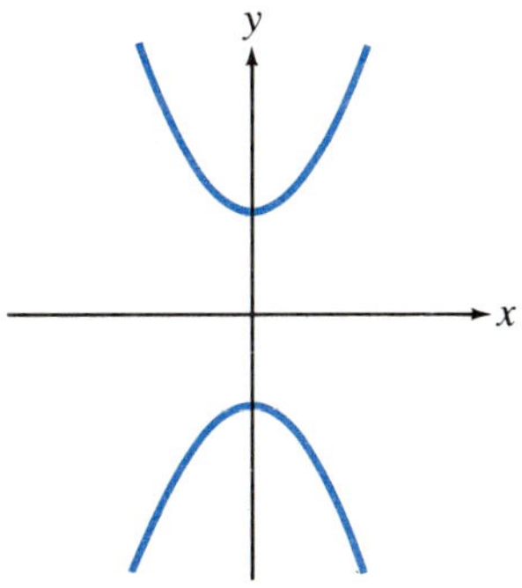

23.

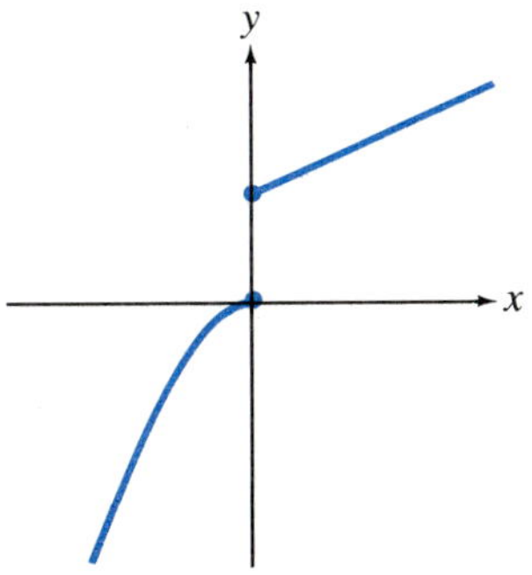

24.

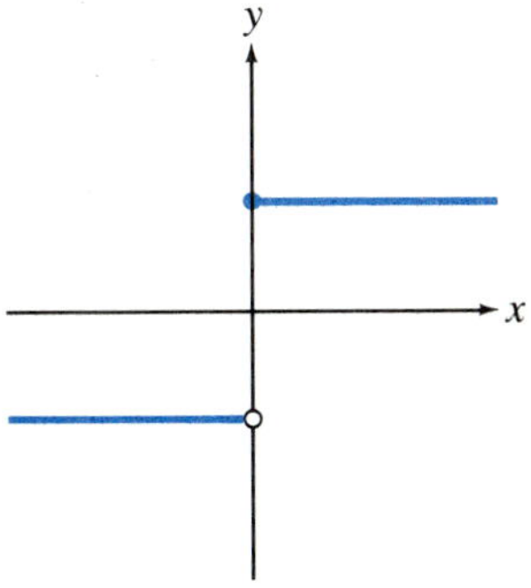

In Problems 25–28, determine the domain of the function f.

25. $f(x) = \dfrac{1}{x - 1}$

26. $f(x) = \sqrt{|2x - 3|}$

27. $f(x) = \sqrt[3]{\dfrac{x + 3}{3}}$

28. $f(x) = \sqrt{3x + 6}$

29. Let $f(x) = \dfrac{1}{\sqrt{x}}$. Determine each of the following:

(a) $f(1)$ (b) $f(4)$ (c) $f(a^2)$ (d) $f(f(16))$

30. Let $f(x) = x^2 - 2x + 5$. Determine each of the following:

(a) $\dfrac{f(x+h) - f(x)}{h}$ (b) $f(x^2)$ (c) $f(f(x))$

In Problems 31 and 32, let f be the function defined by the given equation and find $f(x)$.

31. $5x - 2y + 3 = 0$

32. $\dfrac{2y - 3}{5} = x^2$

In Problems 33 and 34, sketch the graph of the function after plotting points with the given x-values.

33. $f(x) = 2 - 3x$; $x = -2, 0, 2, 4$

34. $f(x) = 1 - x^2$; $x = -2, -1, 0, 1, 2$

35. Let f be the function defined in the following figure. Determine each of the following:

(a) $f(-1)$ (b) $f(2)$ (c) $f(1)$
(d) domain (e) range

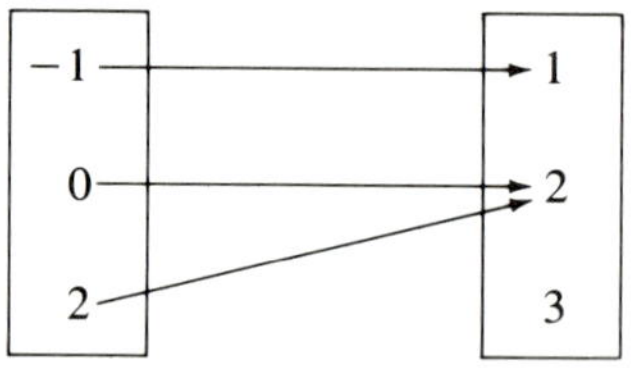

36. Let f be the function graphed in the following figure.

(a) What is the domain of f? (b) What is the range of f?
(c) $f(1) = ?$ (d) $f(0) = ?$

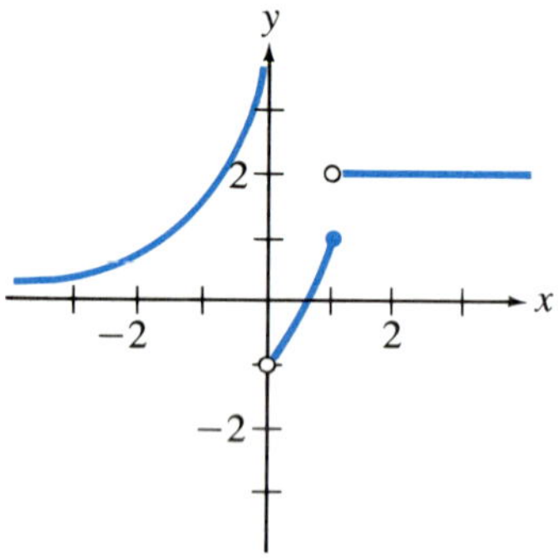

In Problems 37–39, determine whether the function is even, odd, or neither.

37. $f(x) = 2 - x^2$ *38.* $f(x) = \dfrac{1}{x}$ *39.* $f(x) = 5^x$

In Problems 40 and 41, determine whether the function graphed is even, odd, or neither.

40.

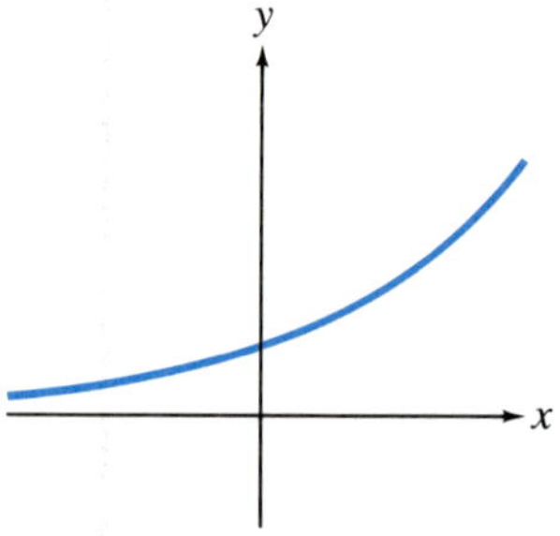

41.

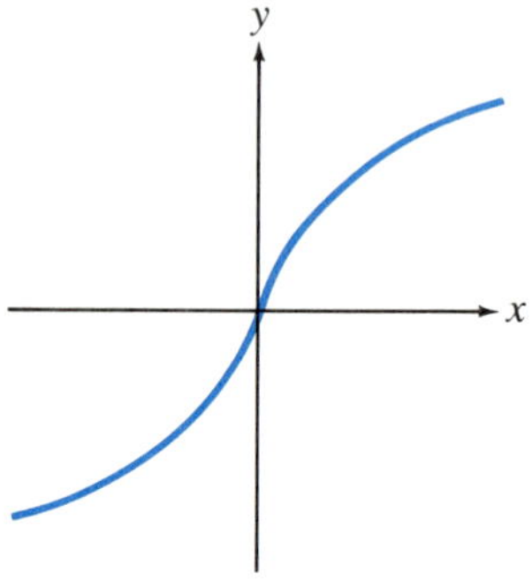

In Problems 42–44, determine whether the function is one-to-one.

42. $f(x) = \dfrac{1}{x^2}$ *43.* $f(x) = -(x-3)^3$ *44.* $f(x) = 5 - 3x$

In Problems 45 and 46, determine whether the function graphed is one-to-one.

45.

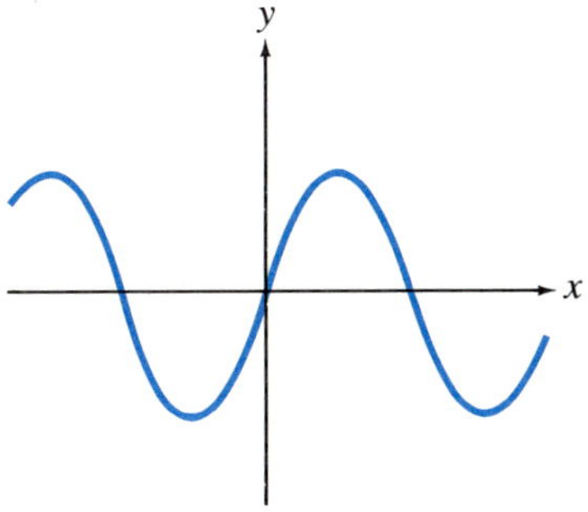

46.

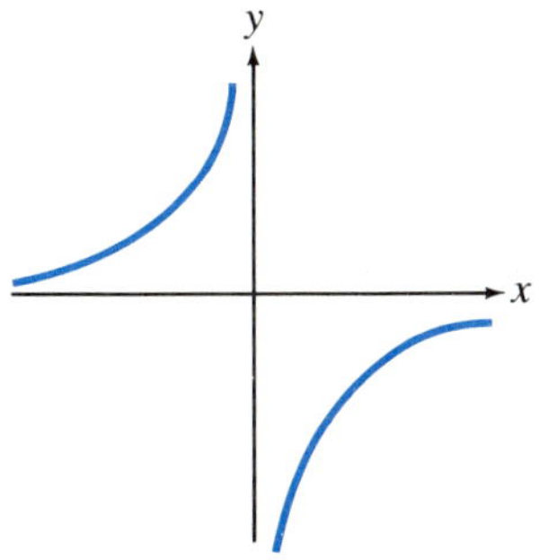

In Problems 47 and 48, find the inverse and specify whether the inverse defines a function.

47. $y = 3 - x^2$ *48.* $y = x^3 + 2$

In Problems 49 and 50, determine f^{-1} for the one-to-one function.

49. $f(x) = 2x - 6$ *50.* $f(x) = x^{1/3} - 1$

2
RIGHT TRIANGLE TRIGONOMETRY

CHAPTER OUTLINE

Trigonometry began as a computational branch of geometry. Its origin is the study of the relationships between the sides and angles of triangles. Literally "trigonometry" means triangle measure. This chapter introduces us to the basic definitions and applications of trigonometry within the framework of right triangles.

2.1

TRIGONOMETRIC RATIOS

We begin our study of trigonometry with the right triangle. If triangle ABC is a right triangle, with $\angle C$ the right angle, then $\overline{AC}$ and $\overline{BC}$, the legs of the triangle, are called, respectively, the **side adjacent** to $\angle A$ and the **side opposite** $\angle A$. Of course, $\overline{AB}$ is the hypotenuse. See Figure 2.1.

Figure 2.1

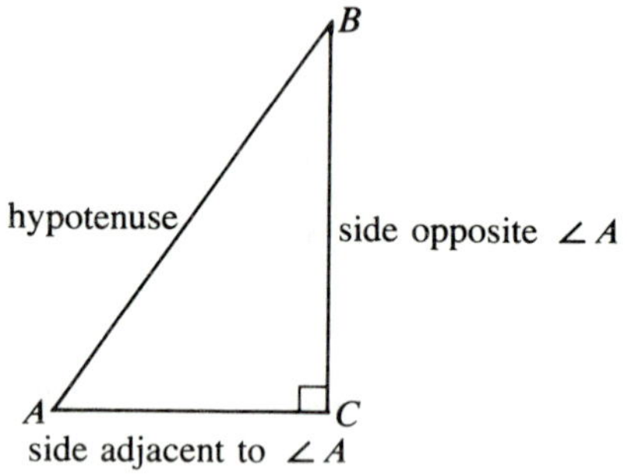

Using the three sides, we find there are six possible ratios, but we shall initially define only three. Letting α (alpha) be the measure of $\angle A$, we define the following **principal trigonometric ratios**.

Name	*Abbreviation*	*Definition*
sine of $\angle A$	$= \sin \alpha$	$= \dfrac{\text{opposite}}{\text{hypotenuse}}$
cosine of $\angle A$	$= \cos \alpha$	$= \dfrac{\text{adjacent}}{\text{hypotenuse}}$
tangent of $\angle A$	$= \tan \alpha$	$= \dfrac{\text{opposite}}{\text{adjacent}}$

Let's consider the right triangle of Figure 2.2 and calculate these three trigonometric ratios for $\angle A$.

Figure 2.2

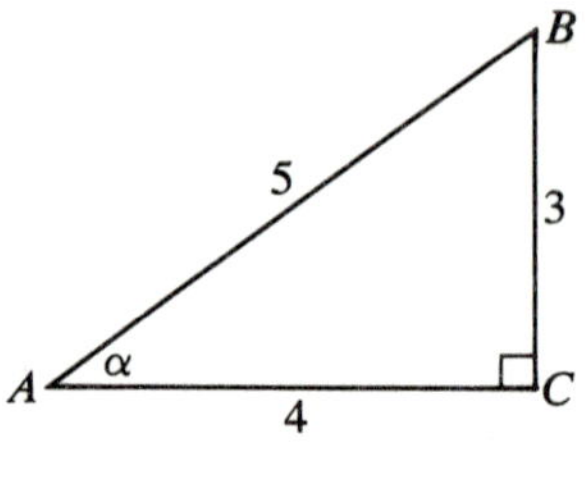

Using the definitions, we have

$$\sin \alpha = \frac{\text{opposite}}{\text{hypotenuse}} = \frac{3}{5},$$

$$\cos \alpha = \frac{\text{adjacent}}{\text{hypotenuse}} = \frac{4}{5},$$

$$\tan \alpha = \frac{\text{opposite}}{\text{adjacent}} = \frac{3}{4}.$$

▲ ***Exercise*** Calculate the three principal trigonometric ratios for $\angle A$ in Figure 2.3.

Figure 2.3

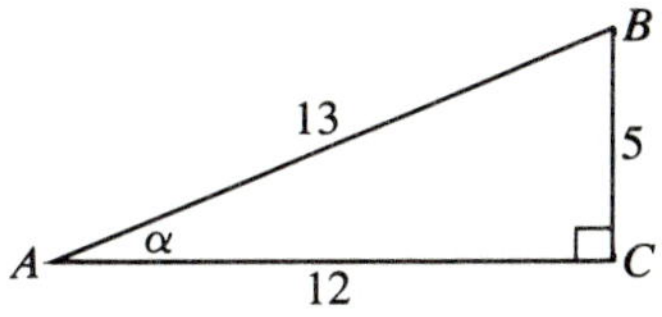

Answer $\sin \alpha = \frac{5}{13}$, $\cos \alpha = \frac{12}{13}$, $\tan \alpha = \frac{5}{12}$

The values of the trigonometric ratios for a given acute angle are independent of the *particular* right triangle in which the angle may exist. That is, given two right triangles ABC and $A'B'C'$ in which angles A and A' are equal (have the same measure α), the ratios for $\angle A$ will be identical to those for $\angle A'$. With respect to Figure 2.4,

$$\frac{BC}{AB} = \frac{B'C'}{A'B'},$$

$$\frac{AC}{AB} = \frac{A'C'}{A'B'},$$

$$\frac{BC}{AC} = \frac{B'C'}{A'C'}.$$

Figure 2.4

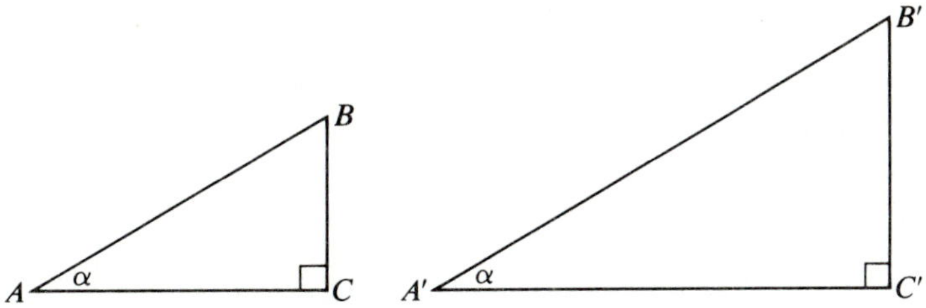

We obtain this result because the two triangles are similar (by the Angle-Angle Similarity Theorem), and thus the ratios of corresponding sides are equal. Hence the values of the trigonometric ratios depend only on the measure of the angle.

Also note that because each leg of a right triangle is shorter than the hypotenuse,

$$\sin \alpha < 1 \quad \text{and} \quad \cos \alpha < 1 .$$

The tangent ratio, however, may assume any positive value.

EXAMPLE 1 Find cos 45° and tan 45°.

SOLUTION We draw a 45°-45°-90° triangle and label the angles and sides, as seen in Figure 2.5.

Figure 2.5

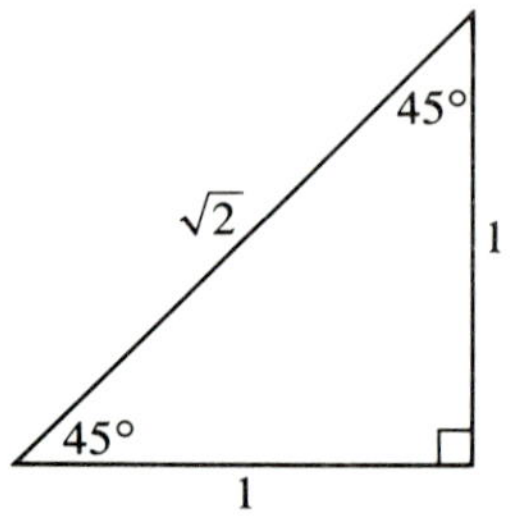

From the definitions we have

$$\cos 45^\circ = \frac{\text{adjacent}}{\text{hypotenuse}} = \frac{1}{\sqrt{2}},$$

$$\tan 45^\circ = \frac{\text{opposite}}{\text{adjacent}} = \frac{1}{1} = 1 .$$

▲

If we choose to rationalize the denominator in computing cos 45°, we have

$$\cos 45^\circ = \frac{1}{\sqrt{2}} \cdot \frac{\sqrt{2}}{\sqrt{2}} = \frac{\sqrt{2}}{2} .$$

Although we defined the ratios using $\angle A$, they may be applied to the other acute angle, $\angle B$. (The ratios are not valid for the right angle, however.) See Figure 2.6.

Figure 2.6

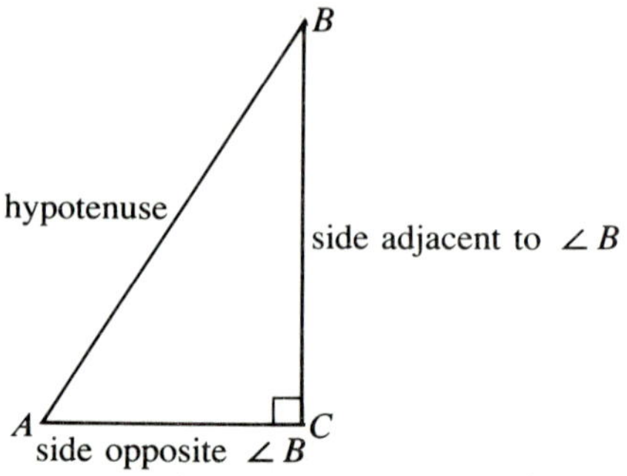

EXAMPLE 2 Find sin 30° and tan 60°.

SOLUTION We first draw a 30°-60°-90° triangle and label the angles and sides, as shown in Figure 2.7.

Figure 2.7

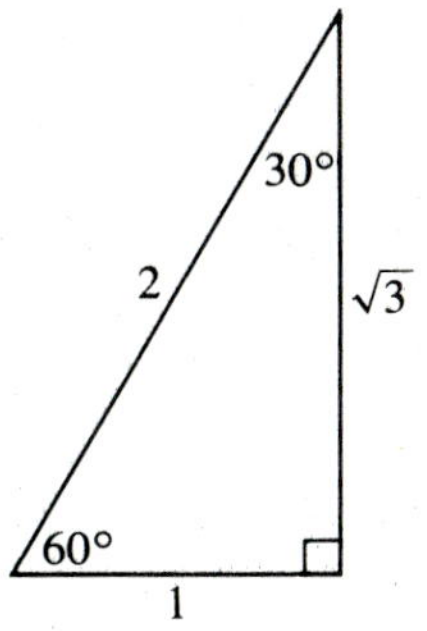

Using the definitions from the perspective of the 30° angle, we get

$$\sin 30^\circ = \frac{\text{opposite}}{\text{hypotenuse}} = \frac{1}{2}.$$

From the perspective of the 60° angle we have

$$\tan 60^\circ = \frac{\text{opposite}}{\text{adjacent}} = \frac{\sqrt{3}}{1} = \sqrt{3}.$$

▲

▲ ***Exercise*** Find each of the following, exactly:

(a) sin 45° (b) tan 30° (c) cos 60°

Answer (a) $\frac{1}{\sqrt{2}}\left(\text{or } \frac{\sqrt{2}}{2}\right)$ (b) $\frac{1}{\sqrt{3}}\left(\text{or } \frac{\sqrt{3}}{3}\right)$ (c) $\frac{1}{2}$

Note that for acute angles A and B of a right triangle, the side opposite $\angle B$ is the side adjacent to $\angle A$, and the side adjacent to $\angle B$ is the side opposite $\angle A$. See Figure 2.8.

Figure 2.8

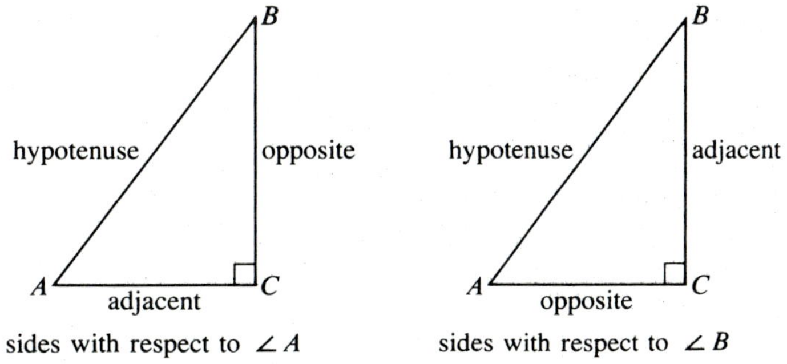

Consequently,

$$\text{sine } \angle B = \frac{\text{side opposite } \angle B}{\text{hypotenuse}} = \frac{\text{side adjacent } \angle A}{\text{hypotenuse}} = \text{cosine } \angle A,$$

$$\text{cosine } \angle B = \frac{\text{side adjacent } \angle B}{\text{hypotenuse}} = \frac{\text{side opposite } \angle A}{\text{hypotenuse}} = \text{sine } \angle A.$$

Angles A and B are complementary (the sum of their measures is 90°). So *the sine of an angle equals the cosine of its complement*, and vice versa. For instance, $\sin 20° = \cos 70°$ and $\cos 15° = \sin 75°$. Because of this, the trigonometric ratios sine and cosine are called **coratios**.

Throughout this chapter we shall use a standard notation for right triangle ABC. Angle C will be the right angle, and the acute angles A and B will have measures denoted by α and β (beta), respectively. The length of the side opposite an angle will be represented by the same letter in lower case. See Figure 2.9.

Figure 2.9

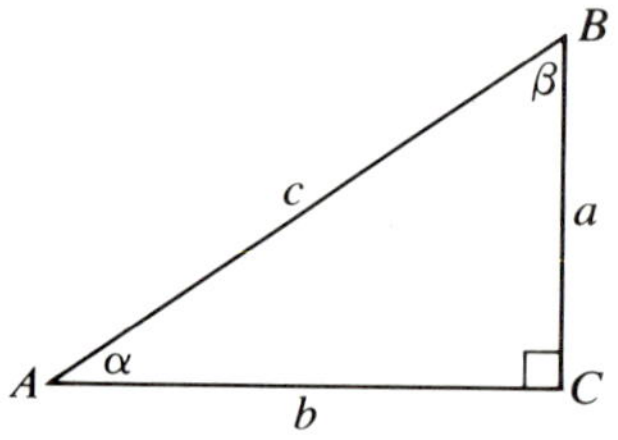

EXAMPLE 3 (a) If $\tan \alpha = 3$, find $\sin \alpha$. (b) If $\cos \alpha = \frac{2}{5}$, find $\tan \alpha$.

SOLUTION (a) We first construct a right triangle with acute angle α such that its tangent is 3. Since $3 = \frac{3}{1}$, we may use the triangle shown in Figure 2.10.

Figure 2.10

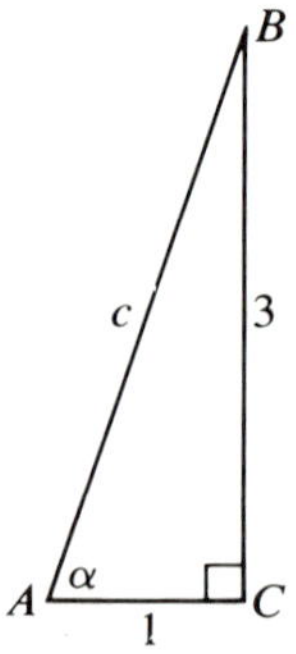

By the Pythagorean Theorem, $c^2 = 1^2 + 3^2 = 10$. Hence $c = \sqrt{10}$. Therefore,

$$\sin \alpha = \frac{3}{\sqrt{10}} \left(\text{or } \frac{3\sqrt{10}}{10}\right).$$

(b) We construct a right triangle with an angle whose cosine is $\frac{2}{5}$, as shown in Figure 2.11.

Figure 2.11

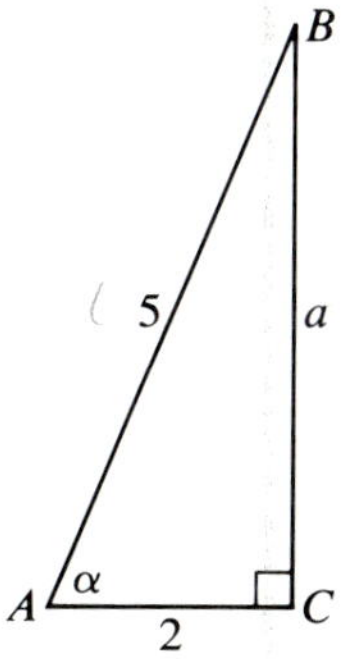

By the Pythagorean Theorem, $a^2 = 5^2 - 2^2 = 21$. Thus $a = \sqrt{21}$. Therefore,

$$\tan \alpha = \frac{\sqrt{21}}{2}.$$

▲

▲ *Exercise*

(a) If $\tan \alpha = \frac{3}{2}$, find $\sin \alpha$, exactly.

(b) If $\cos \alpha = \frac{1}{3}$, find $\tan \alpha$, exactly.

Answer (a) $\frac{3}{\sqrt{13}}\left(\text{or } \frac{3\sqrt{13}}{13}\right)$ (b) $2\sqrt{2}$

EXAMPLE 4 If $\sin \alpha = \frac{3}{4}$, find $\tan \beta$.

SOLUTION We construct a right triangle such that α has a sine of $\frac{3}{4}$, as shown in Figure 2.12.

Figure 2.12

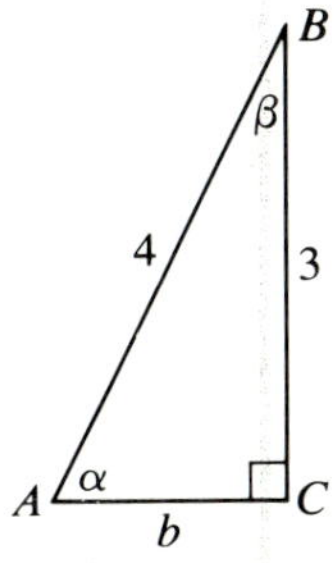

By the Pythagorean Theorem, $b^2 = 4^2 - 3^2 = 7$. So $b = \sqrt{7}$. Thus, using the definitions from the perspective of $\angle B$, we get

$$\tan \beta = \frac{\sqrt{7}}{3}.$$

▲

▲ ***Exercise*** If $\sin \alpha = \frac{5}{6}$, find $\tan \beta$, exactly.

Answer $\frac{\sqrt{11}}{5}$

Problem Set 2.1

In Problems 1–4, calculate the three principal trigonometric ratios for $\angle A$.

1.

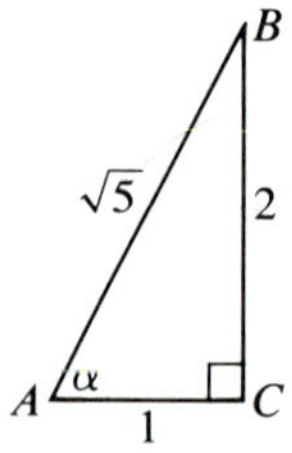

2.

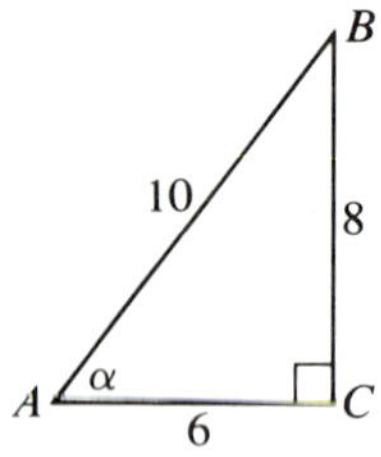

3.

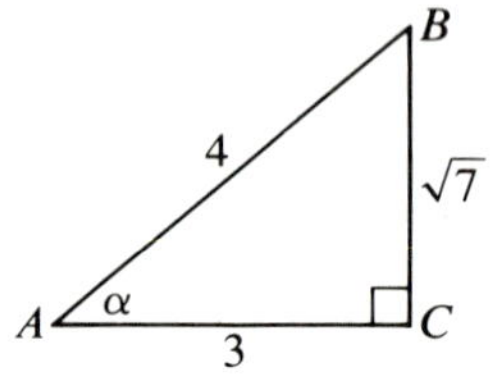

4.

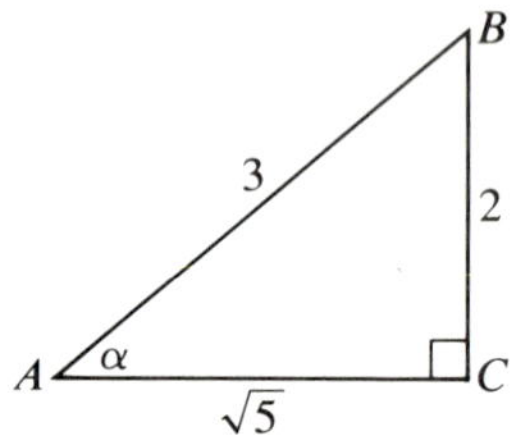

In Problems 5 and 6, calculate the three principal trigonometric ratios for $\angle B$.

5.

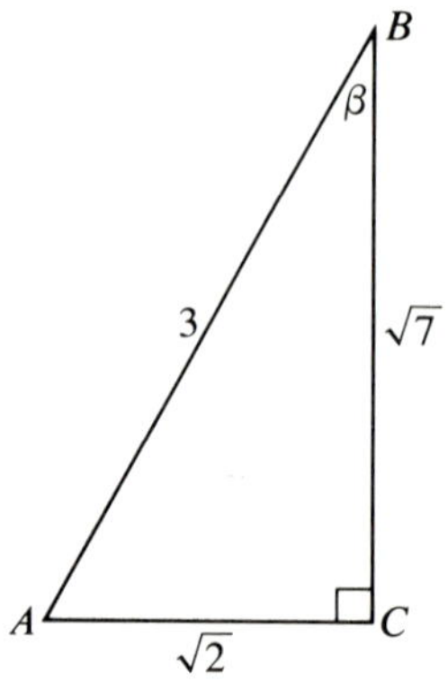

6.

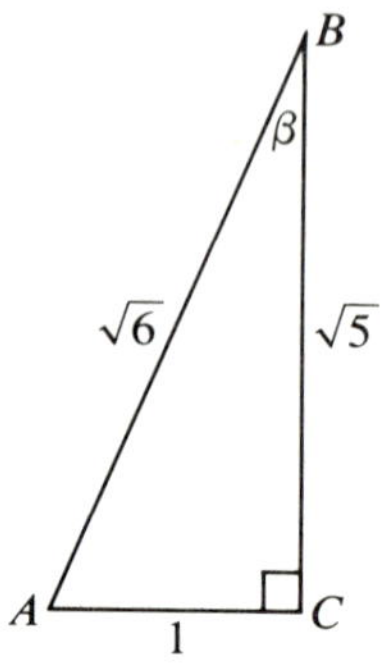

In Problems 7–24, answers are to be exact.

7. If $\tan \alpha = 4$, find $\cos \alpha$.

8. If $\tan \alpha = 5$, find $\cos \alpha$.

9. If $\tan \alpha = \frac{5}{2}$, find $\sin \alpha$.

10. If $\tan \alpha = \frac{2}{3}$, find $\sin \alpha$.

11. If $\cos \alpha = \frac{2\sqrt{2}}{5}$, find $\sin \alpha$.

12. If $\cos \alpha = \frac{1}{4}$, find $\sin \alpha$.

13. If $\cos \alpha = \frac{\sqrt{6}}{3}$, find $\tan \alpha$.

14. If $\cos \alpha = \frac{1}{5}$, find $\tan \alpha$.

15. If $\sin \alpha = \frac{3}{4}$, find $\cos \alpha$.

16. If $\sin \alpha = \frac{4}{5}$, find $\cos \alpha$.

17. If $\sin \alpha = \frac{\sqrt{6}}{6}$, find $\tan \alpha$.

18. If $\sin \alpha = \frac{\sqrt{21}}{7}$, find $\tan \alpha$.

19. If $\sin \alpha = \frac{\sqrt{3}}{5}$, find $\cos \beta$.

20. If $\cos \alpha = \frac{5}{9}$, find $\sin \beta$.

21. If $\sin \alpha = \frac{7}{8}$, find $\tan \beta$.

22. If $\sin \alpha = \frac{\sqrt{5}}{6}$, find $\tan \beta$.

23. If $\tan \alpha = \frac{2}{5}$, find $\sin \beta$.

24. If $\tan \alpha = \frac{\sqrt{3}}{5}$, find $\cos \beta$.

25. Show that $\sin^2 \alpha + \cos^2 \alpha = 1$ for all acute angles α.

26. Show that $\tan \alpha = \frac{\sin \alpha}{\cos \alpha}$ for all acute angles α.

27. Calculate the three principal trigonometric ratios for an angle of 30°, exactly.

28. Calculate the three principal trigonometric ratios for an angle of 60°, exactly.

29. Calculate the three principal trigonometric ratios for an angle of 45°, exactly.

2.2

TRIGONOMETRIC VALUES FOR ACUTE ANGLES

We can find the values of the trigonometric ratios for the special angles of 30°, 60°, and 45° because we know the lengths of the sides of right triangles containing these angles. What about other angles? How do we find sin 42.3° or tan 67.8°?

Advanced mathematical methods enable us to calculate the trigonometric ratio of any acute angle as accurately as we like. Scientific calculators use such methods, and the results of such computations are also presented in tables like the ones found in Appendix A.4. (Instructions for their use are included there and may be used whenever a calculator is unavailable.) Because calculators are faster, easier to use, and generally more accurate, the discussions in the text (excluding A.4) will rely on calculators instead of tables.

Calculators have buttons for sine, cosine, and tangent. They also have a button that allows us to specify the mode of measurement—degrees or radians. Since radian measure of an angle will not be discussed until Chapter 4, we will temporarily confine ourselves to degree mode.

EXAMPLE 1 Find $\sin 42.3^\circ$.

SOLUTION *In the degree mode*, we enter 42.3 and then push the sine button [sin]. An eight digit display shows 0.6730125. Thus

$$\sin 42.3^\circ = 0.6730125\,.$$

▲

Therefore, in a right triangle with an acute angle of 42.3°, the ratio of the side opposite to the hypotenuse is 0.6730125. In other words, if the hypotenuse had length 1, the side opposite the 42.3° angle would have length 0.6730125. See Figure 2.13.

Figure 2.13

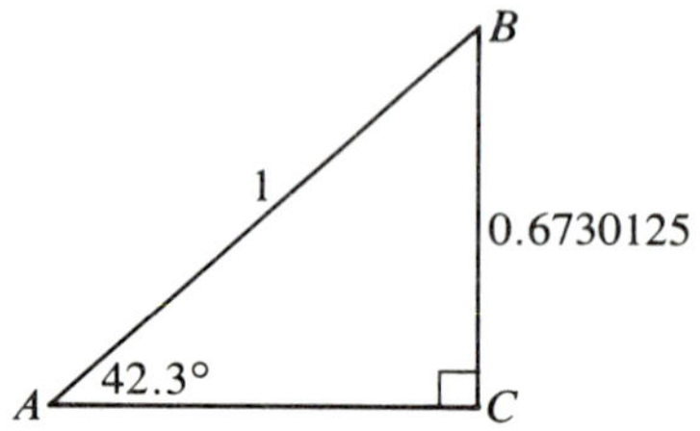

When we use a calculator to find the values of the trigonometric ratios, the displayed result is usually an approximation. Technically, then, we should write

$$\sin 42.3^\circ \approx 0.6730125$$

because the internally stored number (which may also be an approximation) has been rounded to the seventh decimal place for the display. However, we shall continue to use equal marks when we write the number displayed. We shall use approximation marks when *we* round the result, as when rounding to four significant digits:

$$\sin 42.3^\circ \approx 0.6730\,.$$

(For a discussion of approximations and significant digits, see Appendix A.1.)

What is the relationship between the accuracy of an angle measurement and the accuracy of the values of its trigonometric ratios? For instance, if an angle is measured to the nearest degree, what is the accuracy of its sine? As a general guide, we shall use the following table.

Accuracy of degree measure	*Accuracy of trigonometric values*
nearest one	two significant digits
nearest tenth	three significant digits
nearest hundredth	four significant digits

The guidelines for computations with approximations (see Appendix A.2) will be followed for applied problems. However, intermediate results will be carried with any nonsignificant digits so that rounding is delayed until the final step. For nonapplied problems, however, we shall assume that the numbers given are exact, but we shall round results when necessary to provide uniform answers.

EXAMPLE 2 Find the trigonometric value to four significant digits.

(a) tan 67.8° (b) cos 49.18°

SOLUTION (a) In the degree mode we enter 67.8 and then push the tangent button tan. An eight-digit display is 2.4504252. Rounding to four significant digits (which is to the nearest thousandth here), we get

$$\tan 67.8^\circ \approx 2.450\,.$$

(b) In the degree mode we enter 49.18 and then push the cosine button cos. To eight digits we get 0.6536848. Rounding to four significant digits gives

$$\cos 49.18^\circ \approx 0.6537\,.$$

▲

▲ ***Exercise*** Find the trigonometric value to four significant digits.

(a) sin 41.56° (b) cos 22.5° (c) tan 82.5°

Answer (a) 0.6634 (b) 0.9239 (c) 7.596

EXAMPLE 3 If a projectile is fired from ground level at an angle of α to the horizontal, then its maximum vertical height, h, in feet, is given by

$$h = \frac{(v_0 \sin \alpha)^2}{64}, \qquad \textbf{(2.1)}$$

where v_0 is the initial velocity in feet per second. If $\alpha = 40°$ and $v_0 = 55$, what is the maximum height (to two significant digits)?

SOLUTION Substituting 55 for v_0 and 40° for α gives

$$h = \frac{(55 \sin 40°)^2}{64} = \frac{(55(0.6427876))^2}{64} = 19.529018 \approx 20.$$

The maximum height is approximately 20 feet.

▲

▲ *Exercise* Use equation (2.1) to determine the maximum height (to two significant digits) of a projectile fired from ground level with an initial velocity of 75 feet per second at an angle of 70° to the horizontal.

Answer 78 ft

For a given trigonometric ratio and a given trigonometric value, there is only one acute angle for which the specified ratio will yield the specified value. For example, there is only one acute angle whose sine (a given ratio) will equal 0.825 (a given value). We may determine that angle by using the calculator. However, the method of doing so varies from calculator to calculator. On some we push [inv] or [arc] followed by the sine, cosine, or tangent button, as the case may be. Other calculators have a single button: [$\sin^{-1}$], [$\cos^{-1}$], and [$\tan^{-1}$]. The mathematics behind this concept will be discussed in Chapter 5.

EXAMPLE 4 Find angle α (in degrees) if $\cos \alpha = 0.7976$.

SOLUTION First we insure that we are in degree mode. Next we enter .7976 and then push [inv] [cos] (or its equivalent, which will be [arc] [cos] or [$\cos^{-1}$]). An eight-digit display reads 37.098473. Thus

$$\alpha = 37.098473°.$$

▲

The result of Example 4 is in fact an approximation, but we shall continue the practice of using approximation symbols when *we* (not the calculator) do the rounding.

EXAMPLE 5 Find angle α to the nearest hundredth of a degree, given the trigonometric value.

(a) $\tan \alpha = 2.5$ (b) $\sin \alpha = \dfrac{57}{130}$ (c) $\cos \alpha = 1.5$

SOLUTION (a) In degree mode we enter 2.5, followed by [inv] [tan] (or its equivalent). An eight-digit display shows 68.198591. Rounding to the nearest hundredth, we obtain

$$\alpha \approx 68.20^\circ.$$

(b) First we need to divide 57 by 130. We use the calculator to get this result. Next (in degree mode) we push [inv] [sin] (or its equivalent). We see 26.005762 (eight-digit display). Rounding to the nearest hundredth, we get

$$\alpha \approx 26.01^\circ.$$

(c) We enter 1.5, followed by [inv] [cos] (or its equivalent). We get an error message! As noted earlier, the cosine and sine of an acute angle are always less than one. There is no angle α such that $\cos \alpha = 1.5$.

▲

▲ ***Exercise*** Find angle α to the nearest hundredth of a degree, given the trigonometric value.

(a) $\sin \alpha = 0.1516$ (b) $\tan \alpha = 3.1$ (c) $\cos \alpha = \dfrac{30}{57}$ (d) $\sin \alpha = 2.5$

Answer (a) 8.72° (b) 72.12° (c) 58.24° (d) no such α

EXAMPLE 6 Use equation (2.1) to determine a projectile's angle to the horizontal (to the nearest degree) if an initial velocity of 125 feet per second results in a maximum height of 85 feet.

SOLUTION Equation (2.1) is

$$h = \frac{(v_0 \sin \alpha)^2}{64}.$$

Since $h = 85$ when $v_0 = 125$, we have

$$85 = \frac{(125 \sin \alpha)^2}{64}.$$

Solving for $\sin \alpha$ gives

$$\begin{aligned} (125 \sin \alpha)^2 &= 64(85) \\ 125 \sin \alpha &= \sqrt{64(85)} \\ \sin \alpha &= \frac{\sqrt{64(85)}}{125} = 0.5900509\,. \end{aligned}$$

The angle with this sine is 36.160616°. Thus

$$\alpha \approx 36^\circ.$$

▲

▲ *Exercise* Use equation (2.1) to determine a projectile's angle to the horizontal (to the nearest degree) if an initial velocity of 115 feet per second results in a maximum height of 95 feet.

Answer 43°

Problem Set 2.2

In Problems 1–6, find the trigonometric value to four significant digits.

1. sin 22.6°

2. sin 5°

3. cos 41°

4. cos 75.2°

5. tan 81.56°

6. tan 31.2°

In Problems 7–18, find angle α to the nearest hundredth of a degree, given the trigonometric value.

7. $\sin \alpha = 0.24$

8. $\sin \alpha = 0.6$

9. $\cos \alpha = 0.8$

10. $\cos \alpha = 0.125$

11. $\tan \alpha = 4.5$

12. $\tan \alpha = 0.65$

13. $\sin \alpha = \dfrac{15}{38}$

14. $\cos \alpha = \dfrac{27}{31}$

15. $\tan \alpha = \dfrac{131}{5}$

16. $\tan \alpha = \dfrac{98}{105}$

17. $\cos \alpha = 1.5$

18. $\sin \alpha = 2.17$

19. Find $\sin \alpha$ to four significant digits if $\cos \alpha = 0.82$.

20. Find $\tan \alpha$ to four significant digits if $\sin \alpha = 0.412$.

21. Neglecting friction and air resistance, the acceleration of a ball rolling down an inclined plane from rest is given by

$$a = 32 \sin \alpha,$$

where a represents acceleration in feet per second per second and α is the angle of inclination. What is the acceleration (to three significant digits) if $\alpha = 61.5°$?

22. If a golf ball is hit at an angle of α to the horizontal, then the horizontal distance traveled is given by

$$k = \frac{v_0^2 \sin \alpha \cos \alpha}{16},$$

where k represents the horizontal distance traveled in feet and v_0 is the initial velocity in feet per second. If $\alpha = 35°$ and $v_0 = 165$, what is the horizontal distance traveled (to two significant digits)?

23. Neglecting friction and air resistance, the velocity of a ball rolling down an inclined plane from rest is given by

$$v = (32 \sin \alpha)t,$$

where v represents velocity in feet per second, t represents elapsed time in seconds since the release of the ball, and α is the angle of inclination. What is the angle of inclination (to the nearest degree) if the ball is rolling 48 feet per second after 2 seconds?

24. If a projectile is fired from ground level at an angle α to the horizontal, then its distance above ground, s, in feet, after t seconds, is given by

$$s = -16t^2 + (\sin \alpha)v_0 t,$$

where v_0 is the initial velocity in feet per second. If the initial velocity is 98 feet per second, what is the angle α (to the nearest degree) if the object is 130 feet above the ground after 2.5 seconds?

As light passes from one medium to another, it bends because its speed changes. Suppose that there are two mediums, 1 and 2, separated by a plane boundary. Let α_1 be the angle that a ray of light makes with a line perpendicular to the boundary as it passes through medium 1. Let α_2 be the angle that the ray makes with a line perpendicular to the boundary when it exits through medium 2. Further, let v_1 = light speed in medium 1 and v_2 = light speed in medium 2. See Figure 2.14.

Figure 2.14

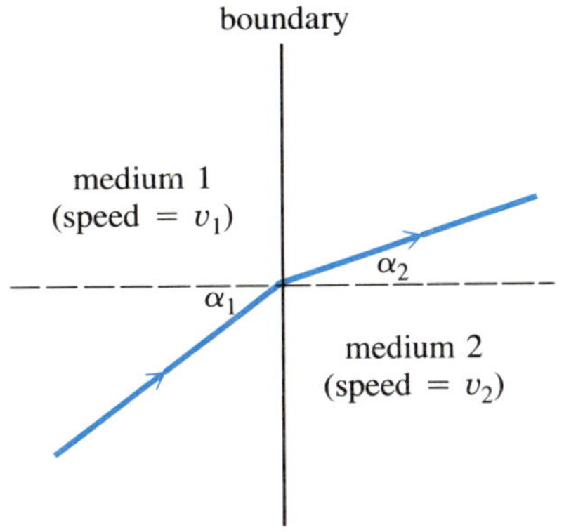

According to Snell's Law, these values are related as follows:

$$\frac{\sin \alpha_1}{\sin \alpha_2} = \frac{v_1}{v_2}.$$

This information is needed in Problems 25–28.

25. If light is traveling 1.7 times faster in medium 1 than in medium 2 and $\alpha_1 = 20°$, find α_2 (to the nearest degree).

26. If light is traveling 1.7 times faster in medium 2 than in medium 1 and $\alpha_1 = 20°$, find α_2 (to the nearest degree).

27. If light travels half as fast in medium 1 as in medium 2 and $\alpha_1 = 55°$, find α_2 (to the nearest degree).

28. Consider a light ray passing through a pane of glass as illustrated in the following figure. Show that the departing ray is parallel to the incoming ray; i.e., show that $\alpha_1 = \alpha_3$.

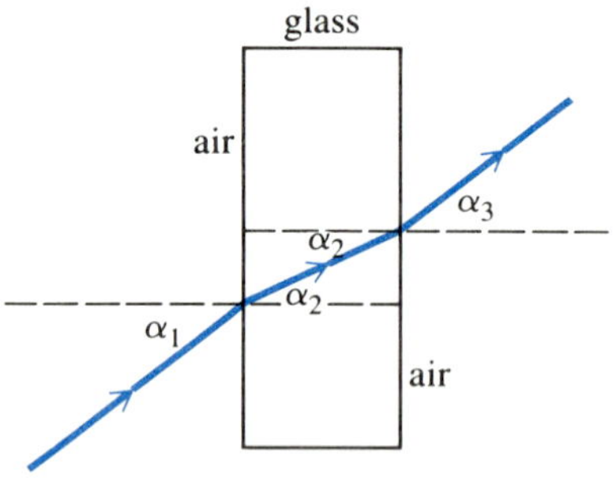

In Problems 29–32, find the length of the specified side to four significant digits.

29. side opposite 35° angle if hypotenuse has length 5

30. side opposite 68° angle if hypotenuse has length 4.5

31. side adjacent to 75.2° angle if hypotenuse has length 8

32. side adjacent to 24° angle if hypotenuse has length 5.6

2.3 SOLVING RIGHT TRIANGLES

The lengths of the three sides and the measures of the three angles of a triangle are called its parts. The process of finding the remaining parts using those given is called **solving a triangle**. It is possible to solve a right triangle if we know the lengths of two sides or if we know the length of one side and the measure of one of the acute angles. We shall use the notation introduced in Section 2.1. Given right triangle ABC, α and β will represent the measures of angles A and B respectively, C will be the right angle, while a, b, and c will denote the lengths of the opposite sides. See Figure 2.15.

Figure 2.15

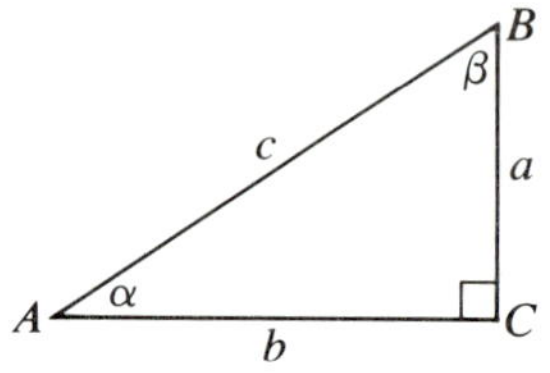

EXAMPLE 1 Solve the right triangle ABC, given that $b = 5$ and $\alpha = 40°$.

SOLUTION We have the triangle shown in Figure 2.16, and we must find a, c, and β.

Figure 2.16

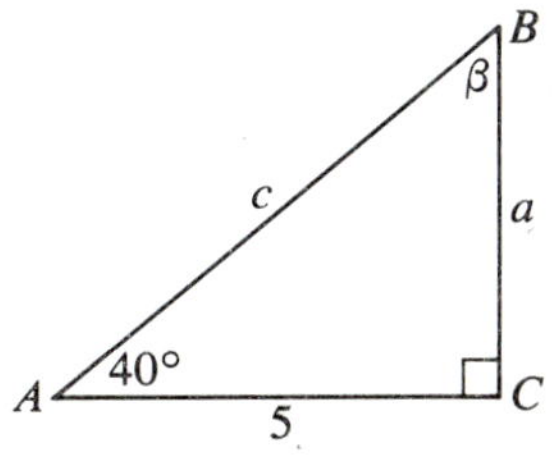

Since angles A and B are complements,

$$\beta = 90° - \alpha = 90° - 40° = 50°.$$

To find a note that, with respect to the 40° angle, we are looking for the side opposite, and we know the side adjacent; so we must use a ratio involving opposite and adjacent. Using the tangent, then, we have

$$\tan 40° = \frac{a}{5}.$$

Thus

$$a = 5 \tan 40°.$$

Computing tan 40° and multiplying by 5 gives us

$$a = 4.1954982 \approx 4.195.$$

To find c we note that

$$\cos 40° = \frac{5}{c}.$$

Hence

$$c \cos 40° = 5$$

$$c = \frac{5}{\cos 40°} = 6.5270364 \approx 6.527.$$

▲

In Example 1 note that, after determining a, we could have found c by using the Pythagorean Theorem. However, since calculated parts may be approximate, solving an unknown part by using given parts only is generally more accurate. Nonetheless, the Pythagorean Theorem serves as a useful check:

$$c = \sqrt{5^2 + (4.1954982)^2} \approx 6.527,$$

which agrees with the previous calculation.

In the next section we will construct triangles that represent physical situations. In such applied problems the lengths of sides and measures of angles are considered approximate because they are determined with mechanical devices. In this section, though, the lengths and measures are taken to be exact. However, to provide uniform answers, we shall round off lengths to four significant digits and angle measures to the nearest hundredth of a degree.

EXAMPLE 2 Solve the right triangle ABC given that $c = 12$ and $\beta = 67°$.

SOLUTION We have the triangle shown in Figure 2.17 and must find a, b, and α.

Figure 2.17

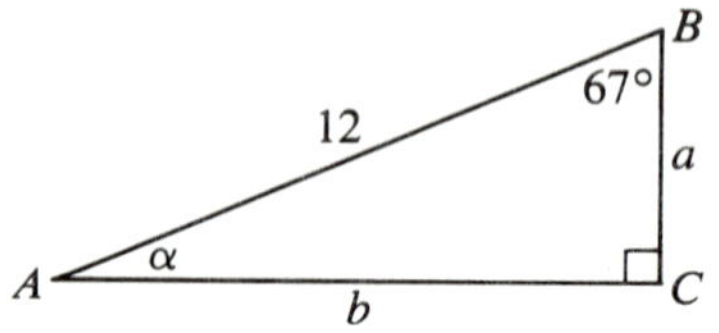

We have

$$\alpha = 90° - \beta = 90° - 67° = 23°.$$

To determine a we note that

$$\cos 67° = \frac{a}{12}.$$

Hence

$$a = 12 \cos 67° \approx 4.689,$$

and

$$\sin 67° = \frac{b}{12},$$

so

$$b = 12 \sin 67° \approx 11.05.$$

▲

▲ *Exercise* Solve the right triangle ABC for the given conditions.

(a) $b = 7$ and $\alpha = 20°$ (b) $c = 8$ and $\beta = 72°$

Answer (a) $\beta = 70°$, $a \approx 2.548$, $c \approx 7.449$ (b) $\alpha = 18°$, $a \approx 2.472$, $b \approx 7.608$

EXAMPLE 3 Solve the right triangle ABC given that $a = 4$ and $b = 10$.

SOLUTION We have the triangle shown in Figure 2.18 and must find c, α, and β.

Figure 2.18

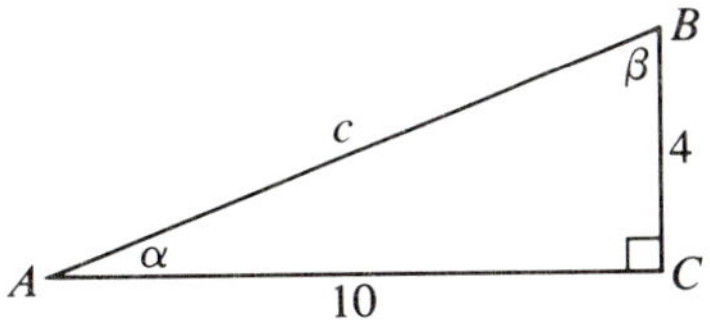

We see that

$$\tan \alpha = \frac{4}{10}.$$

Using our calculators to determine the angle whose tangent is $\frac{4}{10}$, we get

$$\alpha = 21.801409^\circ \approx 21.80^\circ.$$

Thus

$$\beta \approx 90^\circ - 21.80^\circ = 68.20^\circ.$$

To determine c let's use the Pythagorean Theorem. We have

$$c = \sqrt{4^2 + 10^2} \approx 10.77.$$

▲

In Example 3 note that we could have computed c by using one of the acute angles and a trigonometric ratio. However, we continued our practice of calculating unknown parts from *given* parts. Nonetheless, this alternative is a useful check. For instance, we see that

$$\cos 21.801409^\circ = \frac{10}{c}.$$

So

$$c \cos 21.801409^\circ = 10$$

$$c = \frac{10}{\cos 21.801409^\circ} \approx 10.77,$$

our previous result.

EXAMPLE 4 Solve the right triangle ABC given that $a = 5$ and $c = 12$.

SOLUTION We have the triangle shown in Figure 2.19 and must find b, α, and β.

Figure 2.19

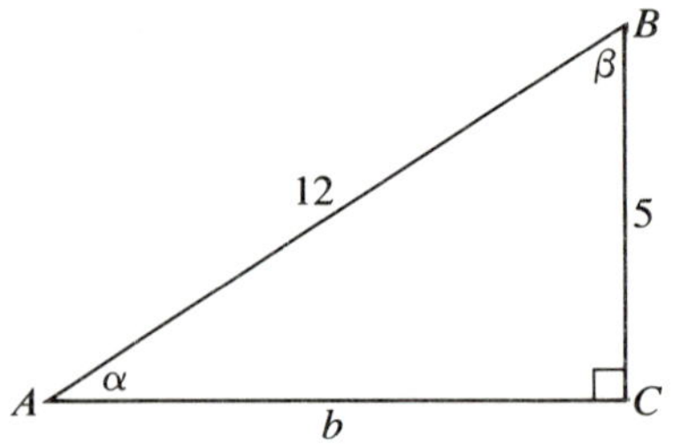

We have

$$\sin \alpha = \frac{5}{12}.$$

Thus

$$\alpha \approx 24.62°.$$

Hence

$$\beta \approx 90° - 24.62° = 65.38°.$$

To determine b let's use the Pythagorean Theorem. We get

$$b = \sqrt{12^2 - 5^2} \approx 10.91.$$

▲

▲ ***Exercise*** Solve the right triangle ABC for the given conditions.

(a) $a = 6$ and $b = 9$ (b) $a = 3$ and $c = 10$

Answer (a) $\alpha \approx 33.69°$, $\beta \approx 56.31°$, $c \approx 10.82$
(b) $\alpha \approx 17.46°$, $\beta \approx 72.54°$, $b \approx 9.539$

Problem Set 2.3

In Problems 1–24, solve the right triangle *ABC* for the given conditions.

1. $\alpha = 42°$, $a = 10$

2. $\alpha = 25°$, $a = 8.5$

3. $\alpha = 68°$, $b = 5$

4. $\alpha = 32.5°$, $b = 7$

5. $\alpha = 51°$, $c = 5$

6. $\alpha = 35°$, $c = 6.2$

7. $\beta = 71.2°$, $a = 8$

8. $\beta = 31.7°$, $a = 9$

9. $\beta = 27.81°$, $b = 6.5$

10. $\beta = 14.52°$, $b = 7.25$

11. $\beta = 42.5°$, $c = 6$

12. $\beta = 29.1°$, $c = 9.05$

13. $a = 6$, $b = 10$

14. $a = 12$, $b = 8$

15. $a = 18.5$, $b = 13$

16. $a = 5.3$, $b = 7.1$

17. $a = 10, c = 14$

18. $a = 12.1, c = 15.05$

19. $a = 12.2, c = 17$

20. $a = 3.05, c = 5.28$

21. $b = 5, c = 8$

22. $b = 6.2, c = 9.5$

23. $b = 11, c = 14$

24. $b = 2.5, c = 5.5$

2.4 APPLICATIONS

We are now in a position to use the techniques learned in the preceding section to solve certain physical problems that involve right triangles.

EXAMPLE 1 A surveyor wishes to determine the distance across the Great Meteor Crater in Arizona. She places two landmarks at the edges of the crater, on opposite sides. She then places a stake 150 feet from one landmark so that a triangle is formed by the stake and the two landmarks, with a right angle at the landmark that is 150 feet from the stake. If the angle at the stake is 87.9°, what is the distance (to three significant digits) across the crater?

SOLUTION We have the triangle shown in Figure 2.20, where

$$x = \text{distance across crater (in feet)}.$$

Figure 2.20

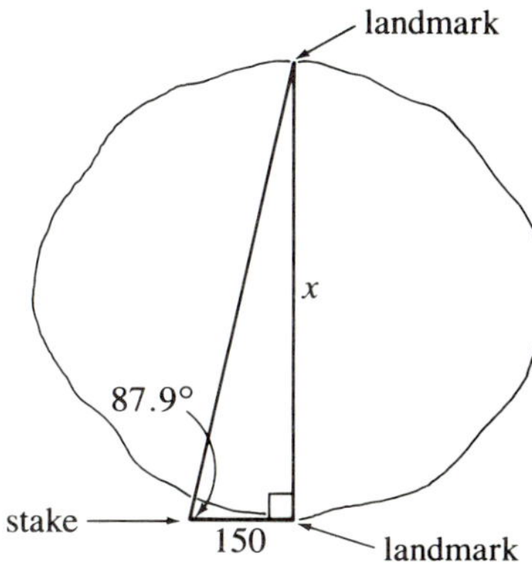

We have

$$\tan 87.9^\circ = \frac{x}{150}$$

$$x = 150 \tan 87.9^\circ \approx 4090 \text{ (nearest ten)}.$$

Thus, to the nearest 10 feet, the distance across is 4090 feet.

▲

As previously noted, the lengths of sides and measures of angles are considered approximate because they are determined with mechanical devices. Consequently, for the answer, each problem will specify a degree of accuracy or level of accuracy consistent with the rules for operations with approximations (see Appendix A.2).

EXAMPLE 2 The grade of a highway is the vertical rise divided by the corresponding road length. If a highway has a grade of 15%, what angle (to the nearest degree) does the highway make with the horizontal?

SOLUTION We have the situation shown in Figure 2.21, where

α = measure of angle between highway and horizontal (in degrees).

Figure 2.21

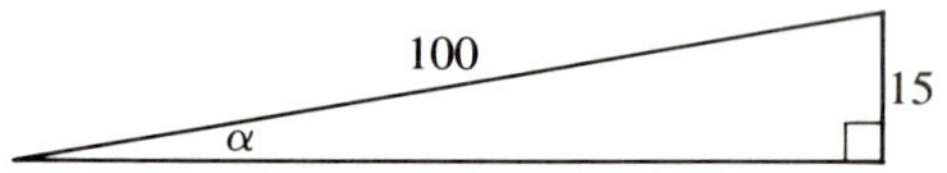

To find α we note that

$$\sin \alpha = \frac{15}{100}.$$

Using our calculators, we see that

$$\alpha \approx 9^\circ.$$

▲

▲ *Exercise*

(a) A surveyor wishes to determine the distance across Lake Tahoe. She places two landmarks at the edges of the lake, on opposite sides. She then places a stake 1.5 miles from one landmark so that a triangle is formed by the stake and the two landmarks, with a right angle at the landmark that is 1.5 miles from the stake. If the angle at the stake is 82.9°, what is the distance (to two significant digits) across the lake?

(b) If a highway has a grade of 20% (grade is vertical rise divided by corresponding road length), what angle (to the nearest degree) does the highway make with the horizontal?

Answer (a) 12 mi (b) 12°

The **angle of elevation** or **angle of depression** from a point A to a point B is the angle formed by the ray $\overrightarrow{AB}$ and the horizontal line passing through A. If the ray points upward, it is called an angle of elevation. If it points downward, it is called an angle of depression (see Figure 2.22). These terms are needed for Examples 3 and 4.

Figure 2.22

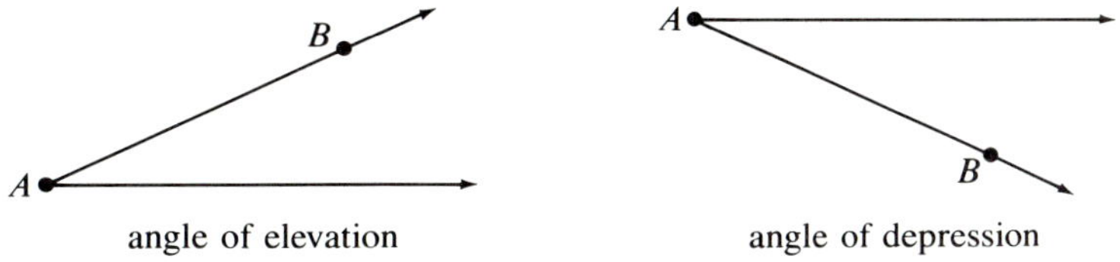

EXAMPLE 3 The angle of elevation from a point at ground level 50.0 feet from the base of a radio tower to its top is 70.0°. How long (to three significant digits) must a guy wire be to stretch from this point to the top?

SOLUTION We have the triangle shown in Figure 2.23, where

$$x = \text{length of wire (in feet)}.$$

Figure 2.23

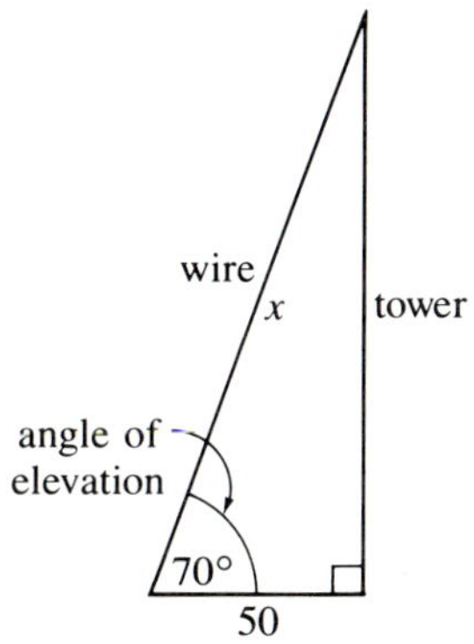

Thus

$$\cos 70^\circ = \frac{50}{x}$$

$$x = \frac{50}{\cos 70^\circ} \approx 146.$$

Hence, the wire must be approximately 146 feet along.

▲

EXAMPLE 4 From a point 160 feet above the water, the angle of depression from a lighthouse to a boat is 65°. How far (to two significant digits) is the boat from the lighthouse?

SOLUTION We have the triangle shown in Figure 2.24, where

$$x = \text{distance from boat to lighthouse (in feet)}.$$

Figure 2.24

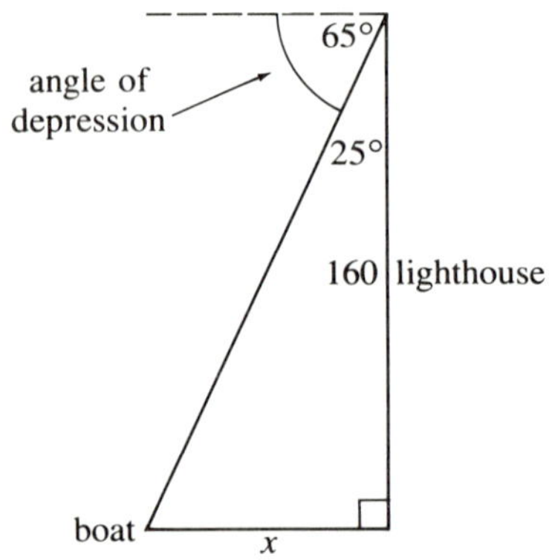

Note that in this particular problem the angle of depression is not part of the triangle. The angle of depression has one side *horizontal*, not vertical. Continuing,

$$\tan 25^\circ = \frac{x}{160}$$

$$x = 160 \tan 25^\circ \approx 75.$$

Thus the boat is approximately 75 feet from the lighthouse.

▲

▲ ***Exercise***

(a) The angle of elevation from a point at ground level 85.0 feet from the base of a TV tower to its top is 68.0°. How long (to three significant digits) must a guy wire be to stretch from this point to the top?

(b) From a point 125 feet above the water, the angle of depression from a lighthouse to a boat is 52°. How far (to two significant digits) is the boat from the lighthouse?

Answer (a) 227 ft (b) 98 ft

Problem Set 2.4

1. A surveyor wishes to determine the distance between a rock and a tree on opposite sides of a gully. He places a stake 50 meters from the tree, so that a triangle is formed by the stake, tree, and rock, with a right angle at the tree. If the angle at the stake is 75°, what is the distance (to two significant digits) between the rock and the tree?

2. A surveyor wishes to determine the distance between a rock and a cliff on opposite sides of a river. He places a stake 75 feet from the rock so that a triangle is formed by the stake, cliff, and rock, with a right angle at the rock. If the angle at the stake is 62°, what is the distance (to two significant digits) between the rock and the cliff?

3. The top of a ladder 20 feet long is leaning against the top of a vertical wall. If the ladder makes an angle of 75° with the ground, how high (to two significant digits) is the wall?

4. A 10-foot ladder is leaning across the top of an 8.0 foot wall. How much of the ladder (to the nearest inch) must overhang to make an angle of 70° between the ladder and the ground?

5. Suppose that a commerical airliner is flying at an altitude of 2.5 miles. If the pilot wishes to descend to an airport such that his path makes an angle of 5° with the ground, how far from the airport (as measured along the ground, to two significant digits) must he begin his descent?

6. A ramp from ground level into a bookstore is to be built. If the entrance to the bookstore is 4.8 feet above the ground, and the inclination must not exceed 25°, what is the shortest ramp (to two significant digits) that could be built?

7. The pitch of a roof is the vertical rise over the horizontal run. If a roof has pitch $\frac{5}{12}$, how steep (to the nearest degree) is the roof, as measured between roof and horizontal?

8. A walkway between two buildings spans 45 feet. If one end is 5.0 feet higher than the other, what is the angle of inclination (to the nearest degree)?

9. A 200-foot building is casting a shadow of 120 feet. What is the angle of elevation (to the nearest degree) of the sun at this time?

10. An observation balloon is hovering 250 meters above the ground. What is the angle of depression (to the nearest degree) to a point that is 850 meters from a point directly beneath the balloon?

11. From a point at ground level the angle of elevation to the top of the World Trade Center in New York City is 77.5°. If the point is 300 feet from the base of the building, how tall (to three significant digits) is it?

12. From a point at ground level the angle of elevation to the top of the Sears Tower in Chicago is 80.2°. If the point is 250 feet from the base of the building, how tall (to three significant digits) is it?

13. The tallest known tree in the world is a redwood in northern California. From a point 150 feet from the base of the tree, the angle of elevation to the top is 66.7°. If the observation point is 20 feet above the horizontal line through the base of the tree, how tall (to the nearest foot) is the tree?

14. The horizontal distance from an observer to a point directly below the highest point of the Rock of Gilbraltar is 8000 feet. If the angle of elevation to the top is 9.22°, and the observation is made 100 feet above sea level, how high (to the nearest 10 feet) is the Rock of Gilbraltar?

15. The angle of elevation from a point at ground level to the top of a tower is 78°. If the tower is 120 feet high, how far (to two significant digits) is the observation point from the base of the tower?

16. From a helicopter hovering 100 feet (nearest 10 feet) above a lifeboat, the angle of depression to the recovery ship is 5°. How far from the lifeboat (to two significant digits) is the recovery ship?

17. From an observation point the angle of elevation to the top of Yosemite Falls is 73°, and the angle of depression to the bottom is 68°. If the observation point is 250 feet from the falls, what is the height (to the nearest 10 feet) of Yosemite Falls?

18. A plane flying at an altitude of 2.5 miles is approaching the Strait of Dover. If the angle of depression to the near shore directly in front is 68° and the angle of depression to the far shore is 6.5°, how wide (to the nearest mile) is the Strait of Dover? (Assume that the Strait is level.)

19. A plane is flying with a ground speed of 450 miles per hour (nearest 10 mph). At what angle (to the nearest degree) must it ascend to gain altitude at a rate of 25 miles per hour if it maintains the same ground speed?

20. A plane is flying with a ground speed of 380 miles per hour (nearest 10 mph). At what angle (to the nearest degree) must it descend to lose altitude at a rate of 0.5 miles per minute if it maintains the same ground speed?

21. A submarine on the surface begins a dive, its path making an angle of 10° with the surface. If its speed along this path is 12 kilometers per hour, what is its depth (to two significant digits) after 2 minutes?

22. A plane is cruising upward at a speed of 485 miles per hour along a path that makes an 8.9° angle with the horizontal. How long (to the nearest second) does it take the plane to gain 1 mile in altitude?

23. Suppose that a plane is flying a level course at an altitude of 3200 feet (nearest 100 feet). The pilot observes an object straight ahead on the ground and notes that the angle of depression is 15°. After 10 seconds the same object has an angle of depression of 30°. What is the speed of the plane (to the nearest 10 mph)?

24. Suppose that a balloon is flying a level course at an altitude of 500 feet (nearest 10 feet). An observer on the ground eyes the balloon and notes that the angle of elevation is 50°. After 10 seconds the balloon (in the same direction with respect to the observer) has an angle of elevation of 35°. What is the speed of the balloon (to the nearest mph)?

25. A satellite is orbiting a planet at an altitude of 105 miles. If the angle formed by the lines from the satellite to the horizons is 82.2°, what is the radius of the planet (to three significant digits)?

26. Assume that the distance to the moon is 384,000 kilometers. An astronomer determines that in moving the telescopic sights from one side of the moon to the other, the telescope rotates through an angle of 0.525°. What is the radius of the moon (to three significant digits)?

27. The radius of the earth is approximately 3960 miles. What is the altitude of a satellite (to three significant digits) if the angle formed by the lines from the satellite to the horizons is 137.6°?

28. What is the distance (to three significant digits) from earth to moon if the angle formed by the lines from the closest point on the moon's surface to the earth's horizons is 1.9°? (Use 3960 miles as the earth's radius.)

29. A flagpole 30 feet high is casting an afternoon shadow of 12.4 feet. What time is it (to the nearest minute) if the sun rose at 7:15 and there were no shadows at exactly 12:15?

2.5 ADDITIONAL TRIGONOMETRIC RATIOS

In this section we define the remaining three trigonometric ratios. Letting α be the measure of $\angle A$, we define the **reciprocal ratios** in the following table.

Name	*Abbreviation*	*Definition*
cosecant of $\angle A$	$= \csc \alpha$	$= \dfrac{1}{\sin \alpha}$
secant of $\angle A$	$= \sec \alpha$	$= \dfrac{1}{\cos \alpha}$
cotangent of $\angle A$	$= \cot \alpha$	$= \dfrac{1}{\tan \alpha}$

The cosecant is the reciprocal of the sine, the secant is the reciprocal of the cosine, and the cotangent is the reciprocal of the tangent. Thus, for example, if

$$\sin \alpha = \frac{2}{3},$$

then

$$\csc \alpha = \frac{1}{\frac{2}{3}} = \frac{3}{2}.$$

▲ *Exercise* Find $\sec \alpha$ if $\cos \alpha = \dfrac{3}{5}$.

Answer $\dfrac{5}{3}$

Although our calculators don't have buttons for these trigonometric ratios, we can still calculate their values by using the reciprocal button [1/x], as will be seen in Example 1.

EXAMPLE 1 Find the trigonometric value to four significant digits.

(a) csc 28° (b) sec 72.5° (c) cot 21.25°

SOLUTION (a) The cosecant is the reciprocal of the sine. Thus

$$\csc 28^\circ = \frac{1}{\sin 28^\circ}.$$

So we begin by finding sin 28°. An eight-digit display shows this as 0.4694716. By taking the reciprocal of this number, we will have csc 28°. Pushing the reciprocal button [1/x] displays 2.1300545. Thus

$$\csc 28^\circ = 2.1300545 \approx 2.130.$$

(b) The secant is the reciprocal of the cosine. So

$$\sec 72.5^\circ = \frac{1}{\cos 72.5^\circ}.$$

We begin by finding cos 72.5°. We get 0.3007058. Taking the reciprocal of this gives 3.3255095. Hence

$$\sec 72.5^\circ \approx 3.326.$$

(c) The cotangent is the reciprocal of the tangent. Hence

$$\cot 21.25^\circ = \frac{1}{\tan 21.25^\circ}.$$

So we find tan 21.25° and then take its reciprocal. We get

$$\cot 21.25^\circ \approx 2.571.$$

▲

Note that in the computations of Example 1, we are taking the reciprocal of a trigonometric value. We are *not* taking the reciprocal of an angle. Thus, in part (a), entering 28, followed by [1/x], followed by [sin] would be *incorrect*.

▲ ***Exercise*** Find the trigonometric value to four significant digits.

(a) csc 68° (b) sec 17.1° (c) cot 78.35°

Answer (a) 1.079 (b) 1.046 (c) 0.2062

Let's make a few observations about the reciprocal ratios. For an acute angle α we know that

$$\sin\alpha < 1 \quad \text{and} \quad \cos\alpha < 1.$$

Thus

$$\csc\alpha = \frac{1}{\sin\alpha} > 1 \quad \text{and} \quad \sec\alpha = \frac{1}{\cos\alpha} > 1.$$

The cotangent ratio, however, may assume any positive value since the tangent may assume any positive value.

Furthermore, the secant and cosecant ratios behave like sine and cosine with respect to complementary angles. That is, *the secant of an angle equals the cosecant of its complement*, and vice versa. The reason is that if α and β are complementary acute angles, then

$$\sec\alpha = \frac{1}{\cos\alpha} = \frac{1}{\sin\beta} = \csc\beta,$$

and

$$\csc\alpha = \frac{1}{\sin\alpha} = \frac{1}{\cos\beta} = \sec\beta.$$

Thus, for example, $\sec 15^\circ = \csc 75^\circ$ and $\csc 50^\circ = \sec 40^\circ$. Like sine and cosine, then, the secant and cosecant ratios are called coratios. Similarly, the tangent and cotangent are coratios.

Suppose that for one of these reciprocal ratios we are given its value and asked to determine the angle. We may do so by using the calculator, utilizing the reciprocal button and the corresponding principal ratio. See Example 2.

EXAMPLE 2 Find angle α to the nearest hundredth of a degree, given the trigonometric value.

(a) $\sec\alpha = 1.5$ (b) $\csc\alpha = 2.3$ (c) $\cot\alpha = 0.895$

SOLUTION (a) Since the cosine is the reciprocal of the secant,

$$\cos\alpha = \frac{1}{\sec\alpha} = \frac{1}{1.5}.$$

That is, $\cos\alpha$ is the reciprocal of 1.5. So we are looking for the angle whose cosine is the reciprocal of 1.5. To determine this we enter 1.5 and then push the reciprocal button [1/x]. An eight-digit display shows 0.6666667. Next we push [inv] [cos] (or its equivalent [arc] [cos] or [$\cos^{-1}$]), displaying 48.189685. Thus

$$\alpha = 48.189685^\circ \approx 48.19^\circ.$$

(b) Since the sine is the reciprocal of the cosecant,

$$\sin \alpha = \frac{1}{\csc \alpha} = \frac{1}{2.3}.$$

So we are looking for the angle whose sine is the reciprocal of 2.3. We begin by entering 2.3 and taking its reciprocal, getting 0.4347826. Pushing [inv] [sin] gives an angle with this sine. We get 25.771462. Thus

$$\alpha \approx 25.77^\circ.$$

(c) Since the cotangent is the reciprocal of the tangent,

$$\tan \alpha = \frac{1}{\cot \alpha} = \frac{1}{0.895}.$$

We are looking for an angle whose tangent is the reciprocal of 0.895. So we enter .895, take its reciprocal, and then push [inv] [tan]. We get

$$\alpha \approx 48.17^\circ.$$

▲

Like Example 1, the computations of Example 2 are such that we are taking the reciprocal of a trigonometric value. We are *not* taking the reciprocal of an angle. In part (c), entering .895, following by [inv] [tan], followed by [1/x] is *not* correct.

▲ ***Exercise*** Find angle α to the nearest hundredth of a degree, given the trigonometric value.

(a) $\sec \alpha = 1.7$ (b) $\csc \alpha = 2.1$ (c) $\cot \alpha = 0.791$

Answer (a) 53.97° (b) 28.44° (c) 51.66°

The reciprocal ratios may also be viewed in terms of the sides of a right triangle such as the one shown in Figure 2.25.

Figure 2.25

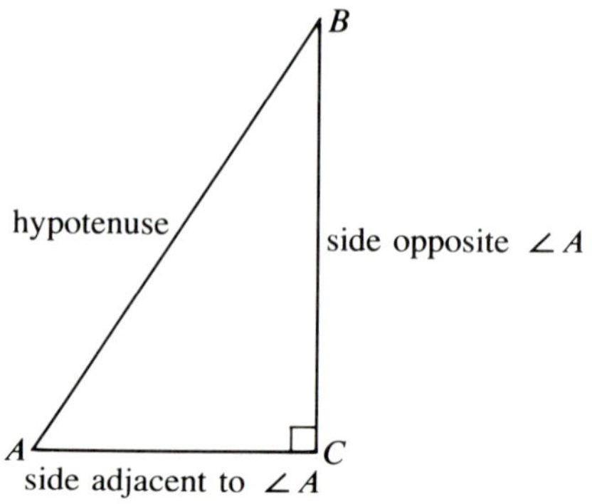

Indeed,

$$\text{cosecant of } \angle A = \frac{1}{\text{sine of } \angle A} = \frac{1}{\frac{\text{opposite}}{\text{hypotenuse}}} = \frac{\text{hypotenuse}}{\text{opposite}}.$$

Similarly,

$$\text{secant of } \angle A = \frac{\text{hypotenuse}}{\text{adjacent}},$$

and

$$\text{cotangent of } \angle A = \frac{\text{adjacent}}{\text{opposite}}.$$

Let's use this alternative view of the reciprocal ratios in Examples 3 and 4.

EXAMPLE 3 Find sec 30°, csc 30°, and cot 30°, exactly.

SOLUTION We first draw a 30°-60°-90° triangle, labeling the angles and sides as shown in Figure 2.26.

Figure 2.26

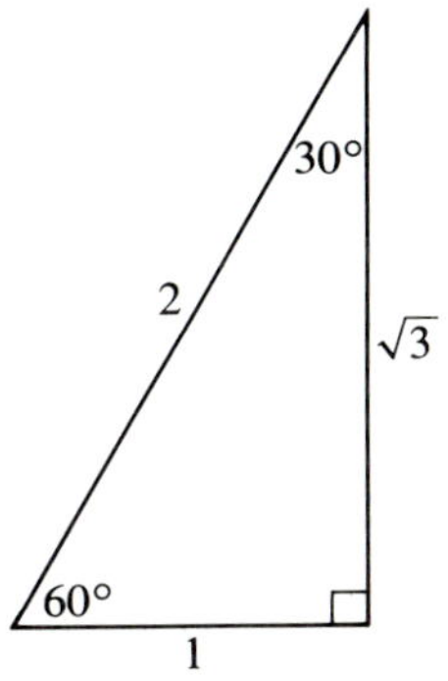

We have

$$\sec 30^\circ = \frac{\text{hypotenuse}}{\text{adjacent}} = \frac{2}{\sqrt{3}} \left(\text{or } \frac{2\sqrt{3}}{3}\right),$$

$$\csc 30^\circ = \frac{\text{hypotenuse}}{\text{opposite}} = \frac{2}{1} = 2,$$

$$\cot 30^\circ = \frac{\text{adjacent}}{\text{opposite}} = \frac{\sqrt{3}}{1} = \sqrt{3}.$$

▲

▲ *Exercise* Find sec 60°, csc 60°, and cot 60°, exactly.

Answer $\sec 60° = 2$, $\csc 60° = \frac{2}{\sqrt{3}}\left(\text{or } \frac{2\sqrt{3}}{3}\right)$, $\cot 60° = \frac{1}{\sqrt{3}}\left(\text{or } \frac{\sqrt{3}}{3}\right)$

EXAMPLE 4 If $\sec \alpha = \frac{7}{3}$, find $\cot \alpha$, exactly.

SOLUTION We construct a right triangle with an angle whose secant is $\frac{7}{3}$, as shown in Figure 2.27.

Figure 2.27

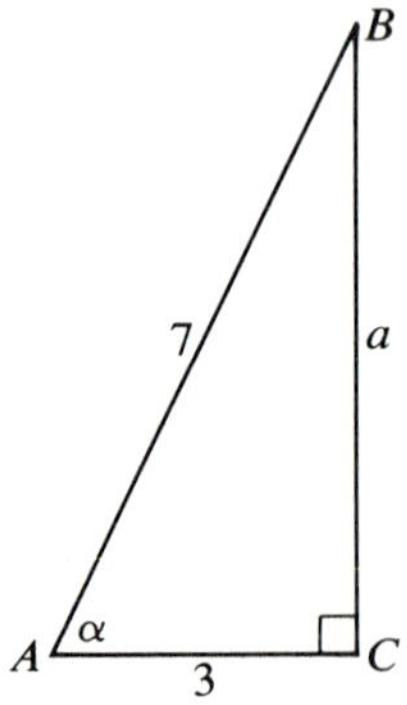

By the Pythagorean Theorem, $a^2 = 7^2 - 3^2 = 40$. Thus $a = \sqrt{40} = 2\sqrt{10}$, and

$$\cot \alpha = \frac{3}{2\sqrt{10}}\left(\text{or } \frac{3\sqrt{10}}{20}\right).$$

▲

▲ *Exercise* If $\csc \alpha = \frac{8}{5}$, find $\cot \alpha$, exactly.

Answer $\frac{\sqrt{39}}{5}$

Problem Set 2.5

In Problems 1–6, answers are to be exact.

1. If $\tan \alpha = \frac{3}{8}$, find $\cot \alpha$.

2. If $\tan \alpha = \sqrt{5}$, find $\cot \alpha$.

3. If $\sin \alpha = \frac{\sqrt{2}}{5}$, find $\csc \alpha$.

4. If $\sin \alpha = \frac{2}{7}$, find $\csc \alpha$.

5. If $\cos \alpha = \frac{5}{6}$, find $\sec \alpha$.

6. If $\cos \alpha = \frac{\sqrt{3}}{6}$, find $\sec \alpha$.

In Problems 7–12, find the trigonometric value to four significant digits.

7. $\csc 15.7^\circ$

8. $\csc 71^\circ$

9. $\sec 84.1^\circ$

10. $\sec 22.05^\circ$

11. $\cot 61.33^\circ$

12. $\cot 28^\circ$

In Problems 13–20, find angle α to the nearest hundredth of a degree, given the trigonometric value.

13. $\csc \alpha = 5$

14. $\csc \alpha = 2.486$

15. $\sec \alpha = 2.4$

16. $\sec \alpha = 7.6$

17. $\cot \alpha = 1.327$

18. $\cot \alpha = 0.15$

19. $\sec \alpha = 0.52$

20. $\csc \alpha = 0.7$

In Problems 21 and 22, calculate the secant, cosecant, and cotangent ratios for $\angle A$.

21.

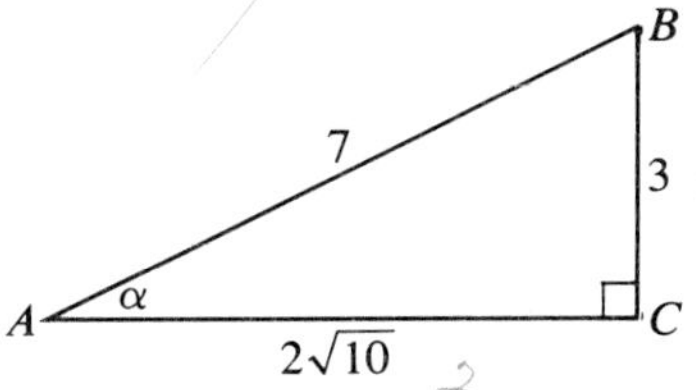

22.

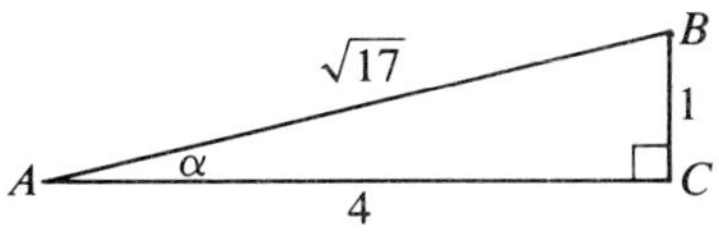

In Problems 23–26, answers are to be exact.

23. If $\cot \alpha = \frac{2}{5}$, find $\csc \alpha$.

24. If $\sin \alpha = \frac{\sqrt{5}}{4}$, find $\cot \alpha$.

25. If $\sec \alpha = \frac{7}{3}$, find $\tan \alpha$.

26. If $\csc \alpha = \frac{8}{5}$, find $\sec \alpha$.

In Problems 27–32, find the exact values of the other trigonometric ratios of α for the given conditions.

27. $\sin \alpha = \frac{\sqrt{3}}{5}$

28. $\cos \alpha = \frac{1}{5}$

29. $\cot \alpha = 4$

30. $\tan \alpha = \frac{2}{\sqrt{7}}$

31. $\sec \alpha = \sqrt{5}$

32. $\csc \alpha = \frac{5}{4}$

33. Find csc α to four significant digits if tan $\alpha = 4.72$.

34. Find cot α to four significant digits if sin $\alpha = 0.791$.

35. Find sec α to four significant digits if cot $\alpha = 3.052$.

36. Find cot α to four significant digits if csc $\alpha = 1.595$.

37. Find the exact values of sec 45°, csc 45°, and cot 45°.

CHAPTER 2 REVIEW

Key Terms

angle of depression, page 68

angle of elevation, page 68

coratios, page 52

principal trigonometric ratios, page 48

$$\text{sine} = \frac{\text{opposite}}{\text{hypotenuse}}$$

$$\text{cosine} = \frac{\text{adjacent}}{\text{hypotenuse}}$$

$$\text{tangent} = \frac{\text{opposite}}{\text{adjacent}}$$

reciprocal ratios, page 73

$$\text{cosecant} = \frac{1}{\text{sine}}$$

$$\text{secant} = \frac{1}{\text{cosine}}$$

$$\text{cotangent} = \frac{1}{\text{tangent}}$$

solving a triangle, page 62

Problem Set

1. Specify "true" or "false":

(a) If sin $\alpha = x$, then cos $(90° - \alpha) = x$.

(b) The sine and the cosine of an acute angle are always less than 1.

(c) The tangent of an acute angle is always less than 1.

(d) It is possible to solve a right triangle if we know only the measures of the two acute angles.

(e) If sec $\alpha = x$, then $\cos \dfrac{1}{\alpha} = x$.

(f) If cot $\alpha = x$, then $\tan (90° - \alpha) = \dfrac{1}{x}$.

(g) The only acute angle that has equal tangent and cotangent ratios is 45°.

(h) If $\sin (90° - \alpha) = \dfrac{1}{x}$, then sec $\alpha = x$.

(i) The secant and the cosecant of an acute angle are always less than 1.

2. Calculate the three principal trigonometric ratios for $\angle A$ using the following figure.

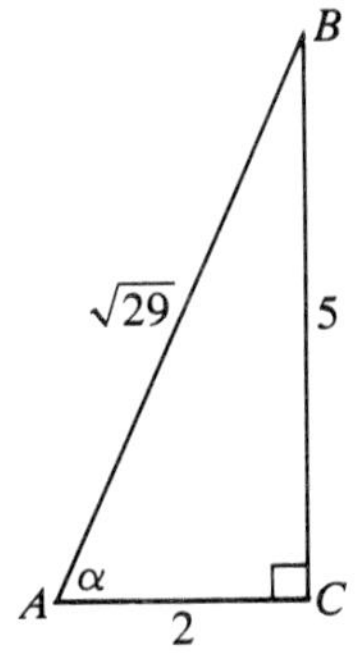

In Problems 3–6, answers are to be exact.

3. If $\sin\alpha = \frac{5}{7}$, find $\cos\alpha$.

4. If $\cos\alpha = \frac{\sqrt{3}}{6}$, find $\tan\alpha$.

5. If $\tan\alpha = \frac{3}{5}$, find $\sec\alpha$.

6. If $\sec\alpha = \frac{3\sqrt{2}}{4}$, find $\tan\alpha$.

In Problems 7–14, find the exact trigonometric value.

7. $\sin 45^\circ$

8. $\cot 45^\circ$

9. $\cos 30^\circ$

10. $\cot 30^\circ$

11. $\sin 60^\circ$

12. $\tan 60^\circ$

13. $\csc 30^\circ$

14. $\sec 60^\circ$

In Problems 15–20, find the trigonometric value to four significant digits.

15. $\cos 81.2^\circ$

16. $\tan 31^\circ$

17. $\sin 54.21^\circ$

18. $\sec 18^\circ$

19. $\csc 28.5^\circ$

20. $\cot 71.215^\circ$

21. If a projectile is fired from ground level at an angle α to the horizontal, then its range, r (horizontal distance traveled), in feet, is given by

$$r = \frac{v_0^2 \sin 2\alpha}{32},$$

where v_0 is the initial velocity in feet per second. If $\alpha = 25^\circ$ and $v_0 = 120$, what is the range (to two significant digits)?

22. If a projectile is fired from ground level at an angle α to the horizontal, then its time in flight, T, in seconds, is given by

$$T = \frac{v_0 \sin \alpha}{16},$$

where v_0 is the initial velocity in feet per second. How long (to two significant digits) will a soccer ball stay in flight if kicked 80 feet per second at an angle of $15°$?

In Problems 23–28, find angle α to the nearest hundredth of a degree, given the trigonometric value.

23. $\sin \alpha = 0.27$

24. $\tan \alpha = 4$

25. $\cos \alpha = 0.1531$

26. $\cot \alpha = 1.39$

27. $\sec \alpha = 5.12$

28. $\csc \alpha = 3.5$

29. Neglecting friction and air resistance, the velocity v of a ball when it reaches the bottom of an inclined plane, having started from rest, is

$$v = 8\sqrt{l \sin \alpha},$$

where v is in feet per second, l is the length of the plane in feet, and α is the angle of inclination. What is the angle (to the nearest degree) if the length of the plane is 3 feet and the velocity at the bottom is 12 feet per second?

30. As light passes from one medium to another, part of the light may be reflected and part refracted. Brewster's angle is the angle of incidence such that the reflected ray is perpendicular to the refracted ray. (At this angle the reflected beam is completely polarized.) See the following figure.

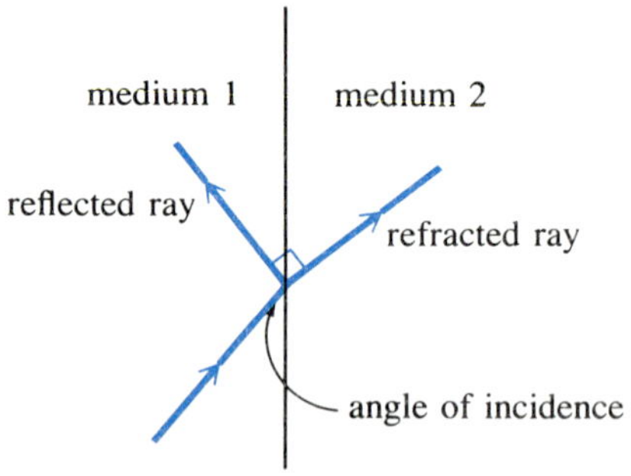

Brewster's angle α is given by

$$\tan \alpha = \frac{n_2}{n_1},$$

where n_2 is the index of refraction for the second medium and n_1 is the index for the first. Find Brewster's angle (to the nearest degree) as light passes from air into water if the index of refraction for air is 1.0003, and the index for water is 1.33.

In Problems 31–38, solve the right triangle ABC for the given conditions.

31. $\alpha = 59°, a = 4$

32. $\beta = 25°, a = 9$

33. $\alpha = 61.5°, b = 20$

34. $\beta = 41.2°, c = 10$

35. $a = 12, c = 17$

36. $b = 5.6, c = 7.9$

37. $a = 9, b = 8.5$

38. $a = 13, b = 15$

39. A guy wire on a telephone pole makes an angle of 35° with the ground. If it is attached to the pole at a point 20 feet above the ground, how long (to two significant digits) is the wire?

40. In the following figure we see that a radar is tilted 8° above horizontal so that its beam strikes the top of a thundercloud. If a cloud is 80 kilometers away, what is the altitude (to two significant digits) of the top of the thundercloud?

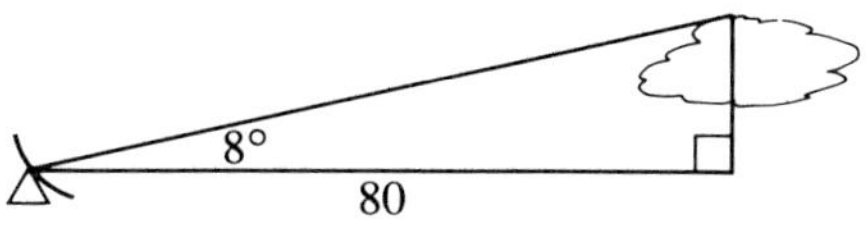

41. The Gateway Arch in St. Louis is 630 feet high. What is the angle of elevation (to the nearest tenth of a degree) to the top from a point on the ground 140 feet from the center of the base?

42. A searchlight at an airport is turned on at night to determine the height of the cloud ceiling. The light illuminates the base of the cloud, which is clearly visible from an observation point. In the following figure we see that the observation point is 850 meters from the searchlight, and the angle of elevation from the observation point to the base of the cloud is 9°. What is the height (to two significant digits) of the cloud ceiling?

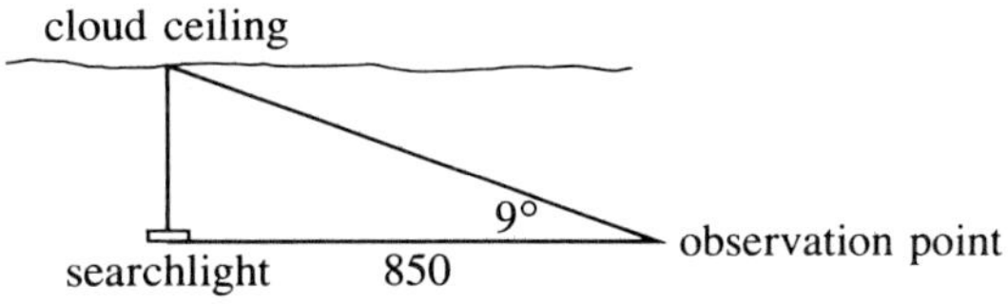

43. The horizontal distance from an observer to a point directly below the top of Pikes Peak is 10,000 feet (to the nearest 100 feet). If the angle of elevation to the top is 41.7°, and the observation is made 5200 feet above sea level, how high (above sea level, to the nearest 10 feet) is Pikes Peak?

44. A surveying team wants to measure the width of a river whose banks are the same distance above the water. A 4-foot surveying transit is placed on one side of the river, and a 10-foot pole is placed on the other. If the angle of elevation to the top of the pole is 2°, how wide (to two significant digits) is the river?

45. The Empire State Building is capped by a TV tower. From ground level, 350 feet from a point directly below the tower, the angle of elevation to the top of the building is $74.4°$, and the angle of elevation to the top of the tower is $76.6°$. What is the height of the building (to three significant digits), and what is the height of the TV tower (to the nearest 10 feet)?

46. When shooting free throws, a basketball player is directly in front of the goal. The goal is 13 feet, 3 inches from the free throw line (measured horizontally), and the diameters of the goal and ball are 18 inches and 9.5 inches, respectively. If a player shoots $5°$ off line, by how much (to the nearest inch) will the ball miss the rim?

47. If a satellite at an altitude of 110 miles subtends an angle of $0.002°$ on the earth's surface, what is the diameter of the satellite (in feet, to two significant digits)?

3
GENERAL TRIGONOMETRY

CHAPTER OUTLINE

In this chapter we change the framework of our study from right triangle to Cartesian plane. With this new framework the concepts of angle measure and trigonometric ratios are extended. The chapter concludes with a study of equations containing trigonometric ratios. The solutions of these equations involve all of the basic ideas presented.

3.1 ANGLE MEASUREMENT EXTENDED

In this section we expand our concept of the measure of an angle. We shall no longer limit the idea to a positive measure between 0° and 180°.

We know that an angle consists of two rays that originate from a common point. The common point is called the vertex of the angle, and each ray is called a side. It is now necessary to identify one of the sides as the *initial side* and the other as the *terminal side*. We may rotate the initial side about the vertex until it coincides with the terminal side. The amount of such rotation is called the **measure** of the angle. The measure is considered positive if the rotation is counterclockwise; it is regarded as negative when the rotation is clockwise.

The measure of an angle BAC with $\overrightarrow{AC}$ the terminal side and $\overrightarrow{AB}$ the initial side may be either positive or negative. To clarify, we must indicate the direction of rotation with a small curved arrow. See Figure 3.1.

Figure 3.1

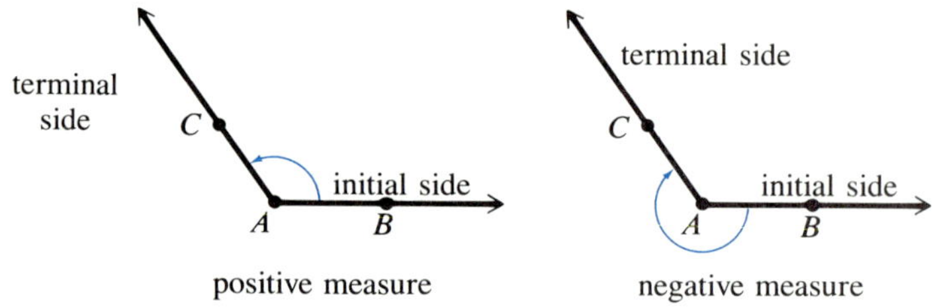

Note that not only may the initial side be made coincident with the terminal side by rotating in either direction, but we may also use more than one complete rotation. See Figure 3.2.

Figure 3.2

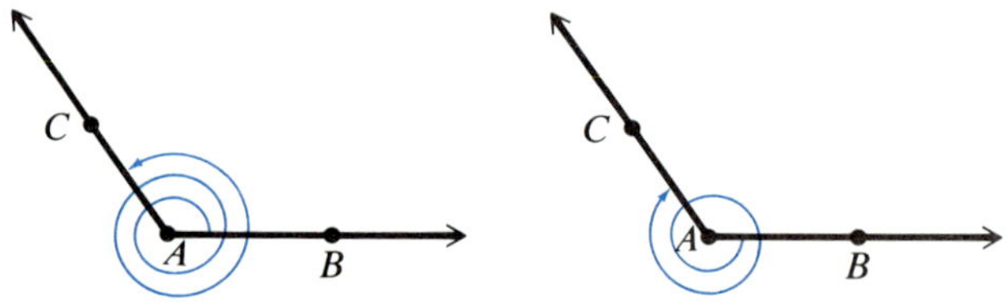

Hence the measure of a given angle depends on both the direction of rotation and the number of rotations indicated. (If the curved arrow is omitted, we will assume that the measure is positive and less than one-half of a complete rotation. Thus the angles in triangles have measures between 0° and 180°.)

We are expanding the meaning of degree. A **degree** is $1/360$ of one complete counterclockwise rotation. Thus

$$360^\circ = 1 \text{ complete counterclockwise rotation}.$$

Figure 3.3

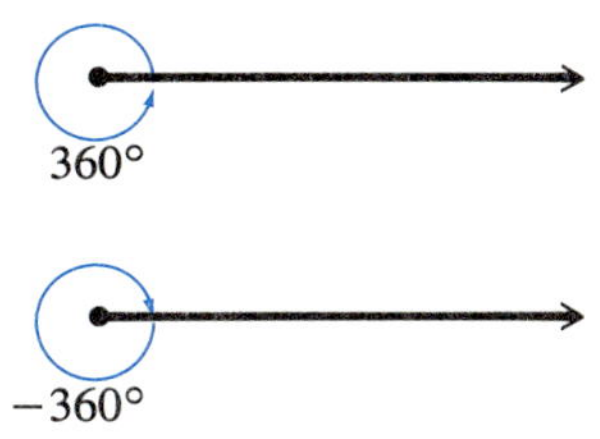

An angle of measure 360°, or of measure −360°, then, will have its initial and terminal sides coincide, as seen in Figure 3.3.

Some other angles and their measures are indicated in Figure 3.4.

Figure 3.4

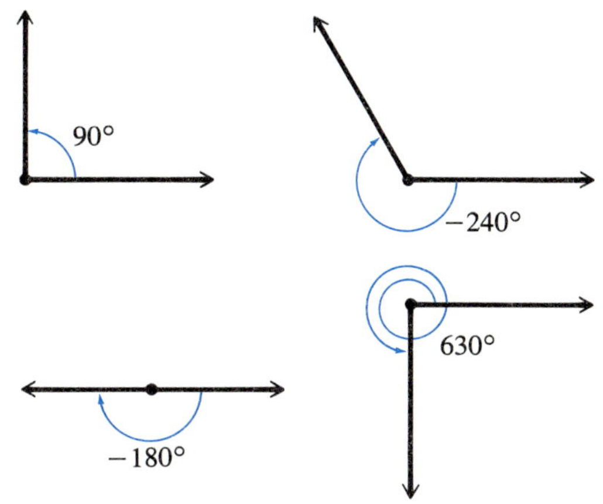

An angle is said to be in **standard position**, with respect to the Cartesian plane, provided that its vertex is at the origin and its initial side coincides with the positive half of the x-axis. In Figure 3.5 we see two angles, one positive and one negative, in standard position.

Figure 3.5

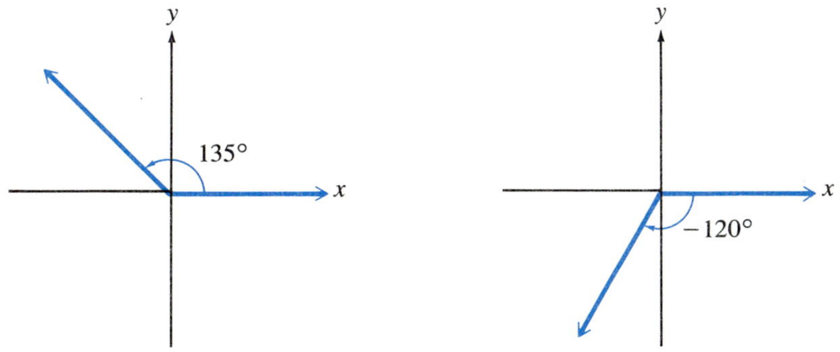

If the terminal side of an angle *in standard position* lies in the first quadrant, we say that the angle is in the first quadrant and call it a **first-quadrant angle**. If the terminal side lies in the second quadrant, the angle is called a **second-quadrant angle**. Third- and fourth-quadrant angles are defined similarly. If the terminal side coincides with an axis, the angle is called **quadrantal**. For example, angles of measure 90°, 180°, and −90° are quadrantal.

Two angles are called **coterminal** provided that their terminal sides coincide when in standard position. Of course, this does not mean that their measures are equal. In Figure 3.6 we see two coterminal angles with measures of 150° and −210°. In fact, given an angle, we can find infinitely many coterminal angles by adding multiples of 360° to or subtracting multiples of 360° from its measures.

Figure 3.6

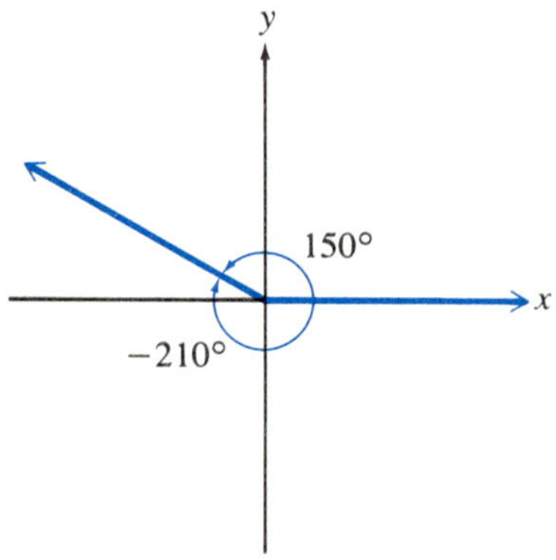

▲ *Exercise* Construct an angle in standard position with the given measure, state the quadrant in which the terminal side lies, and give the measure of a coterminal angle with opposite direction.

(a) 135° (b) −60°

Answer See Figure 3.7.

Figure 3.7 (a) second, −225° (b) fourth, 300°

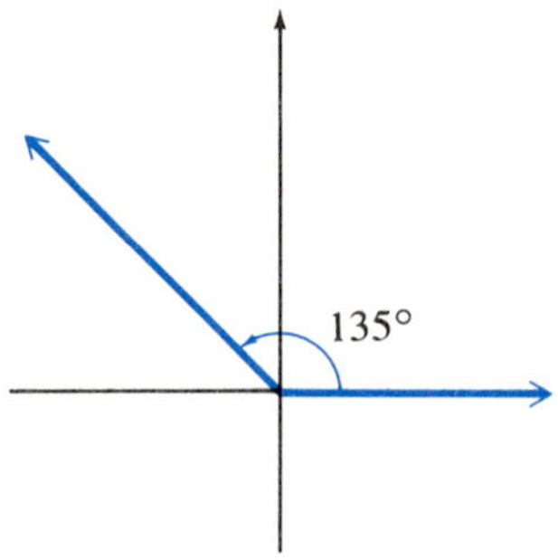

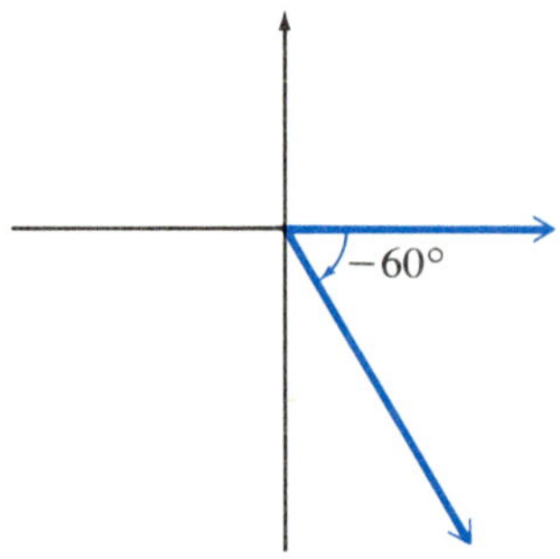

In standard position every nonquadrantal angle is associated with a particular acute angle. The acute angle made by the terminal side of an angle in standard position and the x-axis is called the **reference angle**. In Figure 3.8 we see an angle in each of the four quadrants with the reference angle labeled θ' (theta prime).

Figure 3.8

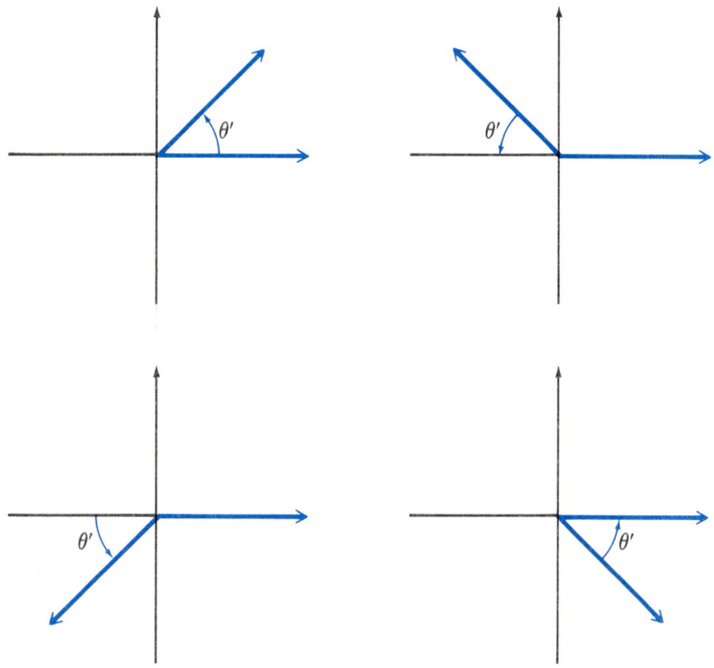

Note that the reference angle is formed with the x-axis. The y-axis is never used. Thus, for an angle of 210° the reference angle is 30°. For an angle of 385° the reference angle is 25°. An angle of −240° has a reference angle of 60°, and an angle of 295° has a reference angle of 65°. See Figure 3.9.

Figure 3.9

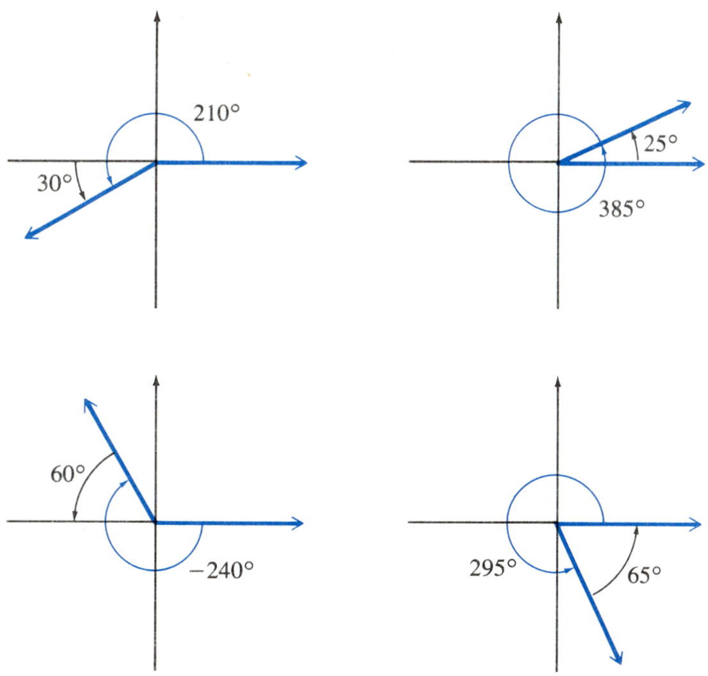

▲ *Exercise* Determine the reference angle.

(a) 120° (b) −135° (c) 392° (d) 297°

Answer (a) 60° (b) 45° (c) 32° (d) 63°

The following chart may also be used in determining the reference angle.

If θ is in quadrant	*For 0° ≤ θ < 360° the reference angle is*
I	θ
II	$180° - \theta$
III	$\theta - 180°$
IV	$360° - \theta$

If $\theta \geq 360°$ or if $\theta < 0°$, we may first find the nonnegative angle less than 360° that is coterminal with it and then find the reference angle. For example, an angle of 500° (being coterminal with one of 140°) has a reference angle of 40°, and an angle of −125° (being coterminal with one of 235°) has a reference angle of 55°.

Most of the time we will be determining θ from its reference angle instead of the reverse.

EXAMPLE 1 Find θ for the given conditions.

(a) $0° \leq \theta < 360°$, θ is in quadrant II, and $\theta' = 70°$
(b) $-180° < \theta < -90°$ and $\theta' = 35°$

SOLUTION (a) We have the situation shown in Figure 3.10.

Figure 3.10

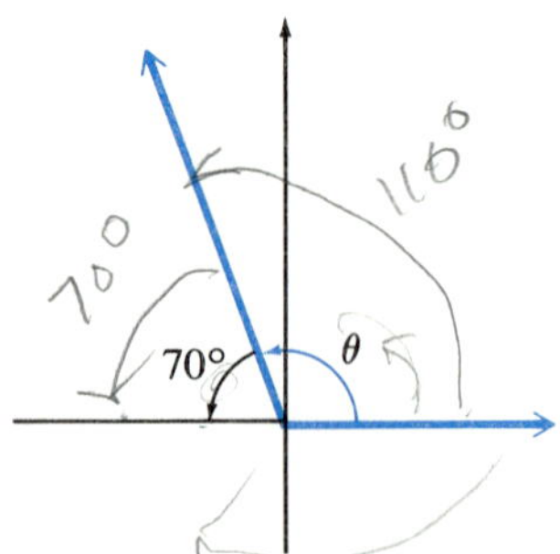

Thus θ is a straight angle, measured positively, less the reference angle. So

$$\theta = 180° - 70° = 110°.$$

(b) We have the situation shown in Figure 3.11.

Figure 3.11

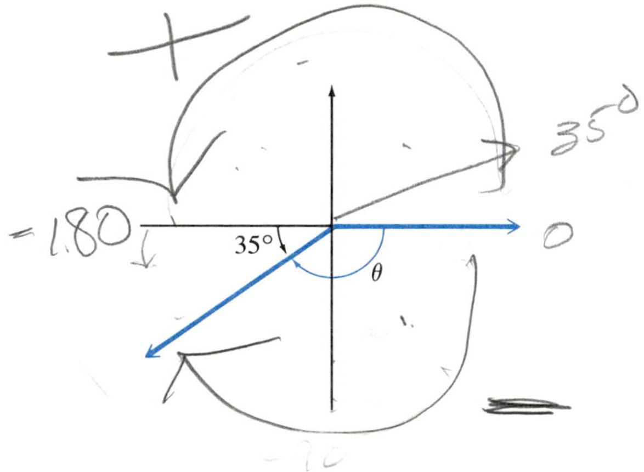

Hence, θ is a straight angle, measured negatively, plus the reference angle. So

$$\theta = -180^\circ + 35^\circ = -145^\circ$$

▲

▲ *Exercise* Find θ for the given conditions.

(a) $0^\circ \leq \theta < 360^\circ$, θ is in quadrant III, and $\theta' = 65^\circ$
(b) $-270^\circ < \theta < -180^\circ$ and $\theta' = 25^\circ$

Answer (a) 245° (b) -205°

Problem Set 3.1

In Problems 1–8, construct an angle in standard position with the given measure, state the quadrant in which the terminal side lies, and give the measure of a coterminal angle with opposite direction.

1. 140°

2. 320°

3. 245°

4. 70°

5. -80°

6. -205°

7. -298°

8. -110°

9. Determine the set of measures of all angles coterminal with an angle of 110°.

10. Determine the set of measures of all angles coterminal with an angle of -20°.

11. Find the point on the terminal side of $\theta = -60^\circ$ that is 2 units from the origin.

12. Find the point on the terminal side of $\theta = 225^\circ$ that is 2 units from the origin.

In Problems 13–22, determine the reference angle.

13. 155°

14. 242°

15. 325°

16. 37°

17. -28°

18. -320°

19. -126°

20. -212°

21. 485°

22. -520°

In Problems 23–26, find θ for the given conditions, where $0^\circ \leq \theta < 360^\circ$.

23. θ is in quadrant III and $\theta' = 10^\circ$.

24. θ is in quadrant II and $\theta' = 75^\circ$.

25. θ is in quadrant IV and $\theta' = 20^\circ$.

26. θ is in quadrant I and $\theta' = 25^\circ$.

In Problems 27–32, find θ for the given conditions.

27. $-180^\circ < \theta < -90^\circ$ and $\theta' = 15^\circ$

28. $-270^\circ < \theta < -180^\circ$ and $\theta' = 50^\circ$

29. $-360^\circ < \theta < -270^\circ$ and $\theta' = 68^\circ$

30. $-90^\circ < \theta < 0^\circ$ and $\theta' = 17^\circ$

31. $-90^\circ < \theta < 90^\circ$ and $\theta' = 40^\circ$

32. $360^\circ < \theta < 540^\circ$ and $\theta' = 5^\circ$

3.2 TRIGONOMETRIC RATIOS EXTENDED

The concept of angle measure was extended in the preceding section. Because we are no longer restricted to acute angles, we must extend the definitions of the trigonometric ratios.

Let θ be the measure of an angle in standard position. Let $P(x, y)$ be any point, excluding the vertex (origin), on the terminal side. Let r be the distance from P to the origin; hence $r = \sqrt{x^2 + y^2}$. If θ is acute, then we have the situation of Figure 3.12.

We have constructed a right triangle by dropping a perpendicular segment from P. Consequently,

$$x = \text{adjacent},$$

$$y = \text{opposite},$$

$$r = \text{hypotenuse}.$$

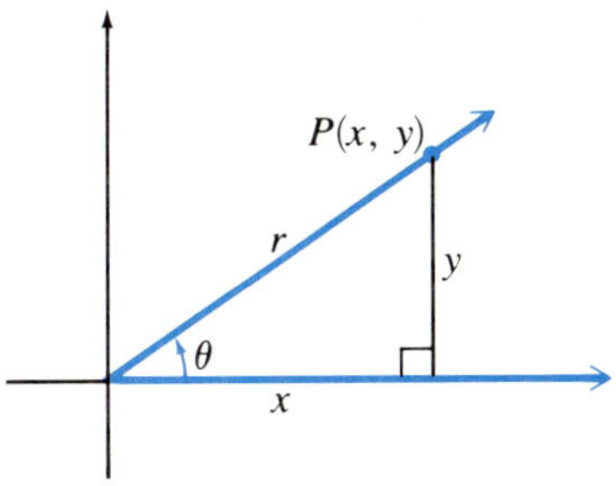

Figure 3.12

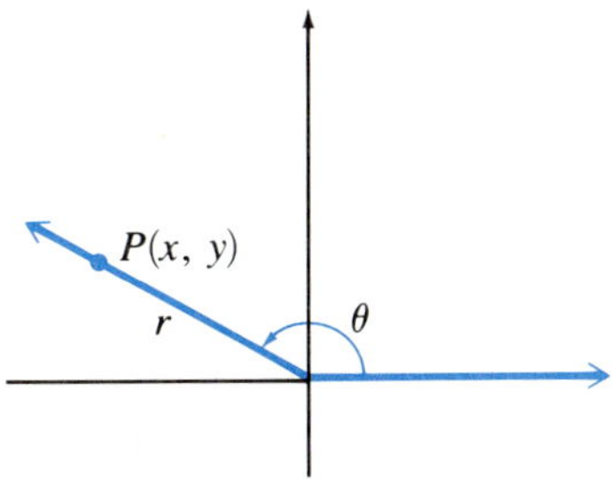

Figure 3.13

Therefore,

$$\sin\theta = \frac{\text{opposite}}{\text{hypotenuse}} = \frac{y}{r},$$

$$\cos\theta = \frac{\text{adjacent}}{\text{hypotenuse}} = \frac{x}{r},$$

$$\tan\theta = \frac{\text{opposite}}{\text{adjacent}} = \frac{y}{x}.$$

The reciprocal ratios, of course, would also be defined in terms of x, y, and r.

However, θ doesn't have to be acute. It might be a second-quadrant angle like that of Figure 3.13.

We can still compute the ratios in terms of x, y, and r, however. This gives us the definitions of the **trigonometric ratios**.

Name	***Abbreviation***	***Definition***
sine θ	$= \sin\theta$	$= \frac{y}{r}$
cosine θ	$= \cos\theta$	$= \frac{x}{r}$
tangent θ	$= \tan\theta$	$= \frac{y}{x}$ (if $x \neq 0$)
cotangent θ	$= \cot\theta$	$= \frac{x}{y}$ (if $y \neq 0$)
secant θ	$= \sec\theta$	$= \frac{r}{x}$ (if $x \neq 0$)
cosecant θ	$= \csc\theta$	$= \frac{r}{y}$ (if $y \neq 0$)

Let's consider an angle θ whose terminal side passes through $(-5, 12)$ and calculate these six ratios. We have the situation shown in Figure 3.14.

Figure 3.14

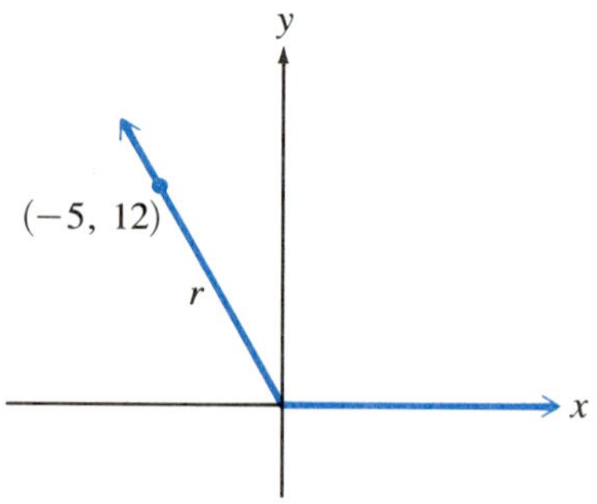

Thus

$$r = \sqrt{(-5)^2 + 12^2} = \sqrt{169} = 13.$$

Using the definitions, then, we have

$$\sin\theta = \frac{y}{r} = \frac{12}{13},$$

$$\cos\theta = \frac{x}{r} = \frac{-5}{13},$$

$$\tan\theta = \frac{y}{x} = \frac{12}{-5},$$

$$\cot\theta = \frac{x}{y} = \frac{-5}{12},$$

$$\sec\theta = \frac{r}{x} = \frac{13}{-5},$$

$$\csc\theta = \frac{r}{y} = \frac{13}{12}.$$

Note that the measure of θ is not indicated in Figure 3.14. It may be positive, negative, or more than one rotation. It doesn't matter as long as we have a point on the terminal side when the angle is in standard position.

▲ ***Exercise*** Find the values of the trigonometric ratios if the terminal side of θ passes through $(3, -4)$.

Answer $\sin\theta = \frac{-4}{5}$, $\cos\theta = \frac{3}{5}$, $\tan\theta = \frac{-4}{3}$, $\cot\theta = \frac{-3}{4}$, $\sec\theta = \frac{5}{3}$, $\csc\theta = \frac{-5}{4}$

There are several observations to be made about these definitions. First, since r is never 0, the sine and cosine ratios will always be defined. But the other four ratios don't always exist because of a possible zero in the denominator. If the angle is quadrantal, either $x = 0$ or $y = 0$. If $x = 0$, the tangent and the secant will not be defined. If $y = 0$, the cotangent and the cosecant will not be defined. Hence if θ is the measure of a quadrantal angle, two of the six ratios will not be defined.

Second, the values of the ratios are independent of the particular point P on the terminal side. If $P'(x', y')$ is another point on the terminal side, and r' is its distance to the origin, we may construct a pair of similar triangles. See Figure 3.15, in which a second-quadrant angle is shown.

Figure 3.15

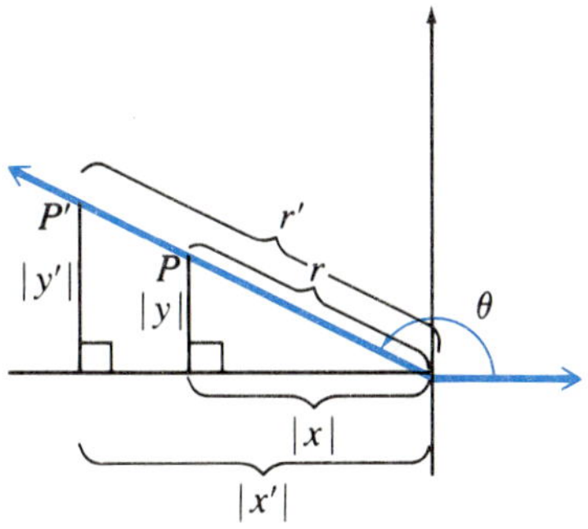

Because corresponding sides are proportional, the six ratios will be the same, regardless of which point is used. For example,

$$\frac{|x|}{r} = \frac{|x'|}{r'}.$$

Therefore,

$$\frac{-x}{r} = \frac{-x'}{r'}$$

$$\frac{x}{r} = \frac{x'}{r'}.$$

The left side of the preceding equation is the cosine ratio using P, while the right side is the cosine ratio using P'. Hence the values of the trigonometric ratios depend only on the position of the terminal side (which in turn depends on the measure of the angle). Coterminal angles will therefore have the same trigonometric values.

If the angle is acute, these ratios are the same as the ones defined in Chapter 2 for right triangles. This is no surprise since we let the triangle definitions motivate the general ones. Of course, the reciprocal relationships observed in Chapter 2 still exist. Since $\sin \theta = y/r$ and $\csc \theta = r/y$ $(y \neq 0)$, we have

$$\sin \theta = \frac{1}{\csc \theta} \quad \text{and} \quad \csc \theta = \frac{1}{\sin \theta}.$$

Similarly,

$$\cos \theta = \frac{1}{\sec \theta} \quad \text{and} \quad \sec \theta = \frac{1}{\cos \theta},$$

$$\tan \theta = \frac{1}{\cot \theta} \quad \text{and} \quad \cot \theta = \frac{1}{\tan \theta}.$$

Further observation reveals that for any angle θ,

$$-1 \leq \sin \theta \leq 1.$$

This may be seen by examining the sine ratio. For any angle θ we have

$$\sin \theta = \frac{y}{r},$$

where (x, y) is a point on the terminal side and $r = \sqrt{x^2 + y^2}$. Since

$$|y| = \sqrt{y^2} \leq \sqrt{x^2 + y^2} = r,$$

$$\frac{|y|}{r} \leq 1.$$

Thus,

$$-1 \leq \frac{y}{r} \leq 1$$

$$-1 \leq \sin \theta \leq 1.$$

A similar analysis shows that for any angle θ,

$$-1 \leq \cos \theta \leq 1,$$

and

$$\sec \theta \geq 1 \quad \text{or} \quad \sec \theta \leq -1,$$

$$\csc \theta \geq 1 \quad \text{or} \quad \csc \theta \leq -1.$$

The tangent and cotangent ratios, on the other hand, may assume any value.

In the definition of the trigonometric ratios, since $r > 0$, the sign of the value of the ratio depends on the signs of x and y. Since $y > 0$ in the first and second quadrants and $y < 0$ in the third and fourth, the sine is positive for first- and second-quadrant angles and is negative for third- and fourth-quadrant angles. Of course, since the cosecant is its reciprocal, it also has the same property. In Figure 3.16, we show the sign of the given ratios by quadrants.

Figure 3.16

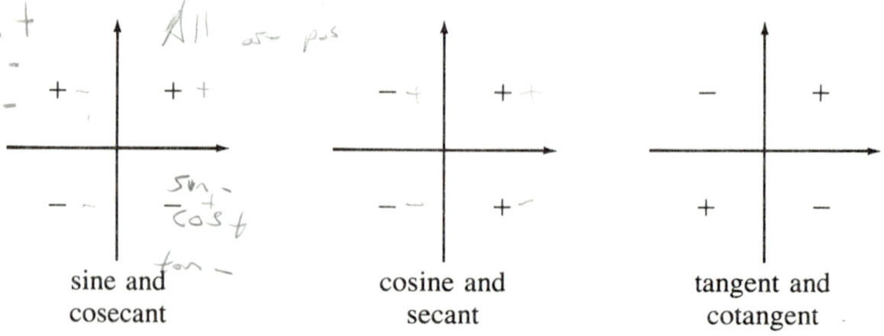

EXAMPLE 1 Find the values of the other trigonometric ratios of θ if $\sec \theta = -3$ and θ is in the third quadrant.

Figure 3.17

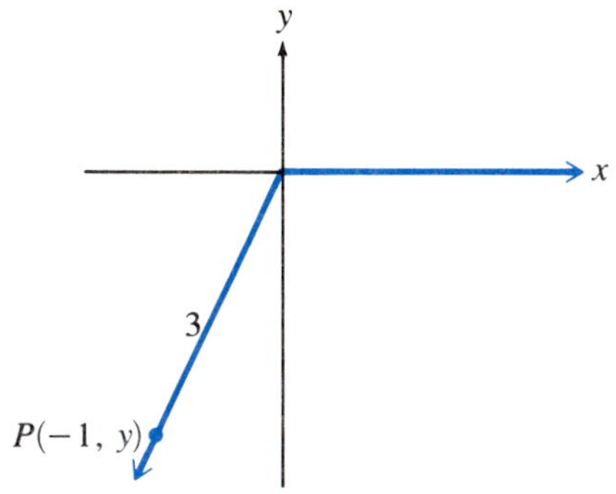

SOLUTION We must locate a point on the terminal side of the angle. Since

$$\sec\theta = \frac{r}{x} = -3,$$

we choose the point P such that $r = 3$ (it must be positive) and $x = -1$. See Figure 3.17.

Since

$$r^2 = x^2 + y^2,$$
$$y = \pm\sqrt{r^2 - x^2} = \pm\sqrt{9-1} = \pm 2\sqrt{2}.$$

Because θ is in the third quadrant, $y < 0$. Hence $y = -2\sqrt{2}$. So the point P is $(-1, -2\sqrt{2})$. We get

$$\sin\theta = \frac{y}{r} = \frac{-2\sqrt{2}}{3},$$

$$\cos\theta = \frac{x}{r} = \frac{-1}{3},$$

$$\tan\theta = \frac{y}{x} = 2\sqrt{2},$$

$$\cot\theta = \frac{1}{\tan\theta} = \frac{1}{2\sqrt{2}} \left(\text{or } \frac{\sqrt{2}}{4}\right),$$

$$\csc\theta = \frac{1}{\sin\theta} = \frac{-3}{2\sqrt{2}} \left(\text{or } \frac{-3\sqrt{2}}{4}\right).$$

▲

▲ ***Exercise*** Find the values of the other trigonometric ratios of θ if $\csc\theta = -4$ and θ is in the fourth quadrant.

Answer $\sin\theta = \frac{-1}{4}$, $\cos\theta = \frac{\sqrt{15}}{4}$, $\tan\theta = \frac{-1}{\sqrt{15}} \left(\text{or } \frac{-\sqrt{15}}{15}\right)$, $\cot\theta = -\sqrt{15}$, $\sec\theta = \frac{4}{\sqrt{15}} \left(\text{or } \frac{4\sqrt{15}}{15}\right)$

EXAMPLE 2 Find the values of the trigonometric ratios if $\theta = 270°$.

SOLUTION The terminal side of θ coincides with the negative half of the y-axis, as shown in Figure 3.18.

Figure 3.18

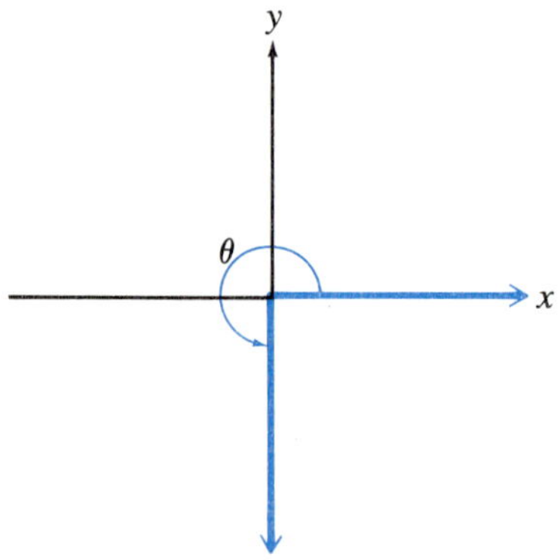

To calculate the ratios, we must choose a point on the terminal side. Since any point will do, let's choose a convenient one like $(0, -1)$. So $r = 1$. Hence, using the definitions,

$$\sin 270° = \frac{y}{r} = -1,$$

$$\cos 270° = \frac{x}{r} = 0,$$

$$\tan 270° = \frac{y}{x} = \text{not defined since } x = 0,$$

$$\cot 270° = \frac{x}{y} = 0,$$

$$\sec 270° = \frac{r}{x} = \text{not defined since } x = 0,$$

$$\csc 270° = \frac{r}{y} = -1.$$

▲

We have already noted that for a quadrantal angle, only four of the trigonometric ratios will be defined. If we choose the point P such that one of the coordinates is ± 1 (the other is 0, of course), we will have $r = 1$. Thus we see that two of the ratios will be 0, and the other two will be either 1 or -1. The two that aren't defined will be the reciprocals of those with values of 0.

▲ ***Exercise*** Find the values of the trigonometric ratios if $\theta = -180°$.

Answer $\sin(-180°) = 0$, $\cos(-180°) = -1$, $\tan(-180°) = 0$, $\cot(-180°) =$ not defined, $\sec(-180°) = -1$, $\csc(-180°) =$ not defined.

EXAMPLE 3 Find the sine, cosine, and tangent of 150°.

SOLUTION We need to choose a point on the terminal side. Since the reference angle is 30°, let's make use of the 30°-60°-90° triangle shown in Figure 3.19.

Figure 3.19

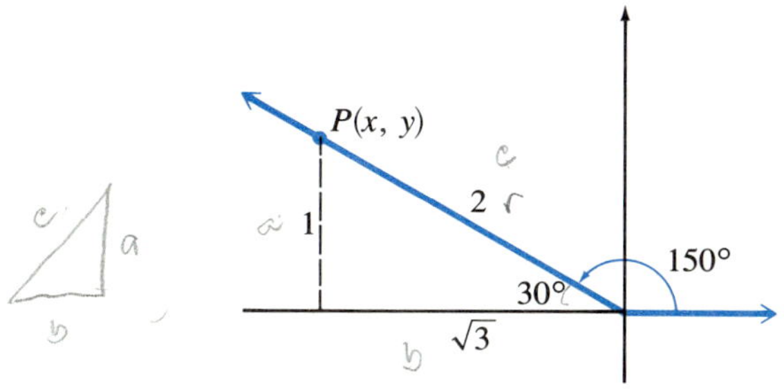

Since the terminal side is in the second quadrant, y is positive and x is negative. Thus $y = 1$ and $x = -\sqrt{3}$. From the definitions, then,

$$\sin 150^\circ = \frac{y}{r} = \frac{1}{2},$$

$$\cos 150^\circ = \frac{x}{r} = \frac{-\sqrt{3}}{2},$$

$$\tan 150^\circ = \frac{y}{x} = \frac{-1}{\sqrt{3}}\left(\text{or } \frac{-\sqrt{3}}{3}\right).$$

▲

▲ ***Exercise*** Find the sine, cosine, and tangent of 120°.

Answer $\sin 120^\circ = \frac{\sqrt{3}}{2}$, $\cos 120^\circ = \frac{-1}{2}$, $\tan 120^\circ = -\sqrt{3}$

EXAMPLE 4 Find the secant, cosecant, and cotangent of 225°.

SOLUTION We need to choose a point on the terminal side. Since the reference angle is 45°, let's make use of the 45°-45°-90° triangle shown in Figure 3.20.

Figure 3.20

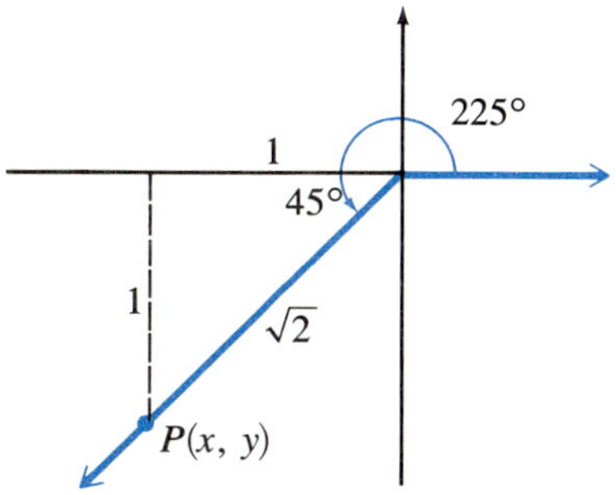

Since the terminal side is in the third quadrant, both x and y are negative. Thus $x = -1$ and $y = -1$. From the definitions, then,

$$\sec 225^\circ = \frac{r}{x} = -\sqrt{2},$$

$$\csc 225^\circ = \frac{r}{y} = -\sqrt{2},$$

$$\cot 225^\circ = \frac{x}{y} = 1.$$

▲

▲ *Exercise* Find the secant, cosecant, and cotangent of 315°.

Answer $\sec 315^\circ = \sqrt{2}$, $\csc 315^\circ = -\sqrt{2}$, $\cot 315^\circ = -1$

Problem Set 3.2

In Problems 1–8, find the values of the trigonometric ratios if the terminal side of θ passes through the given point.

1. $(-3, -4)$

2. $(-2, -3)$

3. $(-1, 2)$

4. $(-3, 1)$

5. $(6, -4)$

6. $(5, -12)$

7. $(1, 3)$

8. $(2, 2)$

In Problems 9–12, find the exact values of the other trigonometric ratios of θ for the given conditions.

9. $\sin\theta = \frac{4}{5}$, θ in the second quadrant

10. $\tan\theta = \frac{12}{5}$, θ in the third quadrant

11. $\sec\theta = -\sqrt{17}$, θ in the third quadrant

12. $\csc\theta = \frac{-8}{5}$, θ in the fourth quadrant

In Problems 13–16, answers are to be exact.

13. If $\cot\theta = \frac{-8}{5}$ and $\sin\theta < 0$, find $\csc\theta$.

14. If $\tan\theta = \dfrac{-5}{2}$ and $\cos\theta < 0$, find $\sin\theta$.

15. If $\sin\theta = \dfrac{-\sqrt{5}}{4}$ and $\tan\theta > 0$, find $\cos\theta$.

16. If $\cos\theta = \dfrac{3}{7}$ and $\sin\theta < 0$, find $\cot\theta$.

In Problems 17–20, find the values of the trigonometric ratios for the given measure.

17. 90° *18.* −90°

19. 180° *20.* 0°

In Problems 21–26, find the sine, cosine, and tangent of the given angle, exactly.

21. 210° *22.* 225°

23. −45° *24.* −240°

25. −210° *26.* 300°

In Problems 27–30, find the secant, cosecant, and cotangent of the given angle, exactly.

27. 330° *28.* 225°

29. −240° *30.* −135°

31. Show that $\sin^2\theta + \cos^2\theta = 1$ for all angles θ.

32. Show that $\tan\theta = \dfrac{\sin\theta}{\cos\theta}$ for all angles θ for which the tangent is defined.

3.3 TRIGONOMETRIC VALUES FOR GENERAL ANGLES

We can find the values of the trigonometric ratios for special angles and for quadrantal angles, but what about for other angles? For example, how do we find sin 198° or tan (−71.5°)? With calculators the process is as easy for general angles as it is for acute angles.

EXAMPLE 1 Find the trigonometric value to four significant digits.

(a) sin 198° (b) cos 322° (c) tan (−71.5°) (d) tan 270°

SOLUTION (a) *In the degree mode*, we enter 198 and then push the sine button [sin]. An eight-digit display shows -0.309017. Thus

$$\sin 198^\circ \approx -0.3090\,.$$

(b) We enter 322 and then push the cosine button [cos]. We get

$$\cos 322^\circ \approx 0.7880\,.$$

(c) We enter -71.5 (make sure you know how to enter a negative number with your calculator) and then push [tan]. We get

$$\tan(-71.5^\circ) \approx -2.989\,.$$

(d) We enter 270 and then push [tan], and we get an error message. This is because the tangent ratio is not defined at 270°. Indeed, all points on the terminal side have $x = 0$, so the ratio y/x is not defined.

▲

What does it mean to say that $\sin 198^\circ \approx -0.3090$? It means that, for any point on the terminal side of an angle of 198°, the ratio of the y-coordinate to the point's distance to the origin is approximately -0.3090.

▲ ***Exercise*** Find the trigonometric value to four significant digits.

(a) $\sin 127^\circ$ (b) $\cos 203^\circ$ (c) $\tan(-35.6^\circ)$ (d) $\tan(-90^\circ)$

Answer (a) 0.7986 (b) -0.9205 (c) -0.7159 (d) not defined

With secant, cosecant, and cotangent, we use the reciprocal relationships:

$$\sec\theta = \frac{1}{\cos\theta},$$

$$\csc\theta = \frac{1}{\sin\theta},$$

$$\cot\theta = \frac{1}{\tan\theta}.$$

EXAMPLE 2 Find the trigonometric value to four significant digits.

(a) $\sec 119^\circ$ (b) $\csc(-41.2^\circ)$ (c) $\cot 281^\circ$

SOLUTION (a) The secant is the reciprocal of the cosine, so

$$\sec 119^\circ = \frac{1}{\cos 119^\circ}.$$

So we begin by finding cos 119°. An eight-digit display shows this as −0.4848096. By taking the reciprocal of this number, we will have sec 119°. Pushing the reciprocal button [1/x] displays −2.0626653. Thus

$$\sec 119^\circ \approx -2.063\,.$$

(b) The cosecant is the reciprocal of the sine. So we find sin (−41.2°) and then take the reciprocal. We get

$$\csc(-41.2^\circ) \approx -1.518\,.$$

(c) The cotangent is the reciprocal of the tangent. So we find tan 281° and then push [1/x]. We get

$$\cot 281^\circ \approx -0.1944\,.$$

▲

▲ *Exercise* Find the trigonometric value to four significant digits.

(a) sec 137° (b) csc (−21.5°) (c) cot 313° (d) csc 180°

Answer (a) −1.367 (b) −2.729 (c) −0.9325 (d) not defined

Suppose now that we know the trigonometric value and wish to find the angle. For instance, suppose we want to find θ given that $\sin\theta = 0.3665012$. In the preceding chapter, when working with acute angles only, we merely entered .3665012 and then pushed [inv] [sin] (or its equivalent). If we do that now we get 21.5° (rounded).

However, θ isn't necessarily acute, and there are other general angles that have the given sine. There is 158.5°, there is −338.5° (try them), and there are infinitely-many more. So the determination with general angles is a little more complicated than with acute angles. We have to make use of the reference angle.

Let's begin by considering a second-quadrant angle of measure θ. Without loss of generality, we may assume that the measure of θ is between 90° and 180°. Let $P(x, y)$ be a point on the terminal side, let r be the distance from P to the origin, and let θ' be the reference angle. Let M be the point $(x, 0)$ and let O be the origin. Let's consider the right triangle OMP. We have $MP = y$ and $OP = r$, and since $x < 0$, $MO = -x$. See Figure 3.21.

Figure 3.21

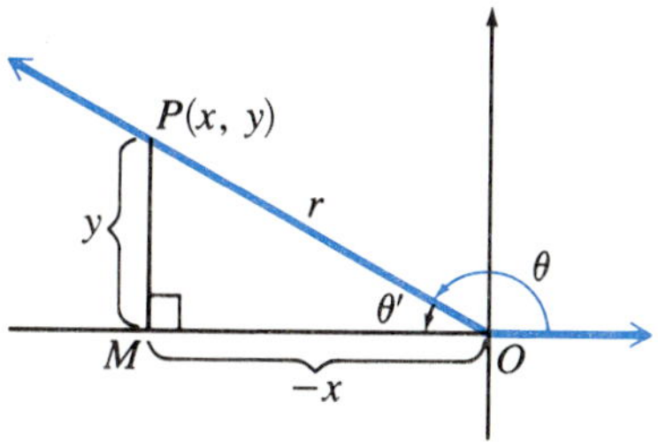

By definition, then,

$$\sin\theta = \frac{y}{r}.$$

Using the right triangle, we have

$$\sin\theta' = \frac{y}{r}.$$

Therefore,

$$\sin\theta = \sin\theta'.$$

From the definition,

$$\cos\theta = \frac{x}{r}.$$

Using the right triangle, we have

$$\cos\theta' = \frac{-x}{r}.$$

Thus

$$\cos\theta = -\cos\theta'.$$

Finally, using the definition,

$$\tan\theta = \frac{y}{x},$$

whereas, using the right triangle,

$$\tan\theta' = \frac{y}{-x}.$$

Hence

$$\tan\theta = -\tan\theta'.$$

We have just proven that if θ is the measure of a second-quadrant angle and θ' is the measure of its reference angle, then

$$\sin\theta = \sin\theta',$$
$$\cos\theta = -\cos\theta',$$
$$\tan\theta = -\tan\theta'.$$

Similar results may be obtained in quadrants III and IV. Together, this gives us the following theorem.

If θ is the measure of an angle and θ' is the measure of its reference angle, then

$$\sin\theta = \pm\sin\theta',$$
$$\cos\theta = \pm\cos\theta',$$
$$\tan\theta = \pm\tan\theta',$$

where the particular sign, $+$ or $-$, depends on the quadrant in which θ lies.

This theorem tells us that, depending on the position of the terminal side, the trigonometric value of a given angle is equal to either the trigonometric value of the reference angle or its negative. To determine the sign, we merely need to recall whether the particular trigonometric ratio is positive or negative for the given quadrant. For example, if θ is a third-quadrant angle, we know that

$$\cos\theta = -\cos\theta',$$

where θ' is the reference angle, because the cosine is negative in the third quadrant.

With regard to the reciprocal ratios,

$$\csc\theta = \pm\csc\theta',$$
$$\sec\theta = \pm\sec\theta',$$
$$\cot\theta = \pm\cot\theta'.$$

Thus, if θ is in quadrant III,

$$\cot\theta = \cot\theta'$$

because the cotangent is positive in the third quadrant.

These theorems, then, are handy in finding the exact values of ratios for special angles. For instance, we may determine sin 315° as follows. We note that this is a fourth-quadrant angle, so the sine is negative. The reference angle is $360° - 315° = 45°$. Thus

$$\sin 315° = -\sin 45°.$$

From the 45°-45°-90° triangle we get sin 45°. Therefore,

$$\sin 315° = -\sin 45° = \frac{-1}{\sqrt{2}}\left(\text{or } \frac{-\sqrt{2}}{2}\right).$$

For our purposes, however, the principal consequence of these three theorems is that the trigonometric value of the reference angle (an acute angle) is equal to the absolute value of the trigonometric value of the general angle.

For any angle θ,

$$\sin\theta' = |\sin\theta|,$$
$$\cos\theta' = |\cos\theta|,$$
$$\tan\theta' = |\tan\theta|,$$

where θ' is the reference angle.

These facts are utilized in Examples 3 through 5.

EXAMPLE 3 Find θ (to the nearest hundredth) for the given conditions.

(a) $90° < \theta < 180°$ and $\tan\theta = -0.6644$
(b) $-180° < \theta < -90°$ and $\cos\theta = -0.215$

SOLUTION (a) We have the situation shown in Figure 3.22, where θ' is the reference angle.

Figure 3.22

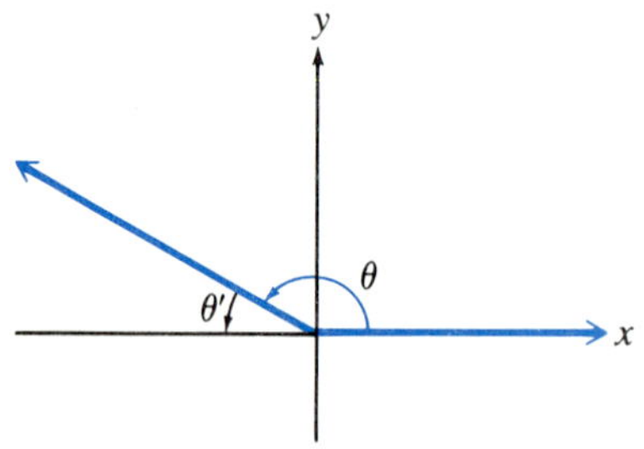

Our first task is to find the reference angle. We have

$$\tan\theta' = |\tan\theta| = 0.6644.$$

So θ' is the acute angle whose tangent is 0.6644. To determine it, we enter .6644 followed by [inv] [tan] (or its equivalent). We get (eight digits)

$$\theta' = 33.600063°.$$

Subtracting this from 180° will give us θ. We have

$$\theta = 180° - 33.600063° \approx 146.40°.$$

(b) We have the situation shown in Figure 3.23. To determine the reference angle we first note that

$$\cos\theta' = |\cos\theta| = 0.215.$$

From our calculations we see that

$$\theta' = 77.584473°.$$

Figure 3.23

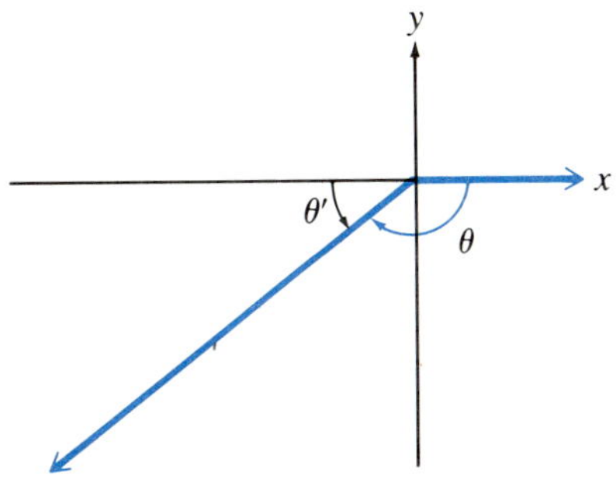

To get θ we must add this reference angle to $-180°$. We have

$$\theta = -180° + 77.584473° \approx -102.42°.$$

▲

▲ ***Exercise*** Find θ (to the nearest hundredth) for the given conditions.

(a) $90° < \theta < 180°$ and $\tan\theta = -0.6088$
(b) $-180° < \theta < -90°$ and $\sin\theta = -0.718$

Answer (a) 148.67° (b) −134.11°

EXAMPLE 4 Find θ (to the nearest hundredth) if $0° \le \theta < 360°$, $\sin\theta < 0$, and $\cos\theta = 0.3153$.

SOLUTION Because the sine is negative and the cosine is positive, θ must be a fourth-quadrant angle, as shown in Figure 3.24.

Figure 3.24

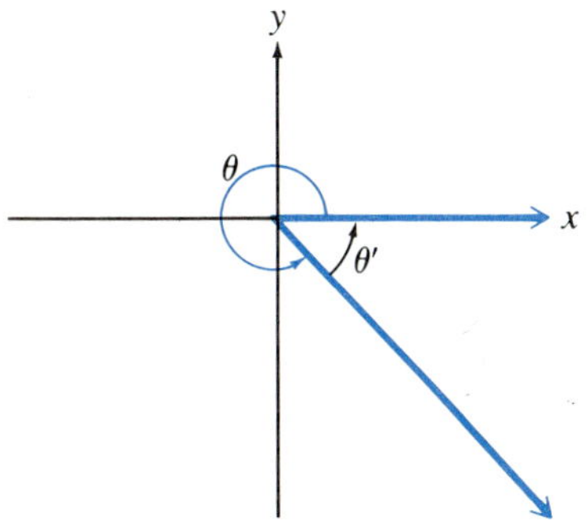

We have

$$\cos\theta' = |\cos\theta| = 0.3153.$$

Thus the reference angle is 71.621075°. Subtracting this from 360° gives

$$\theta \approx 288.38°.$$

▲

▲ ***Exercise*** Find θ (to the nearest hundredth) if $0^\circ < \theta \leq 360^\circ$, $\cos\theta < 0$, and $\tan\theta = 0.7445$.

Answer 216.67°

When using the reciprocal ratios, we may need the relationships

$$\csc\theta' = |\csc\theta|,$$

$$\sec\theta' = |\sec\theta|,$$

$$\cot\theta' = |\cot\theta|.$$

EXAMPLE 5 Find θ (to the nearest hundredth) if $180^\circ < \theta < 270^\circ$ and $\csc\theta = -2.15$.

SOLUTION We have the situation described in Figure 3.25.

Figure 3.25

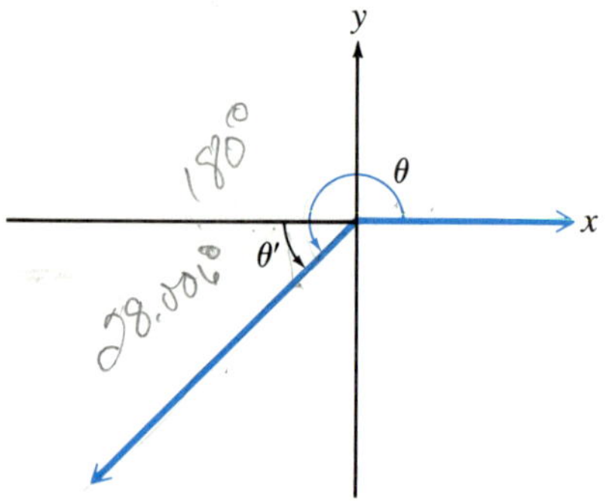

Again we must determine the reference angle first. We note that

$$\csc\theta' = |\csc\theta| = 2.15.$$

To find the acute angle whose cosecant is 2.15, we find the acute angle whose sine is the reciprocal of 2.15. We enter 2.15, push [1/x], followed by [inv] [sin] to get

$$\theta' = 27.717748^\circ.$$

Adding this to 180° gives

$$\theta \approx 207.72^\circ.$$

▲

▲ ***Exercise*** Find θ (to the nearest hundredth) if $180^\circ < \theta < 270^\circ$ and $\sec\theta = -3.56$.

Answer 253.69°

Problem Set 3.3

In Problems 1–12, find the trigonometric value to four significant digits.

1. $\sin 97.2^\circ$

2. $\sin 212^\circ$

3. $\cos(-25^\circ)$

4. $\cos 202^\circ$

5. $\tan 160^\circ$

6. $\tan(-110.2^\circ)$

7. $\sec 205.6^\circ$

8. $\sec 95.1^\circ$

9. $\csc(-227^\circ)$

10. $\csc 446^\circ$

11. $\cot 100.5^\circ$

12. $\cot(-84^\circ)$

In Problems 13–18, write the ratio in terms of the reference angle and then find the exact trigonometric value.

13. $\sin 210^\circ$

14. $\cos 225^\circ$

15. $\tan(-240^\circ)$

16. $\tan 300^\circ$

17. $\sec 330^\circ$

18. $\csc(-135^\circ)$

In Problems 19–32, find θ (to the nearest hundredth) for the given conditions.

19. $90^\circ < \theta < 180^\circ$ and $\tan\theta = -0.95$

20. $180^\circ < \theta < 270^\circ$ and $\sin\theta = -0.83$

21. $270^\circ < \theta < 360^\circ$ and $\tan\theta = -0.384$

22. $90^\circ < \theta < 180^\circ$ and $\tan\theta = -4.5$

23. $-180^\circ < \theta < 0^\circ$ and $\cos\theta = 0.6216$

24. $360^\circ < \theta < 540^\circ$ and $\tan\theta = -9.51$

25. $90^\circ < \theta < 180^\circ$ and $\csc\theta = 2.25$

26. $180^\circ < \theta < 270^\circ$ and $\sec\theta = -1.152$

27. $0^\circ \le \theta < 360^\circ$, $\sin\theta < 0$, and $\cos\theta = -0.8829$

28. $0^\circ \le \theta < 360^\circ$, $\tan\theta < 0$, and $\cos\theta = 0.75$

29. $0^\circ \le \theta < 360^\circ$, $\cos\theta < 0$, and $\tan\theta = 2.23$

30. $0^\circ \le \theta < 360^\circ$, $\sin\theta < 0$, and $\tan\theta = -6.2$

31. $0^\circ \le \theta < 360^\circ$ and $\cos\theta = 0.65$

32. $0^\circ \le \theta < 360^\circ$ and $\cot\theta = -0.875$

3.4 TRIGONOMETRIC EQUATIONS

Equations like

$$\sin 2\theta = \frac{1}{2}, \quad \cos(\theta - 180^\circ) = 0, \quad \text{and} \quad \tan^2\theta - \tan\theta - 3 = 0$$

are called **trigonometric equations**. Because the trigonometric values depend directly on the position of the terminal side of the angle, any solution of a trigonometric equation gives rise to infinitely more solutions. In $[0^\circ, 360^\circ)$ the solutions of

$$\sin\theta = \frac{\sqrt{2}}{2}$$

are in quadrants I and II, where the sine is positive, with a reference angle of 45°. Hence the solutions are 45° and 135°. The general solutions are $45^\circ + 360^\circ \cdot n$ and $135^\circ + 360^\circ \cdot n$, $n = 0, \pm 1, \pm 2, \ldots$.

To simplify matters, we'll restrict the universe of the variable to all values in $[0^\circ, 360^\circ)$ for which the trigonometric ratios are defined.

EXAMPLE 1 Find the exact solutions, where $0^\circ \le \theta < 360^\circ$.

(a) $\sin\theta = \dfrac{-1}{2}$ (b) $\cos 2\theta = \dfrac{\sqrt{3}}{2}$ (c) $\tan(\theta - 270^\circ) = \sqrt{3}$

SOLUTION (a) Let's determine the reference angle first. It is the acute angle whose sine is a positive $\frac{1}{2}$. This is the special angle 30°. Since the sine is negative in quadrants III and IV, we have two solutions. The third-quadrant solution is

$$\theta = 180^\circ + 30^\circ = 210^\circ,$$

and the fourth-quadrant solution is

$$\theta = 360^\circ - 30^\circ = 330^\circ.$$

Hence the solutions are 210° and 330°.

(b) We will begin by finding values of 2θ. Thus we must note the restrictions on this angle. Because $0^\circ \le \theta < 360^\circ$,

$$0^\circ \le 2\theta < 720^\circ.$$

So we must find angles in $[0^\circ, 720^\circ)$ whose cosine is $\sqrt{3}/2$. Since 30° is the acute angle whose cosine is $\sqrt{3}/2$, this is the reference angle for 2θ. The cosine is positive in quadrants I and IV, so we get four possibilities:

$$\begin{array}{cccc} 2\theta = 30^\circ \quad \text{or} & 2\theta = 330^\circ \quad \text{or} & 2\theta = 390^\circ \quad \text{or} & 2\theta = 690^\circ \\ \theta = 15^\circ & \theta = 165^\circ & \theta = 195^\circ & \theta = 345^\circ. \end{array}$$

Thus the solutions are 15°, 165°, 195°, and 345°.

(c) We will begin by finding values of $\theta - 270^\circ$. Since $0^\circ \le \theta < 360^\circ$, the restrictions for $\theta - 270^\circ$ are

$$-270^\circ \le \theta - 270^\circ < 90^\circ.$$

The reference angle for $\theta - 270^\circ$ is 60° (the acute angle whose tangent is $\sqrt{3}$), and the tangent is positive in the first and third quadrants. Between 0° and −270° there is a third-quadrant angle that is $-180^\circ + 60^\circ = -120^\circ$. Between 0° and 90° we get the first-quadrant angle of 60°. Thus

$$\begin{array}{cc} \theta - 270^\circ = -120^\circ \quad \text{or} & \theta - 270^\circ = 60^\circ \\ \theta = 150^\circ & \theta = 330^\circ. \end{array}$$

Hence, the solutions are 150° and 330°.

▲

▲ *Exercise* Find the exact solutions, where $0^\circ \le \theta < 360^\circ$.

(a) $\sin\theta = \dfrac{-\sqrt{3}}{2}$ (b) $\cos 2\theta = \dfrac{\sqrt{2}}{2}$ (c) $\tan(\theta - 270^\circ) = \dfrac{1}{\sqrt{3}}$

Answer (a) 240°, 300° (b) 22.5°, 157.5°, 202.5°, 337.5° (c) 120°, 300°

EXAMPLE 2 Find the approximate solutions to the nearest hundredth, where $0^\circ \le \theta < 360^\circ$.

(a) $\tan\dfrac{\theta}{2} = 5$ (b) $\cos(\theta + 70^\circ) = 0.45$.

SOLUTION (a) We will begin by finding the values of $\theta/2$. Because $0^\circ \le \theta < 360^\circ$, the restrictions for $\theta/2$ are

$$0^\circ \le \frac{\theta}{2} < 180^\circ.$$

We must find angles in [0°, 180°) whose tangent is 5. The reference angle for $\theta/2$ is 78.690068°. Since the tangent is negative in the second quadrant, we have a first-quadrant angle only. Thus,

$$\frac{\theta}{2} = 78.690068^\circ$$

$$\theta = 157.38014^\circ \approx 157.38^\circ.$$

(b) We first note that

$$70^\circ \le \theta + 70^\circ < 430^\circ.$$

So we are looking for the angles in $[70^\circ, 430^\circ)$ that have a cosine of 0.45. The reference angle is 63.256316°. Thus, in the first and fourth quadrants we have

$$\theta + 70^\circ = 360^\circ + 63.256316^\circ \quad \text{or} \quad \theta + 70^\circ = 360^\circ - 63.256316^\circ$$
$$\theta \approx 353.26^\circ \qquad\qquad \theta \approx 226.74^\circ.$$

▲

▲ ***Exercise*** Find the approximate solutions to the nearest hundredth, where $0^\circ \le \theta < 360^\circ$.

(a) $\tan \dfrac{\theta}{2} = -7$ (b) $\sin(\theta + 70^\circ) = 0.68$

Answer (a) 196.26° (b) 67.16°, 332.84°

EXAMPLE 3 Find the approximate solutions, to the nearest hundredth, of $\cot \theta = \dfrac{-4}{5}$, where $0^\circ \le \theta < 360^\circ$.

SOLUTION The reference angle is the acute angle whose cotangent is $\frac{4}{5}$. From our calculators we get 51.340192°. Since the cotangent is negative in the second and fourth quadrants,

$$\theta = 180^\circ - 51.340192^\circ \approx 128.66^\circ \quad \text{or} \quad \theta = 360^\circ - 51.340192^\circ \approx 308.66^\circ.$$

Hence the solutions are 128.66° and 308.66°.

▲

▲ ***Exercise*** Find the approximate solutions, to the nearest hundredth, of $\csc \theta = \dfrac{-3}{2}$, where $0^\circ \le \theta < 360^\circ$.

Answer 221.81°, 318.19°

Sometimes algebraic manipulation is required to get the trigonometric equation into the proper form. This occurs in the next example.

EXAMPLE 4 Find the approximate solutions, to the nearest hundredth, of $\tan^2 \theta = 2 \tan \theta + 3$, where $0^\circ \le \theta < 360^\circ$. [Note: The notation $\tan^2 \theta$ means $(\tan \theta)^2$.]

SOLUTION This equation is quadratic in $\tan \theta$. That is, if we let $u = \tan \theta$, the resulting equation

$$u^2 = 2u + 3$$

is quadratic. Getting one side to zero and factoring, we have

$$u^2 - 2u - 3 = 0$$

$$(u + 1)(u - 3) = 0$$

$$u = -1 \quad \text{or} \quad u = 3.$$

Substituting for u, we have

$$\tan \theta = -1 \quad \text{or} \quad \tan \theta = 3.$$

The solutions to the first equation are

$$135^\circ \quad \text{and} \quad 315^\circ.$$

The approximate solutions to the second are

$$71.57^\circ \quad \text{and} \quad 251.57^\circ.$$

Hence the solutions are 135°, 315°, 71.57°, and 251.57°.

▲

▲ ***Exercise*** Find the approximate solutions, to the nearest hundredth, of $3 \cos^2 \theta = 4 \cos \theta - 1$, where $0^\circ \leq \theta < 360^\circ$.

Answer 0°, 70.53°, 289.47°

Problem Set 3.4

In Problems 1–26, find the exact solutions, where $0^\circ \leq \theta < 360^\circ$.

1. $\sin \theta = \dfrac{\sqrt{2}}{2}$

2. $\sin \theta = \dfrac{-\sqrt{3}}{2}$

3. $\cos \theta = \dfrac{-1}{2}$

4. $\cos \theta = \dfrac{\sqrt{3}}{2}$

5. $\tan \theta = -1$

6. $\tan \theta = \sqrt{3}$

7. $\sin 2\theta = \dfrac{1}{2}$

8. $\cos 2\theta = \dfrac{-\sqrt{2}}{2}$

9. $\cos \dfrac{\theta}{2} = \dfrac{-1}{2}$

10. $\sin \dfrac{\theta}{2} = \dfrac{\sqrt{2}}{2}$

11. $\tan \dfrac{\theta}{2} = 1$

12. $\tan 3\theta = -\sqrt{3}$

13. $\sec \theta = 2$

14. $\csc \theta = -\sqrt{2}$

15. $\cot 2\theta = 1$

16. $\sec \dfrac{\theta}{4} = -\sqrt{2}$

17. $\sin(\theta + 180°) = \dfrac{-\sqrt{2}}{2}$

18. $\cos(\theta - 180°) = \dfrac{\sqrt{3}}{2}$

19. $\tan(180° - \theta) = 1$

20. $\sec(\theta - 90°) = -2$

21. $\sin(2\theta - 180°) = \dfrac{1}{2}$

22. $\cos(180° - 2\theta) = \dfrac{-1}{2}$

23. $2\sin^2\theta - 3\sin\theta = 2$

24. $2\cos^2\theta + 5\cos\theta = 3$

25. $\dfrac{3\cos\theta}{2\cos\theta + 1} - 1 = \dfrac{-\cos\theta - 2}{2\cos\theta + 1}$

26. $\dfrac{\tan\theta}{\tan\theta - 1} + 2 = \dfrac{2\tan\theta - 1}{\tan\theta - 1}$

In Problems 27–40, find the approximate solutions to the nearest hundredth, where $0° \le \theta < 360°$.

27. $\sin\theta = 0.35$

28. $\cos\theta = 0.612$

29. $\tan\theta = -2.53$

30. $\cot\theta = 0.46$

31. $\cos 2\theta = 0.85$

32. $\tan 2\theta = -4$

33. $\sin\dfrac{\theta}{2} = 0.67$

34. $\cos\dfrac{\theta}{2} = -0.414$

35. $\sin(\theta + 80°) = 0.51$

36. $\tan(\theta - 50°) = -2.5$

37. $\tan^2\theta = 3\tan\theta - 2$

38. $12\sin^2\theta = 1 + \sin\theta$

39. $\sin^2\theta = 1 + 2\sin\theta$

40. $\cos^2\theta = 2 - 2\cos\theta$

CHAPTER 3 REVIEW

Key Terms and Formulas

coterminal angles, page 88

degree, page 86

first-, second-, third-, or fourth-quadrant angle, page 87

initial side, page 86

measure of an angle, page 86

quadrantal angle, page 88

reference angle, page 88

standard position, page 87

terminal side, page 86

trigonometric ratios, page 93

sine $= \dfrac{y}{r}$

cosine $= \dfrac{x}{r}$

tangent $= \dfrac{y}{x}$

cotangent $= \dfrac{x}{y}$

secant $= \dfrac{r}{x}$

cosecant $= \dfrac{r}{y}$

Problem Set

1. Specify "true" or "false."

(a) A right angle is considered a first-quadrant angle.

(b) If $|\tan \theta_1| = |\tan \theta_2|$, then $\theta'_1 = \theta'_2$.

(c) Given any quadrantal angle, two of the six trigonometric ratios will equal 0, and two of the six will be undefined.

(d) For some quadrants, the reference angle may be formed with the y-axis.

(e) For all angles θ, $-1 < \sin \theta < 1$.

(f) The sine, cosine, and tangent ratios are all positive in the first quadrant only.

(g) The cotangent ratio is the reciprocal of the tangent ratio.

(h) For a second-quadrant angle, $\cos \theta = \cos \theta'$.

(i) If the sines of two angles are equal, then the angles must be coterminal.

(j) If the value of a trigonometric ratio is zero, then the value of the reciprocal ratio is also zero.

In Problems 2 and 3, construct an angle in standard position with the given measure, state the quadrant in which the terminal side lies, and give the measure of a coterminal angle with opposite direction.

2. $-120°$

3. $312°$

In Problems 4 and 5, determine the reference angle.

4. $168°$

5. $341°$

In Problems 6–8, find θ for the given conditions.

6. $-270° < \theta < -180°$ and $\theta' = 20°$

7. $-180° < \theta < -90°$ and $\theta' = 82°$

8. $270° < \theta < 360°$ and $\theta' = 15°$

In Problems 9 and 10, find the exact values of the trigonometric ratios if the terminal side of θ passes through the given point.

9. $(-4, 3)$

10. $(1, -3)$

In Problems 11 and 12, find the exact values of the other trigonometric ratios of θ for the given conditions.

11. $\cot \theta = \dfrac{-12}{5}$, θ in the second quadrant

12. $\tan \theta = \dfrac{1}{2}$, θ in the third quadrant

In Problems 13–16, answers are to be exact.

13. If $\sin\theta = \dfrac{-\sqrt{17}}{5}$ and $\tan\theta < 0$, find $\sec\theta$.

14. If $\csc\theta = \dfrac{7}{2}$ and $\cot\theta < 0$, find $\cos\theta$.

15. If $\tan\theta = \dfrac{-1}{4}$ and $\cos\theta < 0$, find $\sin\theta$.

16. If $\tan\theta = \dfrac{2\sqrt{10}}{3}$ and $\sin\theta < 0$, find $\sec\theta$.

In Problems 17 and 18, find the values of the trigonometric ratios for the given measure.

17. -270°

18. 360°

In Problems 19–22, find the sine, cosine, and tangent of the given angle, exactly.

19. 135°

20. 330°

21. -60°

22. -135°

In Problems 23 and 24, find the secant, cosecant, and cotangent of the given angle, exactly.

23. 210°

24. -225°

In Problems 25–30, find the trigonometric value to four significant digits.

25. $\sin 215^\circ$

26. $\cos 131.2^\circ$

27. $\tan(-40.5^\circ)$

28. $\cot 218^\circ$

29. $\sec 97.6^\circ$

30. $\csc(-112^\circ)$

In Problems 31–36, find θ (to the nearest hundredth) for the given conditions.

31. $180^\circ < \theta < 270^\circ$ and $\cos\theta = -0.718$

32. $270^\circ < \theta < 360^\circ$ and $\sin\theta = -0.1582$

33. $0^\circ < \theta < 180^\circ$ and $\tan\theta = -3.92$

34. $180^\circ < \theta < 360^\circ$ and $\cot\theta = 0.05$

35. $0^\circ \le \theta < 360^\circ$, $\cos\theta < 0$, and $\tan\theta = -1.557$

36. $0^\circ \le \theta < 360^\circ$, $\csc\theta < 0$, and $\sec\theta = -1.76$

In Problems 37–44, find the exact solutions, where $0^\circ \le \theta < 360^\circ$.

37. $\sin\theta = \dfrac{-1}{2}$

38. $\sec\theta = 2$

39. $\tan\theta = -1$

40. $\cot\frac{\theta}{2} = \sqrt{3}$

41. $\sin 2\theta = \frac{\sqrt{2}}{2}$

42. $\cos\frac{\theta}{2} = \frac{-\sqrt{3}}{2}$

43. $\tan(2\theta - 180^\circ) = 1$

44. $\cos(\theta - 180^\circ) = \frac{-\sqrt{2}}{2}$

In Problems 45–50, find the approximate solutions to the nearest hundredth, where $0^\circ \leq \theta < 360^\circ$.

45. $\tan\theta = 6$

46. $\sec\theta = -2.5$

47. $\cos\frac{\theta}{2} = 0.98$

48. $\sin 2\theta = 0.99$

49. $\tan(\theta + 70^\circ) = 2$

50. $\cos(\theta - 20^\circ) = 0.35$

4
RADIANS AND CIRCULAR FUNCTIONS

CHAPTER OUTLINE

For applications to branches of advanced mathematics such as calculus, as well as to physics and engineering, trigonometric *functions* rather than ratios are required. This chapter will bridge the gap between ratios and functions. We begin by introducing another unit of angle measure, the radian, and conclude by examining the connection between the ratios and points on a circle.

4.1 RADIAN MEASURE

In addition to the degree there is another unit of angle measure called the radian. Given $\angle BAC$, we construct a circle with center at the vertex. Let r be the circle's radius, and let s be the length of the arc intercepted. See Figure 4.1.

Figure 4.1

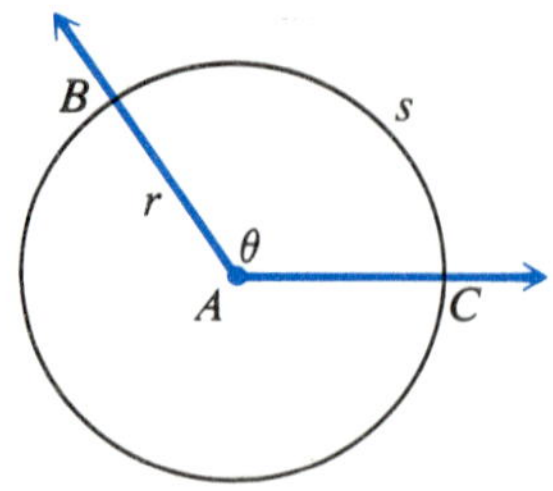

The **radian** measure of this angle is the ratio s/r. Thus an angle of measure 1 radian is one that intercepts an arc equal to the radius of the circle. Furthermore, since the arc length of a semicircle is πr, the ratio $\pi r/r = \pi$ is the radian measure of a straight angle. See Figure 4.2.

Figure 4.2

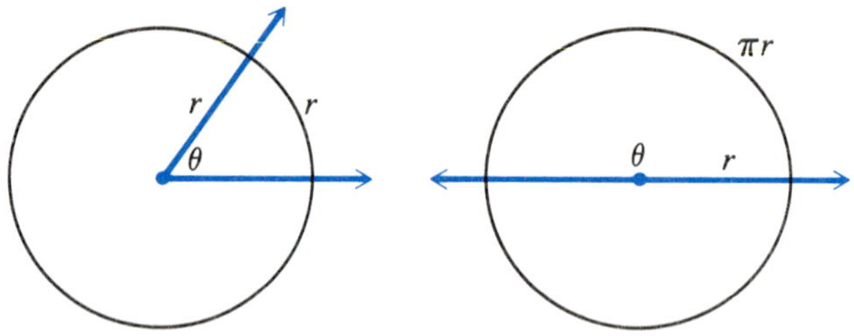

Sometimes we shall need to convert from one unit of measurement to the other. We note that since a straight angle measures π radians and also 180°,

$$\pi \text{ radians} = 180°. \tag{4.1}$$

To convert from one unit to another, we may multiply by a special form of 1. To convert from radians to degrees, we divide both sides of equation (4.1) by π radians to get

$$1 = \frac{180°}{\pi \text{ radians}}. \tag{4.2}$$

Hence, when we multiply a radian measure by 1 in this form, the units "radians" will cancel and give us degrees. See Example 1.

EXAMPLE 1 Convert the radian measure to degrees, exactly.

(a) $\frac{\pi}{3}$ radians (b) $\frac{3\pi}{4}$ radians (c) 2 radians

SOLUTION We multiply the given number of radians by the special form of 1 in equation (4.2).

(a) $\frac{\pi}{3}$ radians $= \frac{\pi}{3}$ radians $\cdot \frac{180°}{\pi \text{ radians}} = 60°$

(b) $\frac{3\pi}{4}$ radians $= \frac{3\pi}{4}$ radians $\cdot \frac{180°}{\pi \text{ radians}} = 135°$

(c) 2 radians = 2 radians $\cdot \frac{180°}{\pi \text{ radians}} = \left(\frac{360}{\pi}\right)°$

▲

▲ ***Exercise*** Convert the radian measure to degrees, exactly.

(a) $\frac{\pi}{4}$ radians (b) $\frac{5\pi}{6}$ radians (c) $\frac{1}{3}$ radians

Answer (a) 45° (b) 150° (c) $\left(\frac{60}{\pi}\right)°$

The measure of a straight angle is 180° or π radians. Thus, if we think of angle measurement in terms of "parts" of a straight angle, some conversions are simplified. For example, if we think of $\pi/3$ radians as being $\frac{1}{3}$ of a straight angle, then we know that the corresponding degree measure is $\frac{1}{3}$ of 180° (i.e., 60°).

In fact, this idea may be seen in a different form of the formula. We multiply both sides of equation (4.2) by x radians to get

$$x \text{ radians} = x\left(\frac{180}{\pi}\right)° = \frac{x}{\pi} \cdot 180°. \quad \textbf{(4.3)}$$

With this view we see that to convert x radians to degrees, we first calculate x/π, which gives us the "part" of a straight angle, and then we multiply by the degree measure of a straight angle.

EXAMPLE 2 Convert the radian measure to degrees, approximating to the nearest hundredth.

(a) 2 radians (b) 0.55 radians (c) 1.5 radians

SOLUTION We'll use equation (4.3), and then use the $\boxed{\pi}$ button on our calculators to approximate.

(a) 2 radians $= \frac{2}{\pi} \cdot 180° \approx 114.59°$

(b) 0.55 radians $= \frac{0.55}{\pi} \cdot 180° \approx 31.51°$

(c) 1.5 radians $= \frac{1.5}{\pi} \cdot 180° \approx 85.94°$

▲

▲ ***Exercise*** Convert the radian measure to degrees, approximating to the nearest hundredth.

(a) 2.5 radians (b) 0.75 radians (c) 1 radian

Answer (a) 143.24° (b) 42.97° (c) 57.30°

To convert from degrees to radians, we may again multiply by 1. We divide both sides of equation (4.1) by 180° to get

$$1 = \frac{\pi \text{ radians}}{180^\circ}. \qquad \textbf{(4.4)}$$

In this way the "degrees" will cancel and give us radians. See Example 3.

EXAMPLE 3 Convert the degree measure to radians, exactly.

(a) 90° (b) 150° (c) 131°

SOLUTION We multiply the given number of degrees by the special form of 1 in equation (4.4).

(a) $90^\circ = 90^\circ \cdot \dfrac{\pi \text{ radians}}{180^\circ} = \dfrac{\pi}{2}$ radians

(b) $150^\circ = 150^\circ \cdot \dfrac{\pi \text{ radians}}{180^\circ} = \dfrac{5\pi}{6}$ radians

(c) $131^\circ = 131^\circ \cdot \dfrac{\pi \text{ radians}}{180^\circ} = \dfrac{131\pi}{180}$ radians

▲

▲ ***Exercise*** Convert the degree measure to radians, exactly.

(a) 135° (b) 120° (c) 47°

Answer (a) $\dfrac{3\pi}{4}$ radians (b) $\dfrac{2\pi}{3}$ radians (c) $\dfrac{47\pi}{180}$ radians

If we again think in terms of "parts" of a straight angle, whose measure is 180° or π radians, some conversions are simplified. For example, if we think of 90° as being $\frac{1}{2}$ of a straight angle, then we know that the corresponding radian measure is $\frac{1}{2}$ of π (i.e., $\pi/2$).

This idea is seen in an alternate method of converting x degrees to radians. Multiplying both sides of equation (4.4) by x° gives

$$x^\circ = x \cdot \frac{\pi}{180} \text{ radians} = \frac{x}{180} \cdot \pi \text{ radians}. \qquad \textbf{(4.5)}$$

We first calculate $x/180$, which gives us the "part" of a straight angle, and then we multiply by the radian measure of a straight angle.

EXAMPLE 4 Convert the degree measure to radians, approximating to the nearest thousandth.

(a) 40° (b) 165° (c) 151.2°

SOLUTION We'll use equation (4.5) and then approximate π with our calculators.

(a) $40^\circ = \dfrac{40}{180} \cdot \pi \text{ radians} \approx 0.698 \text{ radians}$

(b) $165^\circ = \dfrac{165}{180} \cdot \pi \text{ radians} \approx 2.880 \text{ radians}$

(c) $151.2^\circ = \dfrac{151.2}{180} \cdot \pi \text{ radians} \approx 2.639 \text{ radians}$

▲

▲ ***Exercise*** Convert the degree measure to radians, approximating to the nearest thousandth.

(a) 55° (b) 125° (c) 118.1°

Answer (a) 0.960 radians (b) 2.182 radians (c) 2.061 radians

In exactly the same way degree measure was extended to include positive measures greater than 180° and negative measures, radian measure may be extended. Some examples appear in Figure 4.3.

Figure 4.3

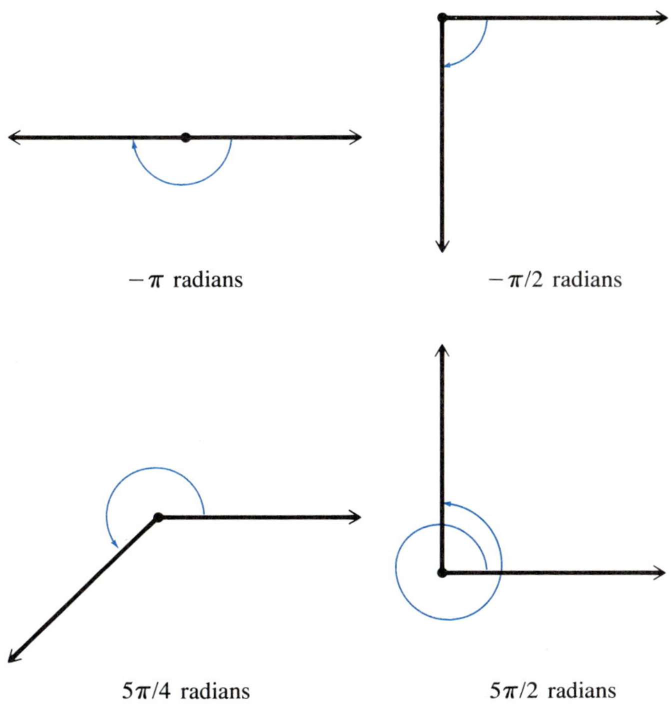

The formulas for converting degrees to radians and vice versa apply for the extended measures as well.

Problem Set 4.1

In Problems 1–6, convert the radian measure to degrees, exactly.

1. $\frac{\pi}{4}$ radians

2. $\frac{2\pi}{3}$ radians

3. $\frac{-5\pi}{6}$ radians

4. $\frac{-3\pi}{2}$ radians

5. -3 radians

6. $\frac{5}{4}$ radians

In Problems 7–10, convert the radian measure to degrees, approximating to the nearest hundredth.

7. 1.25 radians

8. 0.65 radians

9. -0.6 radians

10. -5 radians

In Problems 11–16, convert the degree measure to radians, exactly.

11. 45°

12. 135°

13. $-210°$

14. $-540°$

15. $-67°$

16. 235°

In Problems 17–20, convert the degree measure to radians, approximating to the nearest thousandth.

17. 69°

18. 127.2°

19. $-28°$

20. $-413°$

In Problems 21–24, construct an angle in standard position with the given measure, state the quadrant in which the terminal side lies, determine the reference angle, and give the measure of a coterminal angle with opposite direction.

21. $\frac{5\pi}{4}$ radians

22. 2 radians

23. -0.5 radians

24. $\frac{-11\pi}{6}$ radians

4.2

ARC LENGTH, SECTOR AREA, LINEAR AND ANGULAR SPEED

Radian measure is appropriate to use when we discuss arc length, sector area or linear and angular speed. We already know that, by definition, radian measure θ is

$$\theta = \frac{s}{r}.$$

This means that we may find the length of the arc intercepted by a central angle of measure θ. (A central angle is one whose vertex is at the center of a circle.)

If θ is the radian measure of a central angle, then the length s of the intercepted arc is

$$s = r\theta, \qquad \textbf{(4.6)}$$

where r is the radius.

Note that the formula is valid only when the angle is measured in *radians*.

EXAMPLE 1 Find the length (to four significant digits) of the arc on a circle of radius 15 centimeters that is intercepted by a central angle of 28°.

SOLUTION We must first convert the angle measure to radians. We have

$$28^\circ = \frac{28\pi}{180}\text{ radians}.$$

Using equation (4.6), we get

$$s = 15 \cdot \frac{28\pi}{180} \approx 7.330.$$

The length of the arc is approximately 7.330 centimeters.

▲

▲ ***Exercise*** Find the length (to four significant digits) of the arc on a circle of radius 21 centimeters which is intercepted by a central angle of 121°.

Answer 44.35 cm

Figure 4.4

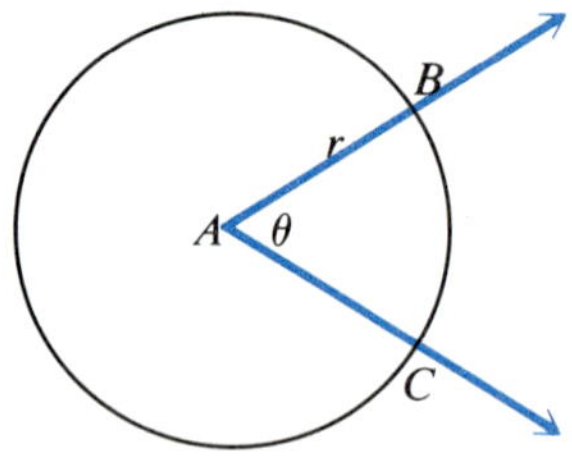

Now consider sector BAC in Figure 4.4. Let r be the radius, and let θ be the radian measure of the central angle. The ratio of the area of the sector to the area of the circle will equal the ratio of θ to 2π, an angle of one complete rotation. Thus, if we let K represent the area of the sector, then

$$\frac{K}{\pi r^2} = \frac{\theta}{2\pi}.$$

Solving this equation for K gives us the following theorem.

If θ is the radian measure of a central angle, then the area K of the corresponding sector is

$$K = \frac{\theta}{2} r^2, \qquad \textbf{(4.7)}$$

where r is the radius.

Note that this formula is valid only when θ is in *radians*.

EXAMPLE 2 Determine the area (to four significant digits) of a sector of radius 1.25 feet with a central angle of 65°.

SOLUTION We first convert the degree measure to radians. We have

$$65^\circ = \frac{65\pi}{180} \text{ radians.}$$

Using equation (4.7) yields

$$K = \frac{65\pi}{180} \cdot \frac{1}{2} \cdot (1.25)^2 \approx 0.8863.$$

The area is approximately 0.8863 square feet.

▲

▲ *Exercise* Determine the area (to four significant digits) of a sector of radius 1.7 feet with a central angle of 200°.

Answer 5.044 ft^2

The final topics to discuss are linear and angular speed. Consider a point P moving at a constant speed on a circle of radius r, centered at O. See Figure 4.5.

Figure 4.5

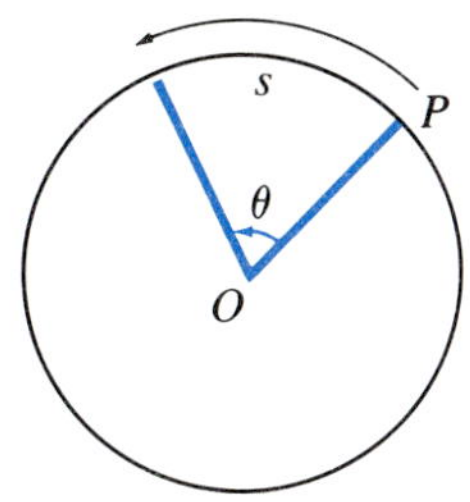

If the point P moves a distance s in time t, then the **linear speed** of P is

$$v = \frac{s}{t}.$$

If the ray $\overrightarrow{OP}$ moves through an angle of measure θ in time t, then the **angular speed** ω (omega) of P is

$$\omega = \frac{\theta}{t}.$$

If θ is measured in radians, then

$$v = \frac{s}{t} = \frac{r\theta}{t} = r \cdot \frac{\theta}{t} = r\omega.$$

That is, the linear speed is the product of the radius and the angular speed. Note that this formula is valid only when ω is measured in *radians* per unit time.

EXAMPLE 3 A pulley of radius 8 centimeters is revolving 40 times per minute (40 rpm). What is the pulley's angular speed, in radians per minute, and what is the speed of a belt driven by the pulley?

SOLUTION Since one revolution equals 2π radians,

$$40 \text{ revolutions} = 80\pi \text{ radians}.$$

Hence the angular speed is 80π radians per minute. The speed of the belt will be the linear speed of a point on the circumference of the pulley. Since $r = 8$, we get

$$v = 8 \cdot 80\pi = 640\pi.$$

The belt is moving 640π centimeters per minute (≈ 20.11 meters per minute). ▲

▲ *Exercise* A pulley of radius 5 inches is revolving 60 times per minute (60 rpm). What is the pulley's angular speed, in radians per minute, and what is the speed of a belt driven by the pulley?

Answer 120π rad/min; 600π in./min (≈ 157.1 ft/min)

Problem Set 4.2

In Problems 1–4, find the length (to four significant digits) of the arc on a circle with the given radius, intercepted by the given central angle.

1. radius 12 centimeters, central angle 0.65 radians

2. radius 25 centimeters, central angle 1.2 radians

3. radius 6 inches, central angle 38°

4. radius 9 feet, central angle 195°

5. If an arc of length 5.2 feet is intercepted by a central angle of 1.68 radians, what is the radius of the circle (to four significant digits)?

6. What is the measure, to the nearest hundredth of a degree, of the central angle of an arc of 12.5 centimeters if the radius of the circle is 6 centimeters?

Let *P* represent a point on the earth's surface, let *N* denote the North Pole and let *S* denote the South Pole. Figure 4.6 shows the intersection of the plane determined by *NPS* with the earth.

Figure 4.6

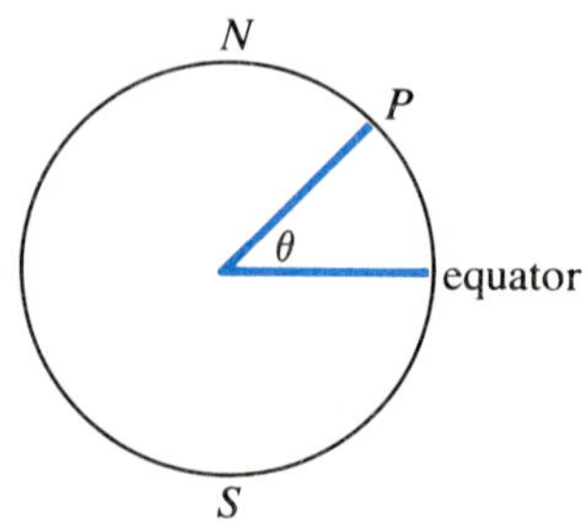

The semicircle *NPS* is called the meridian of *P*. The angle θ, measured in degrees, is called the latitude of *P* and may be either north or south. The arc from *P* to the equator, as measured along the meridian, has a radius of 3960 miles, the approximate radius of the earth. Use this information to answer Problems 7–10.

7. In the third century B.C., Erastothenes was able to make the first reasonable estimate of the earth's radius. On a given day he noted that at high noon, the sun cast no shadow into a well in the city of Syene. At the same time, the sun was 7.2° from vertical in the city of Alexandria, 500 miles north. What was his estimate (to two significant digits) of the earth's radius?

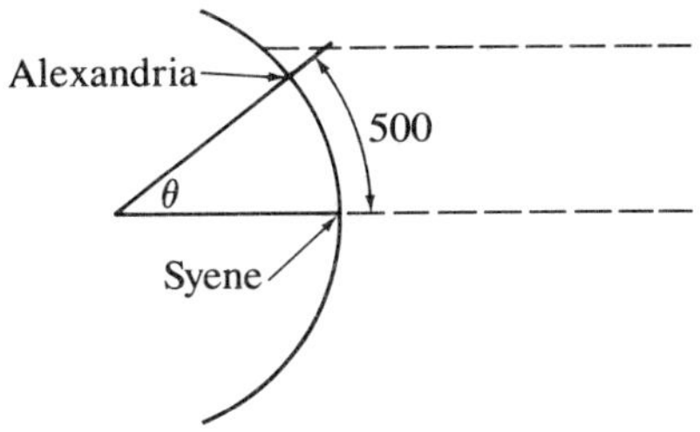

8. Suppose that city A is directly north of city B and that the sun is directly over city B.

(a) If A and B are 420 miles apart, what will be the length (to two significant digits) of the shadow of a 25-foot tree located in city A?

(b) If a 10-foot pole in city A casts a 6-inch shadow, how far apart (to two significant digits) are the two cities?

9. In the 1840s there was a border dispute between the United States and Great Britain over the southern boundary of British Columbia. The British claimed that the boundary was the 49° latitude, while the United States said 54°40′. What was the disputed north-south distance (to two significant digits)?

10. The Tropic of Cancer marks the northern boundary of the tropics. If it is 1600 miles north of the equator, what is its latitude (to the nearest tenth of a degree)?

In Problems 11–14, determine the area (to four significant digits) of the sector described.

11. radius 8.2 centimeters, central angle 2.1 radians

12. radius 7.5 centimeters, central angle 0.85 radians

13. radius 0.5 feet, central angle 87°

14. radius 1.2 feet, central angle 230°

15. If a sector with a central angle of 0.85 radians has an area of 5.6 square feet, what is the radius of the circle (to four significant digits)?

16. What is the measure, to the nearest hundredth of a degree, of the central angle of an arc if the area of the sector is 25 square centimeters and the radius of the circle is 4 centimeters?

17. The blade of a circular saw makes 2100 revolutions per minute. What is the linear speed (in mph, to two significant digits) of a point on the tip of the blade if the diameter is 7.5 inches?

18. Suppose that at the moment of impact with the ball, a tennis racket has an angular speed of 800° per second. What is the speed of the ball (in mph, to two significant digits) if the distance from shoulder to impact point is 3 feet?

19. Suppose that, at the moment of impact, a batted ball leaves the bat at a speed of 110 miles per hour. What is the angular speed of the bat (in radians per second, to two significant digits) if the ball strikes the bat 4 feet from the pivoting point?

20. A turbine has a diameter of 6 feet. If water rushes past its blades at 15 miles per hour, what is the turbine's angular speed (in rpm, to two significant digits)?

21. Two pulleys are connected by a belt. The smaller pulley has a diameter of 6 inches, and the larger has a diameter of 10 inches. If the smaller pulley is making 30 revolutions per minute, what is the angular speed of the larger pulley in radians per minute (to two significant digits)?

22. The front sprocket of a bicycle is 8 inches in diameter, and one of the rear sprockets is 3 inches in diameter. If the front sprocket is making 25 revolutions per minute, what is the speed of the rear sprocket in radians per minute (to two significant digits)?

23. The speed of a jeweler's polishing wheel is given in surface feet per minute, which is the linear speed of a point on the surface. Verify the formula

$$\text{SFPM} = 0.262 \times \text{RPM} \times \text{diameter of wheel in inches.}$$

24. The front sprocket of a bicycle is 8 inches in diameter, and one of the rear sprockets is 2.5 inches in diameter. If the diameter of the wheel is 26 inches, what angular speed (in rpm, to two significant digits) is needed for the front sprocket to maintain a bicycle speed of 15 miles per hour?

25. Let θ be the central angle in the following figure. If θ is close to 0, then the area of sector BAC is close to the area of $\triangle ABC$. Use this to show that for θ near 0, $\theta \approx \sin\theta$.

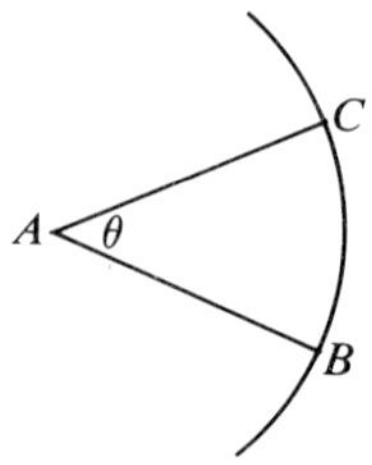

4.3 TRIGONOMETRIC VALUES

To find the values of the trigonometric ratios for angles measured in radians, it is not necessary to convert to degrees. However, it is sometimes helpful to think of the two special triangles in terms of their radian measures. See Figure 4.7.

Figure 4.7

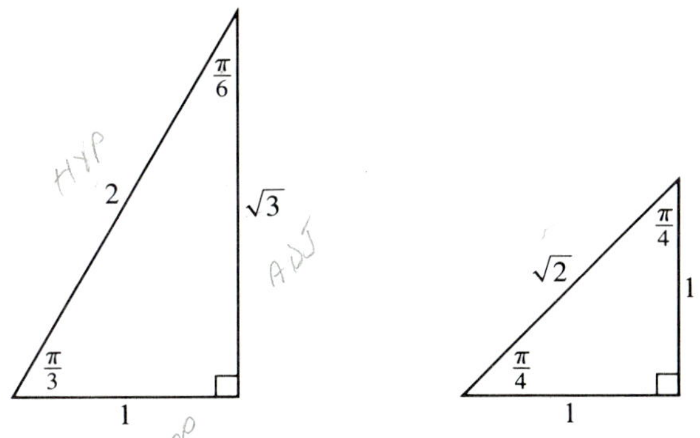

Note that we have omitted the unit name "radians." By convention, when no units are specified, the measure is assumed to be radians. So if $\theta = 3$, $\theta = 3$ *radians, not* $3°$.

Using the triangles in Figure 4.7, we get the exact values of the six trigonometric ratios for the special angles $\pi/6$, $\pi/3$, and $\pi/4$, as shown in the following table.

	$\frac{\pi}{6}$	$\frac{\pi}{3}$	$\frac{\pi}{4}$
sine	$\frac{1}{2}$	$\frac{\sqrt{3}}{2}$	$\frac{1}{\sqrt{2}}\left(\text{or } \frac{\sqrt{2}}{2}\right)$
cosine	$\frac{\sqrt{3}}{2}$	$\frac{1}{2}$	$\frac{1}{\sqrt{2}}\left(\text{or } \frac{\sqrt{2}}{2}\right)$
tangent	$\frac{1}{\sqrt{3}}\left(\text{or } \frac{\sqrt{3}}{3}\right)$	$\sqrt{3}$	1
cosecant	2	$\frac{2}{\sqrt{3}}\left(\text{or } \frac{2\sqrt{3}}{3}\right)$	$\sqrt{2}$
secant	$\frac{2}{\sqrt{3}}\left(\text{or } \frac{2\sqrt{3}}{3}\right)$	2	$\sqrt{2}$
cotangent	$\sqrt{3}$	$\frac{1}{\sqrt{3}}\left(\text{or } \frac{\sqrt{3}}{3}\right)$	1

We may find the exact values then, whenever the angle has a reference angle that is special. We may construct a triangle as we did in Section 3.2 or we may utilize the theorems on reference angles in Section 3.3. For the latter method, we will do the following:

1. Determine the quadrant in which the terminal side lies.
2. Determine the reference angle.
3. Determine the sign of the value of the given trigonometric ratio.
4. Calculate the value.

For working with radians the following chart is useful in determining the reference angle.

If θ is in quadrant	*For $0 \leq \theta < 2\pi$, the reference angle is*
I	θ
II	$\pi - \theta$
III	$\theta - \pi$
IV	$2\pi - \theta$

EXAMPLE 1 Find the sine, cosine, and tangent of $\frac{11\pi}{6}$, exactly.

SOLUTION An angle of $11\pi/6$ lies in the fourth quadrant and has a reference angle of $\pi/6$. In the fourth quadrant, the sine and tangent are negative, and the cosine is positive. Thus we have

$$\sin\frac{11\pi}{6} = -\sin\frac{\pi}{6} = \frac{-1}{2},$$

$$\cos\frac{11\pi}{6} = \cos\frac{\pi}{6} = \frac{\sqrt{3}}{2},$$

$$\tan\frac{11\pi}{6} = -\tan\frac{\pi}{6} = \frac{-1}{\sqrt{3}}\left(\text{or } \frac{-\sqrt{3}}{3}\right).$$

▲

▲ ***Exercise*** Find the sine, cosine, and tangent of $\frac{-3\pi}{4}$, exactly.

Answer $\sin\left(\frac{-3\pi}{4}\right) = \frac{-1}{\sqrt{2}}\left(\text{or } \frac{-\sqrt{2}}{2}\right)$, $\cos\left(\frac{-3\pi}{4}\right) = \frac{-1}{\sqrt{2}}\left(\text{or } \frac{-\sqrt{2}}{2}\right)$, $\tan\left(\frac{-3\pi}{4}\right) = 1$

Using a calculator to approximate the values of trigonometric ratios of radian measures is exactly like doing so with degree measures with one exception: *the calculator must be in the radian mode.*

The relationship between the accuracy of an angle measurement in radians and the accuracy of the values of its trigonometric ratios is noted in the following table.

Accuracy of radian measure	*Accuracy of trigonometric values*
nearest tenth	two significant digits
nearest hundredth	three significant digits
nearest thousandth	four significant digits

We shall continue to follow the guidelines for computations with approximations for applied problems. For nonapplied problems, we continue to round results to provide uniform answers.

EXAMPLE 2 Find the trigonometric value to four significant digits.

(a) sin 1.42 (b) tan (−1.05) (c) sec 3.75 (d) csc 2.17

SOLUTION (a) The measure of the angle is in *radians*. (Recall that when no unit appears, it is assumed to be radians). *In the radian mode*, we enter 1.42 and then push the sine button. An eight-digit display shows 0.9886518. Thus,

$$\sin 1.42 \approx 0.9887.$$

(b) *In the radian mode*, we enter −1.05 and then push the tangent button. We get

$$\tan(-1.05) \approx -1.743.$$

(c) The secant is the reciprocal of the cosine, so we begin by finding cos 3.75. *In the radian mode*, we enter 3.75, push [cos], followed by the reciprocal [1/x]. We get

$$\sec 3.75 \approx -1.219.$$

(d) The cosecant is the reciprocal of the sine, so we first find sin 2.17. *In the radian mode*, we enter 2.17, push [sin], and then [1/x]. We have

$$\csc 2.17 \approx 1.211.$$

▲

▲ ***Exercise*** Find the trigonometric value to four significant digits.

(a) sin 1.81 (b) tan (−1.175) (c) sec 2.11 (d) csc 1.05

Answer (a) 0.9715 (b) −2.393 (c) −1.948 (d) 1.153

The determination of the radian measure of an acute angle with a given trigonometric value is exactly like doing so with degree measures with one exception: *the calculator must be in the radian mode.*

EXAMPLE 3 Find the acute angle θ to the nearest thousandth of a radian, given the trigonometric value.

(a) $\sin\theta = 0.585$ (b) $\cot\theta = 2.21$

SOLUTION (a) *In radian mode*, we enter .585, followed by [inv] [sin] (or its equivalent). An eight-digit display shows 0.6248801. Rounding to the nearest thousandth,

$$\theta \approx 0.625.$$

(b) Since the cotangent is the reciprocal of the tangent, we are looking for an angle whose tangent is the reciprocal of 2.21. So *in the radian mode*, we enter 2.21, take its reciprocal, and then push [inv] [tan]. We get

$$\theta \approx 0.425.$$

▲

▲ ***Exercise*** Find the acute angle θ to the nearest thousandth of a radian, given the trigonometric value.

(a) $\cos\theta = 0.62$ (b) $\csc\theta = 1.507$

Answer (a) 0.902 (b) 0.726

Given the trigonometric values, the determination of the corresponding general angle in radians is analogous to the determination in degrees.

EXAMPLE 4 Find θ (to the nearest thousandth) for the given conditions.

(a) $\frac{\pi}{2} < \theta < \pi$ and $\tan\theta = -1.253$

(b) $-\pi < \theta < \frac{-\pi}{2}$ and $\sec\theta = -1.52$

SOLUTION (a) We have the situation shown in Figure 4.8, where θ' is the reference angle.

Figure 4.8

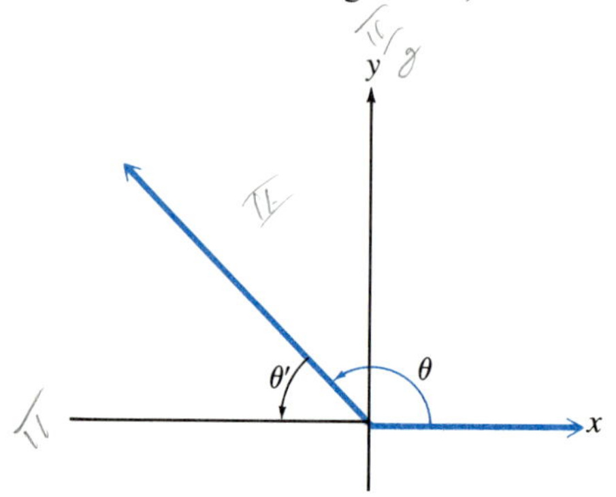

First we must find the reference angle. We note that

$$\tan \theta' = |\tan \theta| = 1.253.$$

So θ' is the acute angle whose tangent is 1.253. To determine it, we put the calculator *in radian mode* and enter 1.253 followed by [inv] [tan] (or its equivalent). We get (eight digits)

$$\theta' = 0.8972244.$$

Subtracting this from π will give us θ. We have

$$\theta = \pi - 0.8972244 \approx 2.244.$$

(b) We have the situation illustrated in Figure 4.9.

Figure 4.9

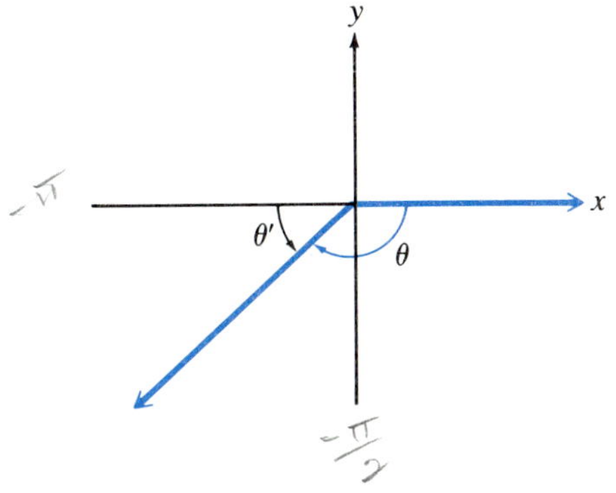

Again we must determine the reference angle first. We note that

$$\sec \theta' = |\sec \theta| = 1.52.$$

To find the acute angle whose secant is 1.52, we find the acute angle whose cosine is the reciprocal of 1.52. *In the radian mode*, we enter 1.52, push [1/x], followed by [inv] [cos] to get

$$\theta' = 0.8527764.$$

Adding this to $-\pi$ gives

$$\theta = -\pi + 0.8527764 \approx -2.289.$$

▲

▲ ***Exercise*** Find θ (to the nearest thousandth) for the given conditions.

(a) $\frac{\pi}{2} < \theta < \pi$ and $\tan \theta = -0.789$

(b) $-\pi < \theta < \frac{-\pi}{2}$ and $\csc \theta = -2.16$

Answer (a) 2.474 (b) -2.660

EXAMPLE 5 Find the approximate solutions, to the nearest thousandth, of $\cos\left(\theta + \frac{\pi}{2}\right) = 0.25$, where $0 \le \theta < 2\pi$.

SOLUTION We begin by finding values of $\theta + (\pi/2)$. We note that since $0 \le \theta < 2\pi$,

$$\frac{\pi}{2} \le \theta + \frac{\pi}{2} < \frac{5\pi}{2}.$$

The reference angle is 1.3181161. Because the cosine is positive in the fourth and first quadrants, we have

$$\theta + \frac{\pi}{2} = 2\pi - 1.3181161 \qquad \text{or} \qquad \theta + \frac{\pi}{2} = 2\pi + 1.3181161$$

$$\theta = \frac{3\pi}{2} - 1.3181161 \approx 3.394 \qquad\qquad \theta = \frac{3\pi}{2} + 1.3181161 \approx 6.031\,.$$

▲

▲ *Exercise* Find the approximate solutions, to the nearest thousandth, of $\cos\left(\theta + \frac{\pi}{2}\right) = 0.85$, where $0 \le \theta < 2\pi$.

Answer 4.158, 5.267

EXAMPLE 6 A tank in the shape of a right circular cylinder with a horizontal axis has a radius of 2.0 feet and is 10 feet long. If it contains oil to a depth of 1.2 feet, what is the volume (to two significant digits) of the oil contained in the tank?

SOLUTION In Figure 4.10 we see the end of the tank.

Figure 4.10

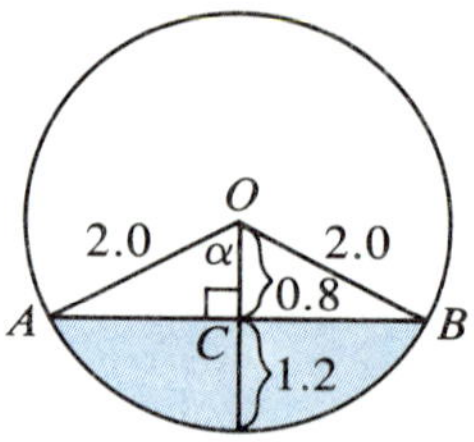

We need to determine the area of the region beneath $\triangle OAB$. To do so we'll compute the area of the sector OAB and subtract the area of $\triangle OAB$. Thus we need the measure of $\angle AOB$. To this end, let's compute the measure of $\angle AOC$. We have

$$\cos\alpha = \frac{0.8}{2.0}.$$

Hence,

$$\alpha = 1.1592795 \text{ radians}\,.$$

Since the measures of $\angle AOC$ and $\angle BOC$ are equal, we have

$$m\,\angle AOB = 2(1.1592795) = 2.318559.$$

So the area of the sector is

$$\frac{m\,\angle AOB}{2} \cdot r^2 = \frac{(2.318559)(2.0)^2}{2} = 4.637118.$$

The area of a triangle is $\frac{1}{2}$ the base times the height. So we must compute AC (or BC). We get

$$AC = \sqrt{(2.0)^2 - (0.8)^2} \approx 1.8330303.$$

So the area of $\triangle OAB$ is

$$(0.8)(1.8330303) = 1.4664242.$$

Hence the area of the desired region is

$$4.637118 - 1.4664242 = 3.1706938.$$

Multiplying this area by the length of the tank will give the volume, which is

$$31.706938.$$

To two significant digits, the volume is 32 cubic feet.

▲

▲ ***Exercise*** A tank in the shape of a right circular cylinder with a horizontal axis has a radius of 2.5 feet and is 12 feet long. If it contains oil to a depth of 1.3 feet, what is the volume (to two significant digits) of the oil contained in the tank?

Answer 49 ft^3

Problem Set 4.3

In Problems 1–4, find the sine, cosine, and tangent of the given angle, exactly.

1. $\frac{13\pi}{6}$

2. $\frac{5\pi}{3}$

3. $\frac{-7\pi}{4}$

4. $\frac{-7\pi}{6}$

In Problems 5 and 6, find the secant, cosecant, and cotangent of the given angle, exactly.

5. $\frac{-2\pi}{3}$

6. $\frac{5\pi}{4}$

In Problems 7–10, find the values of the trigonometric ratios for the given measure.

7. $\frac{\pi}{2}$

8. $\frac{-\pi}{2}$

9. π

10. 2π

In Problems 11–22, find the trigonometric value to four significant digits.

11. sin 1.2

12. sin 3.51

13. cos (−0.72)

14. cos 4.5

15. tan 0.68

16. tan (−1.2)

17. sec 4.25

18. sec 0.21

19. csc (−4.82)

20. csc 9.1

21. cot 3.25

22. cot (−0.812)

In Problems 23–28, find the acute angle θ to the nearest thousandth of a radian, given the trigonometric value.

23. $\sin \theta = 0.95$

24. $\cos \theta = 0.2125$

25. $\tan \theta = 2.11$

26. $\cot \theta = 0.817$

27. $\sec \theta = 3.52$

28. $\csc \theta = 3$

In Problems 29–36, find θ (to the nearest thousandth) for the given conditions.

29. $\pi < \theta < 2\pi$ and $\cos \theta = -0.725$

30. $0 < \theta < \pi$ and $\tan \theta = -0.612$

31. $\pi < \theta < 2\pi$ and $\tan \theta = 3.162$

32. $\pi < \theta < 2\pi$ and $\cot \theta = -2.56$

33. $0 < \theta < 2\pi$, $\cot \theta < 0$, and $\sec \theta = -4.25$

34. $0 < \theta < 2\pi$, $\tan \theta > 0$, and $\sec \theta = -3.59$

35. $0 \le \theta < 2\pi$ and $\csc \theta = 2.5$

36. $0 \le \theta < 2\pi$ and $\tan \theta = -0.516$

In Problems 37–44, find the exact solutions, where $0 \le \theta < 2\pi$.

37. $\sin \theta = \frac{-1}{\sqrt{2}}$

38. $\cos \theta = \frac{1}{2}$

39. $\tan \theta = \sqrt{3}$

40. $\sec \theta = -\sqrt{2}$

41. $\cos 2\theta = \frac{\sqrt{3}}{2}$

42. $\sin \frac{\theta}{2} = \frac{\sqrt{2}}{2}$

43. $\sin (\theta + \pi) = \frac{\sqrt{3}}{2}$

44. $\tan (\theta - \pi) = -1$

In Problems 45–56, find the approximate solutions to the nearest thousandth, where $0 \le \theta < 2\pi$.

45. $\sin \theta = 0.28$

46. $\cos \theta = 0.852$

47. $\tan \theta = 8.3$

48. $\sec \theta = -4$

49. $\sin 2\theta = 0.9$

50. $\cot 2\theta = -1.6$

51. $\cot \frac{\theta}{3} = -0.36$

52. $\tan \frac{\theta}{2} = 14$

53. $\cos (\theta - \pi) = 0.35$

54. $\sin \left(\theta + \frac{\pi}{2}\right) = -0.72$

55. $\sin (\theta - 1) = -0.6816$

56. $\cos (\theta + 1) = 0.98$

57. The cross section of a drainage pipe has the shape of the arc (minor) of a circle. If the pipe is 3.0 feet across the top and the radius of the circle is 2.0 feet, what is the cross-sectional area (to two significant digits)?

4.4 TRIGONOMETRIC FUNCTIONS

So far we have considered trigonometric ratios and their values for angle measures—either degrees or radians. In advanced work it is necessary to think of trigonometric functions and their values at real numbers. The purpose of this section is to develop these concepts.

Consider a circle of radius 1 centered at the origin (a unit circle) and the line parallel and one unit to the right of the y-axis. Let's make this line a number line by letting the point where it crosses the x-axis, (1, 0), be the origin and by choosing a unit distance equal to that on the x- and y-axes. Select the positive direction as upward. See Figure 4.11.

Figure 4.11

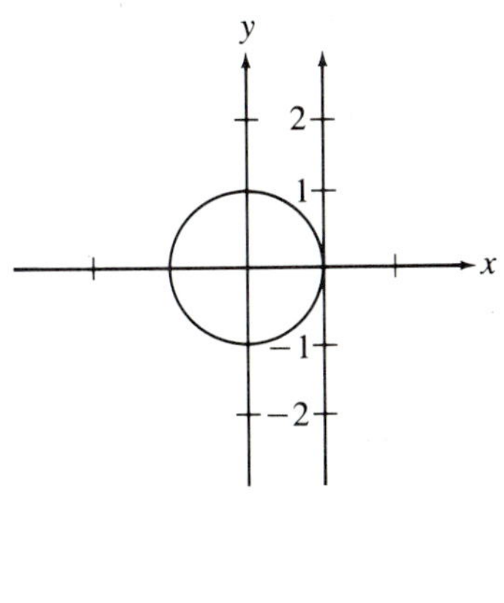

For each real number t on the number line, we will associate with it a point P on the unit circle, determined in the following manner:

1. If $t = 0$, then $P = (1, 0)$.

2. If $t > 0$, then P will be that point on the unit circle which coincides with the point t on the number line when the segment from 0 to t is "wrapped around" the circle in a *counterclockwise* direction.

3. If $t < 0$, then P will be that point on the unit circle which coincides with the point t on the number line when the segment from 0 to t is "wrapped around" the circle in a *clockwise* direction.

Figure 4.12 illustrates the procedure described for a positive and a negative value of t.

Figure 4.12

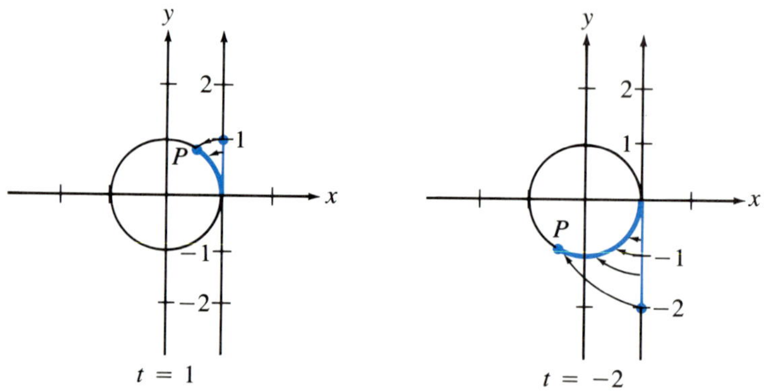

Since the circumference of the circle is 2π, the segment from 0 to 2π would wrap around the circle exactly one time. Hence the point associated with $t = 2\pi$ is $(1, 0)$. See Figure 4.13.

Figure 4.13

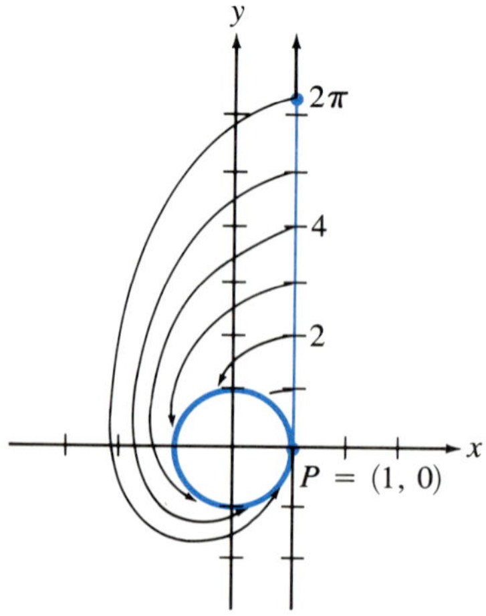

Similarly, the point $(1, 0)$ is associated with $t = -2\pi$. Some other "special" values of t are noted in Figure 4.14.

Figure 4.14

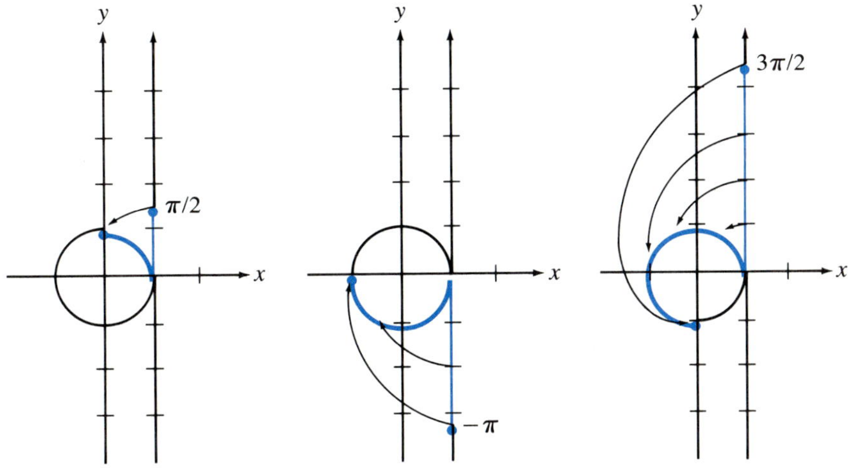

$t = \pi/2$
$P = (0, 1)$

$t = -\pi$
$P = (-1, 0)$

$t = 3\pi/2$
$P = (0, -1)$

Of course, since the circumference is 2π, for any $t > 2\pi$ or $t < -2\pi$, the circle will be wrapped up more than once. For example, for $t = 3\pi$, the segment goes around the circle $1\frac{1}{2}$ times, giving $P = (-1, 0)$.

Every real number t determines a unique point P on the unit circle. Hence this correspondence is a function. It is denoted W and is called the **wrapping function**. Thus W is defined by

$$W(t) = (x, y),$$

where (x, y) is the point determined in the previously described manner. Preceding illustrations indicate that

$$W(0) = (1, 0),$$
$$W(2\pi) = (1, 0),$$
$$W\left(\frac{\pi}{2}\right) = (0, 1),$$
$$W(-\pi) = (-1, 0),$$
$$W\left(\frac{3\pi}{2}\right) = (0, -1).$$

Let's now consider the function that associates each real number t with the second coordinate of $W(t)$. Define f by

$$f(t) = y,$$

where y is the second coordinate of $W(t)$.

How do we find $f(t)$? Let

$$P = (x, y) = W(t).$$

Ray $\overrightarrow{OP}$ may be considered the terminal side of an angle θ in standard position, where the length of the arc intercepted is t. See Figure 4.15.

Figure 4.15

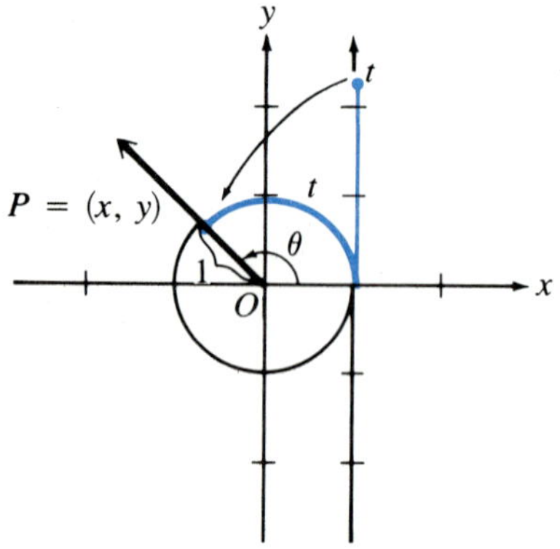

Because the radius of the circle is 1, the measure of θ is t radians. Thus,

$$y = \frac{y}{1} = \sin\theta = \sin t.$$

Therefore, $f(t)$ is the sine of t radians.

So f is a function whose domain is R. Its range is $[-1, 1]$. Because of its direct relationship with the sine ratio, it is called the **sine function** and is abbreviated sin. Hence

$$f(t) = \sin(t).$$

But $\sin(t)$ is generally written $\sin t$. Thus the notation

$$\sin t$$

may be viewed in one of two different ways: as the value of the sine function at the real number t, or as the value of the sine ratio for an angle of t radians. Either way we get the same numerical value.

EXAMPLE 1 A bacteria population is growing in such a way that its weight, w, in milligrams, is given by

$$w = 3^t(2 - \sin t),$$

where t represents elapsed time in hours. Find the weight of the population (to two significant digits) after 2 hours.

SOLUTION We must find the value of w when $t = 2$. We get

$$w = 9(2 - \sin 2).$$

To compute this we must evaluate the sine function at the real number 2. This is numerically equivalent to the value of the sine ratio at 2 radians. We have

$$w = 9(2 - 0.9092974) \approx 10.$$

The weight after 2 hours is approximately 10 milligrams.

Exercise A bacteria population is growing in such a way that its weight, w, in milligrams, is given by

$$w = 2^t(3 - \sin t),$$

where t represents elapsed time in hours. Find the weight of the population (to two significant digits) after 3 hours.

Answer 23 mg

In a similar fashion, we may construct a function g that associates each real number t with the first coordinate of $W(t)$. Thus,

$$g(t) = x,$$

where x is the first coordinate of $W(t)$. Because

$$x = \frac{x}{1} = \cos\theta = \cos t,$$

g is called the **cosine function** and is abbreviated cos. It is a function whose domain is R and whose range is $[-1, 1]$.

We may construct the **tangent function**, abbreviated tan, by defining

$$h(t) = \frac{y}{x},$$

where $W(t) = (x, y)$. This function has domain $R - \{(\pi/2) + n\pi \mid n \in Z\}$ and range R.

The secant, cosecant, and cotangent functions are developed in the same way. These six trigonometric functions are often called **circular functions** because of the association with the unit circle. In the next chapter we study these functions.

Problem Set 4.4

In Problems 1–8, find the exact value, where W is the wrapping function.

1. $W(\pi)$
2. $W(-2\pi)$
3. $W\left(\frac{-\pi}{2}\right)$
4. $W\left(\frac{5\pi}{2}\right)$

5. $W\left(\frac{\pi}{4}\right)$

6. $W\left(\frac{-3\pi}{4}\right)$

7. $W\left(\frac{-2\pi}{3}\right)$

8. $W\left(\frac{5\pi}{6}\right)$

In Problems 9–14, determine the quadrant in which $W(t)$ lies.

9. $W(1.2)$

10. $W(2.7)$

11. $W(-0.56)$

12. $W(-3.2)$

13. $W(20)$

14. $W(-12)$

15. Let $f(x) = \sin x$ and $g(x) = \cos x$. Determine each of the following:
(a) $f(2)$ (b) $f(-\pi)$ (c) $g(3)$ (d) $g(-0.6)$

16. Let $f(t) = \sin t$, $g(t) = \cos t$, and $h(t) = \tan t$. Show each of the following:
(a) $(f(t))^2 + (g(t))^2 = 1$ (b) $h(t) = \dfrac{f(t)}{g(t)}$

17. The voltage e in a certain electrical circuit is given by $e = 100 \sin 40\pi t$, where t is time in seconds. What is the voltage (to the nearest volt) when $t = 2.11$?

18. The amperage i of a certain alternating current is given by $i = 5 \cos 120\pi t$, where t is time in seconds. What is the strength (to the nearest hundredth of an amp) when $t = 4.57$?

19. The number of animals in a certain endangered population varies yearly and is approximated by

$$n = 200 + 100 \sin(5t),$$

where t represents years from the present. What will be the population (to two significant digits) in 4 years?

20. The average number of cases of measles per 10,000 population in North America is approximated by

$$n = 10^{1.5 + 0.7\cos(t/2) + 0.15\sin t},$$

where t is measured in months, and $t = 0$ represents April 1. What is the number of cases per 10,000 for the first of September?

21. What are the domain and range of the secant function?

22. What are the domain and range of the cosecant function?

23. What are the domain and range of the cotangent function?

CHAPTER 4 REVIEW

Key Terms and Formulas

angular speed, page 127

arc length, page 125

$s = r\theta$

area of a sector, page 126

$K = \frac{\theta}{2} r^2$

circular functions, page 143

conversion formulas, pages 120–122

$x \text{ radians} = \frac{x}{\pi} \cdot 180^\circ$

$x^\circ = \frac{x}{180} \cdot \pi \text{ radians}$

linear speed, page 127

$v = r\omega$

radian measure, page 120

wrapping function, page 141

Problem Set

1. Specify "true" or "false."

(a) 2 radians $= \left(\frac{360}{\pi}\right)^\circ$

(b) $10^\circ = \frac{\pi}{18}$ radians

(c) If the arc length exceeds the radius, then the measure of the central angle is less than 1 radian.

(d) If the area of a sector is 75% of the area of the circle, then the central angle has measure 0.75π.

(e) The sine and cosine functions have the same domain and range.

(f) The range of the tangent function is R.

(g) The domain of the cotangent function is R.

(h) The range of the secant function is $(-\infty, -1) \cup (1, \infty)$.

(i) An angle of measure 2 is acute.

2. Construct and angle in standard position with measure $3\pi/4$, state the quadrant in which the terminal side lies, determine the reference angle, and give the measure of a coterminal angle with opposite direction.

In Problems 3 and 4, convert the radian measure to degrees, exactly.

3. $\frac{7\pi}{6}$ radians

4. $\frac{-7\pi}{4}$ radians

In Problems 5 and 6, convert the radian measure to degrees, approximating to the nearest hundredth.

5. -1.79 radians *6.* 3.5 radians

In Problems 7 and 8, convert the degree measure to radians, exactly.

7. 225° *8.* $-270°$

In Problems 9 and 10, convert the degree measure to radians, approximating to the nearest thousandth.

9. $-196°$ *10.* 312°

In Problems 11 and 12, find the length (to four significant digits) of the arc on a circle with the given radius, intercepted by the given central angle.

11. radius 17 centimeters, central angle 1.5 radians

12. radius 3 feet, central angle 115°

In Problems 13 and 14, determine the area (to four significant digits) of the sector described.

13. radius 1.5 feet, central angle 1.05 radians

14. radius 32 centimeters, central angle 250°

15. What is the area (to four significant digits) of a sector whose arc length is 6 inches if the radius is 4.5 inches?

16. What is the arc length (to four significant digits) of a sector whose area is 8.5 square inches if the radius is 3.2 inches?

17. A searchlight with an effective range of 250 feet sweeps an angle of 75°. What is the effective area (to two significant digits) covered by the searchlight?

18. A tank in the shape of a right circular cylinder with horizontal axis has a radius of 3.0 feet and is 12 feet long. What is the volume (to two significant digits) of oil contained in the tank if the depth of the oil is (a) 2.0 feet, (b) 4.2 feet?

19. If the forearm of a pitcher may be rotated about the elbow at 800° per second, what speed (in mph, to two significant digits) can be imparted to a baseball if the distance from ball to elbow is 15 inches?

20. A belt running between two pulleys connects a dentist drill to the motor. The motor pulley has a diameter of 3.0 centimeters, and the drill pulley has a diameter of 0.40 centimeter. If the pulley at the motor has a speed of 20 revolutions per minute, what is the speed of the drill (in rpm, to two significant digits)?

In Problems 21–30, find the exact trigonometric value.

21. $\sin \frac{\pi}{3}$

22. $\cot \frac{\pi}{3}$

23. $\tan \frac{\pi}{4}$

24. $\sec \frac{\pi}{4}$

25. $\cos \frac{\pi}{6}$

26. $\csc \frac{\pi}{6}$

27. $\cos \frac{4\pi}{3}$

28. $\tan \left(\frac{-3\pi}{4} \right)$

29. $\sin \left(\frac{-5\pi}{4} \right)$

30. $\sec \frac{7\pi}{6}$

In Problems 31 and 32, find the values of the trigonometric ratios for the given measures.

31. $\frac{3\pi}{2}$

32. $-\pi$

In Problems 33–38, find the trigonometric value to four significant digits.

33. $\sin 1.9$

34. $\cos 0.77$

35. $\tan (-3.8)$

36. $\cot 1.21$

37. $\sec 4$

38. $\csc (-0.592)$

In Problems 39–44, find the acute angle θ to the nearest thousandth of a radian, given the trigonometric value.

39. $\cos \theta = 0.15$

40. $\sin \theta = 0.8215$

41. $\tan \theta = 3.11$

42. $\cot \theta = 2.9$

43. $\csc \theta = 5$

44. $\sec \theta = 1.425$

In Problems 45–48, find θ (to the nearest thousandth) for the given conditions.

45. $0 < \theta < \pi$ and $\cos \theta = -0.616$

46. $\pi < \theta < 2\pi$ and $\cot \theta = 4.5$

47. $0 < \theta < 2\pi$, $\cot \theta < 0$, and $\csc \theta = 2.55$

48. $0 < \theta < 2\pi$, $\tan \theta < 0$, and $\sec \theta = 1.295$

In Problems 49–54, find the exact solutions, where $0 \le \theta < 2\pi$.

49. $\sin \theta = \frac{1}{2}$

50. $\tan \theta = \frac{-1}{\sqrt{3}}$

51. $\cot \frac{\theta}{2} = -\sqrt{3}$

52. $\sin 2\theta = \frac{\sqrt{2}}{2}$

53. $\sin(\theta - \pi) = \dfrac{-\sqrt{2}}{2}$

54. $\tan(2\theta - \pi) = 1$

In Problems 55–60, find the approximate solutions to the nearest thousandth, where $0 \leq \theta < 2\pi$.

55. $\cos\theta = -0.75$

56. $\csc\theta = 1.096$

57. $\tan\dfrac{\theta}{2} = -2.6$

58. $\sin 2\theta = 0.695$

59. $\cos(\theta - \pi) = -0.62$

60. $\sin(\theta + 1) = 0.287$

In Problems 61 and 62, find the exact value, where W is the wrapping function.

61. $W\left(\dfrac{-9\pi}{2}\right)$

62. $W(3\pi)$

In Problems 63 and 64, determine the quadrant in which $W(t)$ lies.

63. $W(2.8)$

64. $W(-0.87)$

5
TRIGONOMETRIC FUNCTIONS: GRAPHS AND INVERSES

CHAPTER OUTLINE

In this chapter we study the trigonometric functions, whose domains are sets of real numbers. We concentrate on their graphs, which enable us to see properties not readily apparent from the equations. Applications include physical phenomena that are time dependent and repetitive. The chapter concludes with a study of inverses—an important topic in advanced mathematics.

5.1 THE SINE AND COSINE FUNCTIONS

We begin our study of the trigonometric functions by examining their graphs. From Section 4.4 we see that for any trigonometric function T,

$$T(x + 2\pi) = T(x).$$

(In general, if there exists a positive number k such that $f(x+k) = f(x)$ for all x in the domain, we say that f is periodic. The smallest such positive number k is called the **period**.) After x has assumed all the values between 0 and 2π, the values of a trigonometric function will simply repeat themselves. Hence, for each trigonometric function, those portions of the graph on the intervals $[-2\pi, 0]$, $[0, 2\pi]$, $[2\pi, 4\pi]$, and so on, will be identical. Consequently, we will first get a detailed graph over the interval $[0, 2\pi]$, and then use this information to get the entire graph.

We also use two important properties. First, the graphs of the trigonometric functions are smooth. That is, there are no "corners." Second, the trigonometric functions are continuous on their domains. That is, the only "breaks" in the graph occur where the functions are not defined. Hence, if the function is defined on an interval, then the graph will be one piece over this interval.

Beginning with the sine function,

$$y = \sin x,$$

whose domain is R, let's choose values of x in the interval $[0, 2\pi]$. Consider the following table:

x	0	$\frac{\pi}{6}$	$\frac{\pi}{3}$	$\frac{\pi}{2}$	$\frac{2\pi}{3}$	$\frac{5\pi}{6}$	π	$\frac{7\pi}{6}$	$\frac{4\pi}{3}$	$\frac{3\pi}{2}$	$\frac{5\pi}{3}$	$\frac{11\pi}{6}$	2π
y	0	$\frac{1}{2}$	$\frac{\sqrt{3}}{2}$	1	$\frac{\sqrt{3}}{2}$	$\frac{1}{2}$	0	$\frac{-1}{2}$	$\frac{-\sqrt{3}}{2}$	-1	$\frac{-\sqrt{3}}{2}$	$\frac{-1}{2}$	0
$\approx y$	0	0.5	0.9	1	0.9	0.5	0	-0.5	-0.9	-1	-0.9	-0.5	0

Plotting these points and connecting them with a smooth, unbroken curve, we get the graph of $y = \sin x$ on the interval $[0, 2\pi]$. See Figure 5.1. The

Figure 5.1
$y = \sin x$ *on* $[0, 2\pi]$

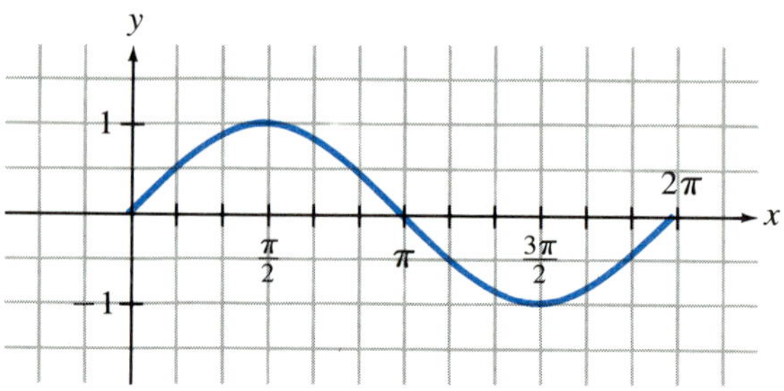

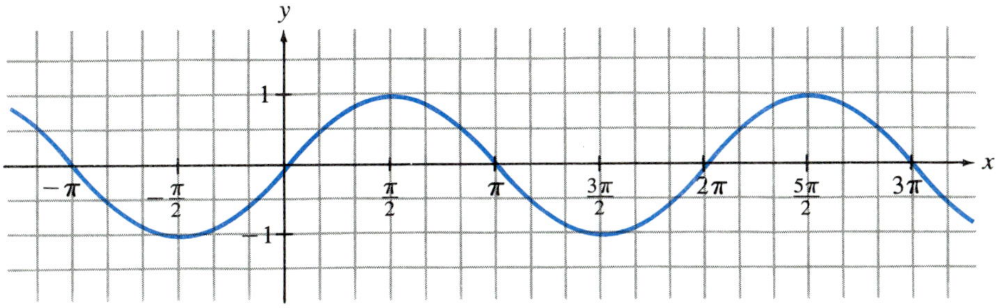

Figure 5.2
$y = \sin x$

"complete" graph, which is a repetition of the graph in Figure 5.1, is shown in Figure 5.2.

From the graph we see that $k = 2\pi$ is the smallest number such that $\sin(x + k) = \sin x$; that is, the values of the sine function repeat themselves every 2π units. Thus the sine function is periodic with period 2π.

Note that the graph is symmetric about the origin, so the sine function is odd. That is,

$$\sin(-x) = -\sin x.$$

Further note that the range of the sine function is $[-1, 1]$. The maximum value of 1 is attained when

$$x = \frac{\pi}{2} + (n \cdot 2\pi), \qquad n = 0, \pm 1, \pm 2, \ldots,$$

and the minimum value of -1 is reached when

$$x = \frac{3\pi}{2} + (n \cdot 2\pi), \qquad n = 0, \pm 1, \pm 2, \ldots.$$

The **amplitude** of a function is $\frac{1}{2}$ of the difference between its maximum and minimum values. Thus the amplitude of the sine function is $\frac{1}{2}(1 - (-1)) = 1$.

In sketching the graph of the cosine function,

$$y = \cos x,$$

whose domain is R, we construct the following table:

x	0	$\frac{\pi}{6}$	$\frac{\pi}{3}$	$\frac{\pi}{2}$	$\frac{2\pi}{3}$	$\frac{5\pi}{6}$	π	$\frac{7\pi}{6}$	$\frac{4\pi}{3}$	$\frac{3\pi}{2}$	$\frac{5\pi}{3}$	$\frac{11\pi}{6}$	2π
y	1	$\frac{\sqrt{3}}{2}$	$\frac{1}{2}$	0	$\frac{-1}{2}$	$\frac{-\sqrt{3}}{2}$	-1	$\frac{-\sqrt{3}}{2}$	$\frac{-1}{2}$	0	$\frac{1}{2}$	$\frac{\sqrt{3}}{2}$	1
$\approx y$	1	0.9	0.5	0	-0.5	-0.9	-1	-0.9	-0.5	0	0.5	0.9	1

Figure 5.3
$y = \cos x$ *on* $[0, 2\pi]$

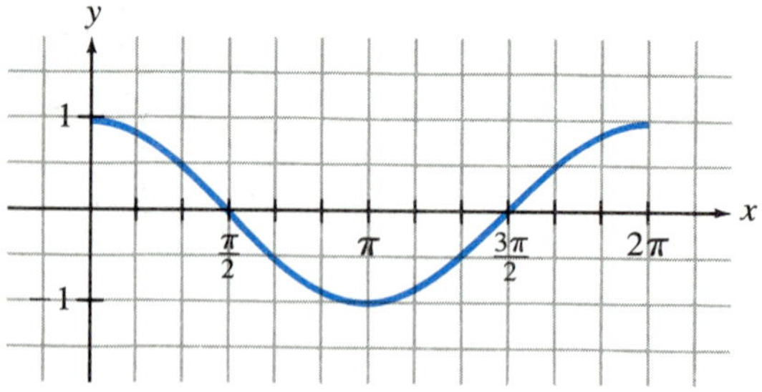

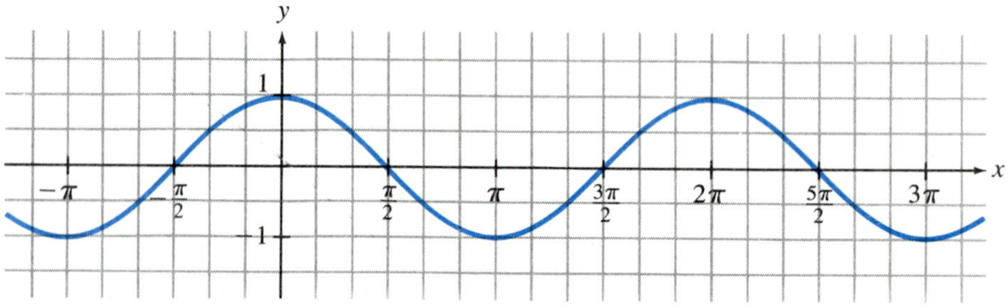

Figure 5.4
$y = \cos x$

Plotting these points leads to the graph of $y = \cos x$ on the interval $[0, 2\pi]$. See Figure 5.3. The "complete" graph is shown in Figure 5.4.

Like the sine function, the cosine function has a period of 2π. Its range is $[-1, 1]$, and its amplitude is 1. Unlike the sine graph, however, the cosine graph is symmetric about the y-axis. So the cosine function is even:

$$\cos(-x) = \cos x .$$

Further, observe that the graphs of the sine and cosine functions have the same shape. They are simply "out of phase." That is, each is a horizontal shift of the other. If the cosine graph were shifted to the right $\pi/2$ units, it would coincide with the sine graph. So the graph of the cosine function on $[-\pi/2, 3\pi/2]$ has the same shape as the graph of the sine function on $[0, 2\pi]$. Since $x - (\pi/2)$ assumes values on $[-\pi/2, 3\pi/2]$ as x assumes values on $[0, 2\pi]$,

$$\cos\left(x - \frac{\pi}{2}\right) = \sin x .$$

This relationship will be verified in Section 6.2.

Henceforth, to sketch a graph, we will first plot some points over one period. By dividing the period into equal fourths, we will plot the maximum and minimum points and the points where the graph crosses the x-axis. This will define the wave. To complete the graph we use the periodic property to extend the shape obtained.

EXAMPLE 1 Sketch the graph of $y = 2 \sin x$.

Solution The period is 2π, so initially we'll sketch our graph over the interval $[0, 2\pi]$. Dividing the interval into equal fourths, we get x-values of $0, \pi/2, \pi, 3\pi/2$, and 2π. Substituting to determine the corresponding y-value gives the table:

x	0	$\frac{\pi}{2}$	π	$\frac{3\pi}{2}$	2π
y	0	2	0	-2	0

Plotting these points and using our knowledge of the general shape of the sine function gives the graph on $[0, 2\pi]$ shown in Figure 5.5. Using the periodic property, we have the graph shown in Figure 5.6.

Figure 5.5
$y = 2 \sin x$ *on* $[0, 2\pi]$

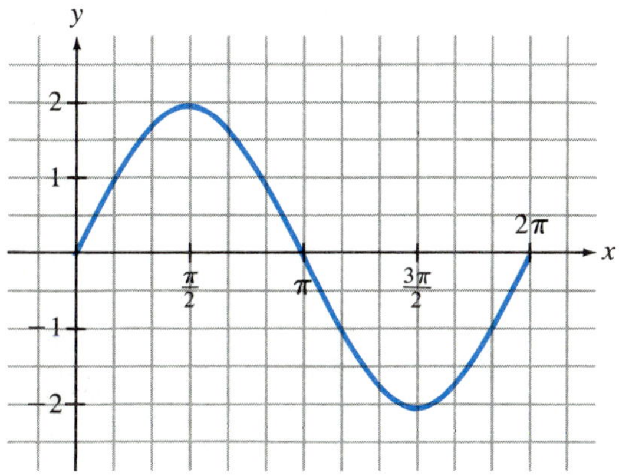

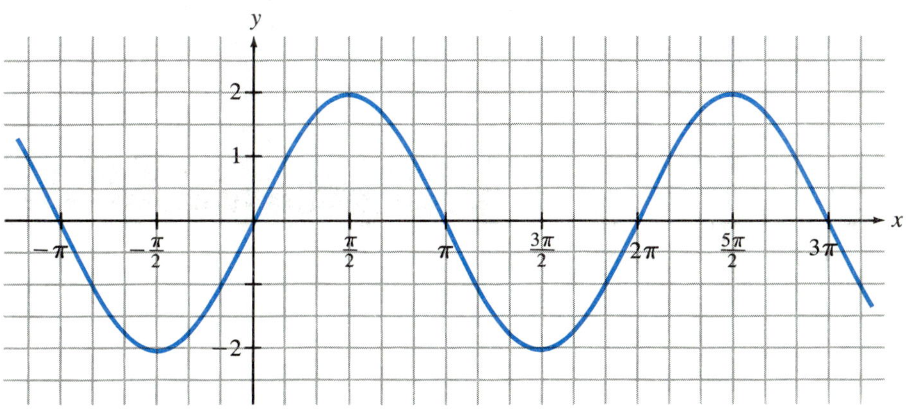

Figure 5.6
$y = 2 \sin x$

▲

If c is a positive constant and f a function, then the graph of $y = c \cdot f(x)$ is a vertical stretching (if $c > 1$) or compressing (if $c < 1$) of the graph of

$y = f(x)$. That is, to get the graph of $y = c \cdot f(x)$ from the graph of $y = f(x)$, each point on the graph of $y = f(x)$ is moved so that it stays on the same side of the x-axis, but its distance from the x-axis is multiplied by c. Thus if each point on the graph of $y = \sin x$ is moved so that its distance from the x-axis is doubled, then we would have the graph of $y = 2 \sin x$, whose amplitude is 2. Thus the c in $y = c \cdot f(x)$ affects the amplitude.

▲ ***Exercise*** Sketch the graph of $y = 2 \cos x$.

Answer See Figure 5.7.

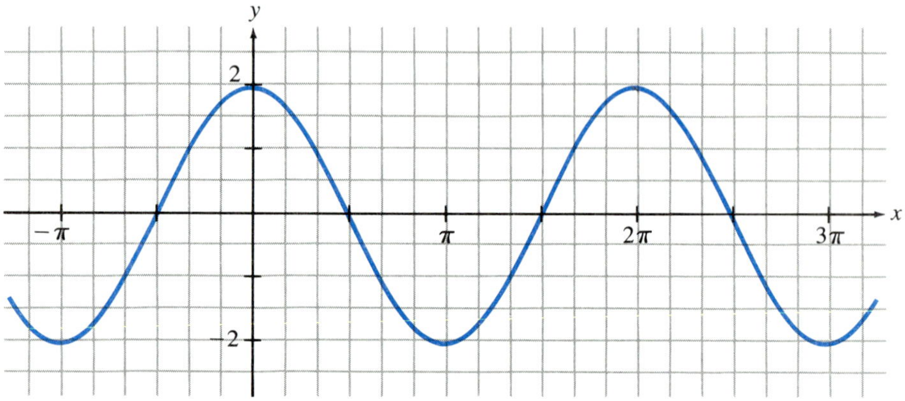

Figure 5.7

EXAMPLE 2 Sketch the graph of $y = \cos 2x$.

SOLUTION Again we plot points over one period, but the period is no longer 2π. All values of $\cos 2x$ will be attained as $2x$ assumes all of the values on $[0, 2\pi]$—that is, as x assumes all of the values on $[0, \pi]$. The period is π. Dividing $[0, \pi]$ into equal fourths gives x-values of $0, \pi/4, \pi/2, 3\pi/4$, and π. We get the following table,

x	0	$\frac{\pi}{4}$	$\frac{\pi}{2}$	$\frac{3\pi}{4}$	π
y	1	0	−1	0	1

which leads to the graph shown in Figure 5.8.

Figure 5.8
$y = \cos 2x$

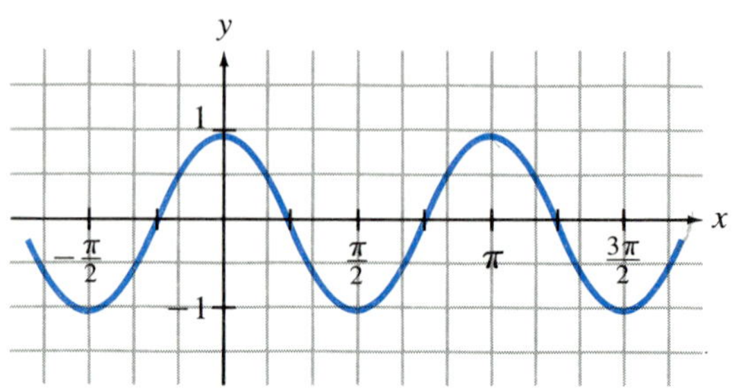

▲

You may find it valuable to have an extra line in the table when the argument of the trigonometric function is something other than x. For instance, in Example 2, we could have written

x	0	$\frac{\pi}{4}$	$\frac{\pi}{2}$	$\frac{3\pi}{4}$	π
$2x$	0	$\frac{\pi}{2}$	π	$\frac{3\pi}{2}$	2π
y					

before computing the y-values. In this way the x-values are all doubled at once, and we use the middle row in computing y.

This optional row will be included in the remaining examples. However, note that when plotting the points on the graph, we refer only to the *top* and *bottom* rows. These are the x and y coordinates.

If c is a positive constant and f a function, then the graph of $y = f(cx)$ is a horizontal compressing (if $c > 1$) or stretching (if $c < 1$) of the graph of $y = f(x)$. That is, to get the graph of $y = f(cx)$ from the graph of $y = f(x)$, each point on the graph of $y = f(x)$ is moved so that it stays on the same side of the y-axis, but its distance from the y-axis is divided by c. Thus, if each point on the graph of $y = \cos x$ is moved so that its distance from the y-axis is halved, then we would have the graph of $y = \cos 2x$, whose period is π. Thus the c in $y = f(cx)$ affects the period.

▲ ***Exercise*** Sketch the graph of $y = \sin 2x$.

Answer See Figure 5.9.

Figure 5.9

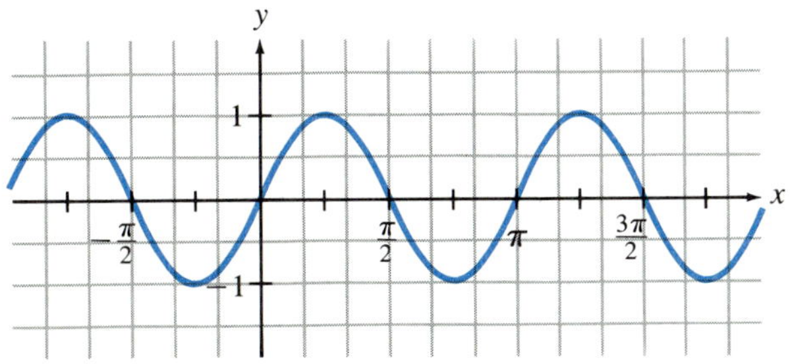

For the functions defined by

$$y = a \sin bx \quad \text{and} \quad y = a \cos bx,$$

where a and b are nonzero constants, the period is $2\pi/|b|$ (which is the usual period divided by the absolute value of the coefficient of x), and the amplitude is $|a|$.

EXAMPLE 3 Sketch the graph.

(a) $y = -3 \sin \dfrac{x}{2}$ (b) $y = \dfrac{1}{2} \cos(-4x)$

SOLUTION (a) The period is $2\pi \div \frac{1}{2} = 4\pi$. Dividing $[0, 4\pi]$ into equal fourths, we get the following table:

x	0	π	2π	3π	4π
$\dfrac{x}{2}$	0	$\dfrac{\pi}{2}$	π	$\dfrac{3\pi}{2}$	2π
y	0	-3	0	3	0

Plotting points and using the periodic property gives us the graph in Figure 5.10.

Figure 5.10
$y = -3 \sin \dfrac{x}{2}$

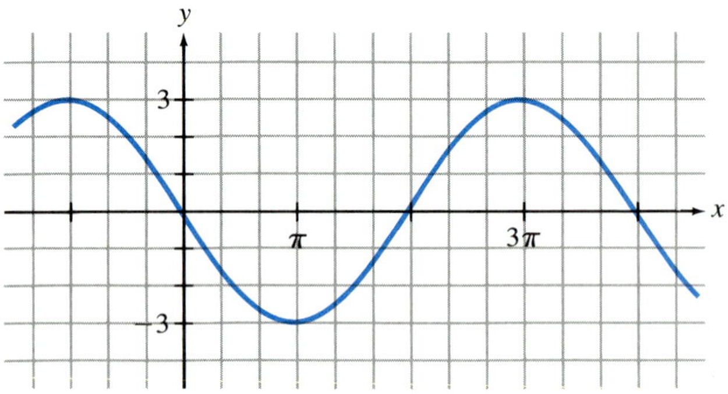

(b) The period is $2\pi \div |-4| = \pi/2$. Dividing $[0, \pi/2]$ into equal fourths gives the following table:

x	0	$\dfrac{\pi}{8}$	$\dfrac{\pi}{4}$	$\dfrac{3\pi}{8}$	$\dfrac{\pi}{2}$
$-4x$	0	$\dfrac{-\pi}{2}$	$-\pi$	$\dfrac{-3\pi}{2}$	-2π
y	$\dfrac{1}{2}$	0	$\dfrac{-1}{2}$	0	$\dfrac{1}{2}$

This gives us the graph in Figure 5.11.

▲

Figure 5.11
$y = \frac{1}{2}\cos(-4x)$

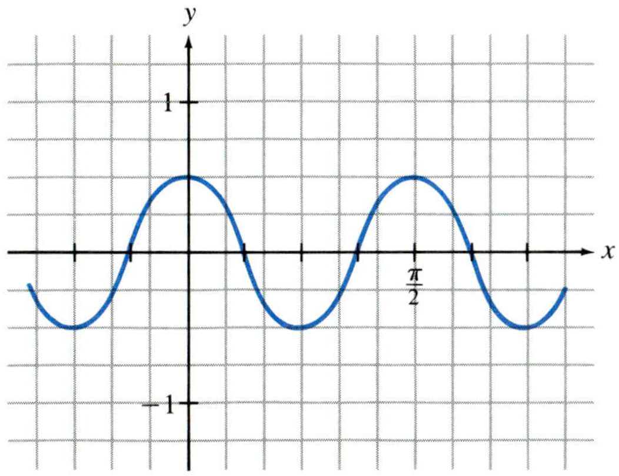

When the coefficient of x in the argument (expression in parentheses) is negative, we may use the even/odd property to simplify. In Example 3(b) we could have written

$$y = \frac{1}{2}\cos(-4x) = \frac{1}{2}\cos 4x\,.$$

This step is possible because the cosine function is even; that is, $\cos(-t) = \cos t$. The benefit of this is that we get familiar values in the optional row. Indeed, we have

x	0	$\frac{\pi}{8}$	$\frac{\pi}{4}$	$\frac{3\pi}{8}$	$\frac{\pi}{2}$
$4x$	0	$\frac{\pi}{2}$	π	$\frac{3\pi}{2}$	2π
y	$\frac{1}{2}$	0	$\frac{-1}{2}$	0	$\frac{1}{2}$

▲ ***Exercise*** Sketch the graph.

(a) $y = 3\cos\frac{x}{2}$ (b) $y = \frac{-1}{2}\sin(-4x)$

Answer See Figure 5.12.

(a)

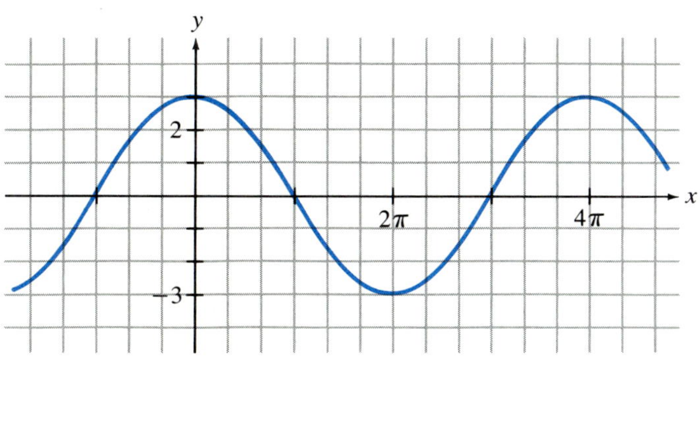

Figure 5.12 (*Continued*)

(b)

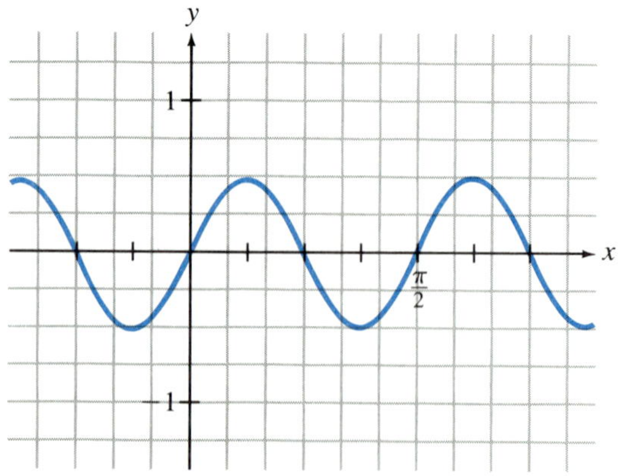

Sound is produced by a vibrating object. The vibration produces a wave of changes in air pressure, which causes the eardrum to vibrate at the same rate as the source. This is then transmitted to the brain as sound.

When struck, a tuning fork will vibrate up and down. Let y represent its directed distance from equilibrium, the position midway between the extremes. See Figure 5.13.

Figure 5.13

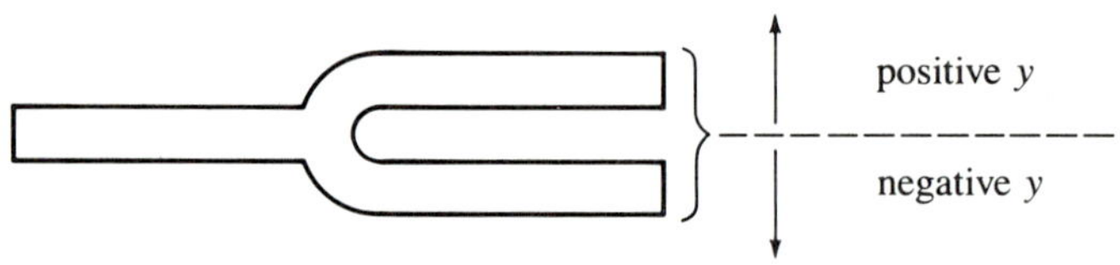

Then the equation of the displacement y as a function of elapsed time t is given by

$$y = a \sin bt,$$

where $a, b > 0$. The number a, the amplitude, is determined by the intensity (loudness) of the sound. The number $2\pi/b$, the period, is determined by the pitch (highness or lowness).

The pitch depends on the frequency of vibration—that is, the number of cycles per unit time. One cycle is a complete "back and forth and back" motion, for example, from equilibrium to maximum displacement to minimum displacement to equilibrium. Since the period is the length, in time, of one cycle, the frequency is its reciprocal. If f represents frequency, then

$$f = \frac{1}{\frac{2\pi}{b}} = \frac{b}{2\pi}.$$

If we solve this equation for b, we get

$$b = 2\pi f,$$

and the displacement equation becomes

$$y = a \sin 2\pi f t.$$

EXAMPLE 4 Suppose a tuning fork vibrates 264 cycles per second, and the maximum displacement is 0.002 inch. Determine the equation of the displacement.

Solution The general equation is

$$y = a \sin 2\pi f t.$$

The maximum displacement equals the amplitude, so $a = 0.002$. The frequency is 264 cycles per second, so $f = 264$. This gives

$$y = 0.002 \sin 528\pi t,$$

where y is the displacement in inches, and t is elapsed time in seconds.

▲

▲ ***Exercise*** Suppose a tuning fork vibrates 330 cycles per second, and the maximum displacement is 0.004 inch. Determine the equation of the displacement.

Answer $y = 0.004 \sin 660\pi t$

Problem Set 5.1

In Problems 1–4, determine the period and amplitude.

1.

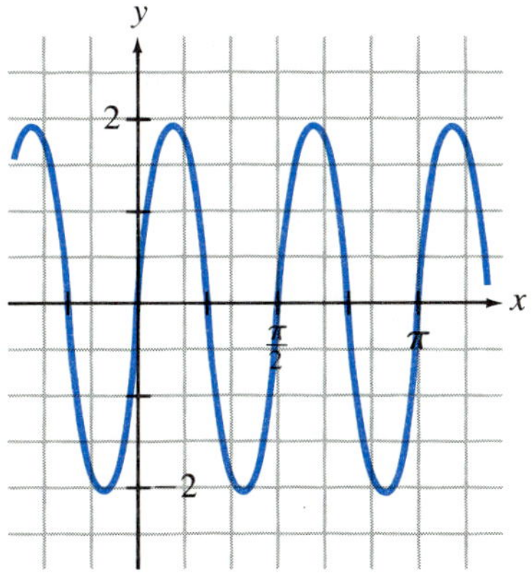

2.

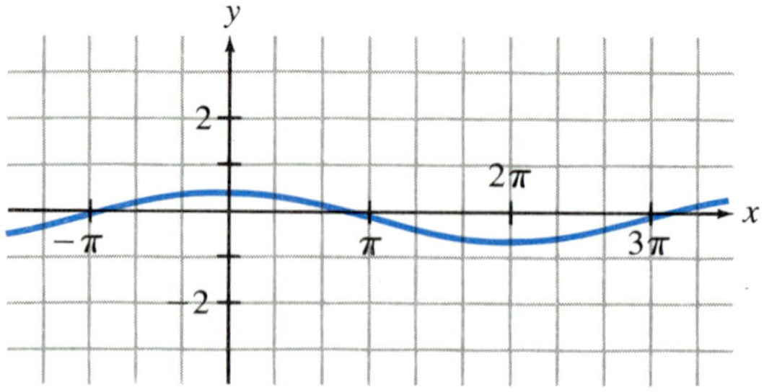

3.

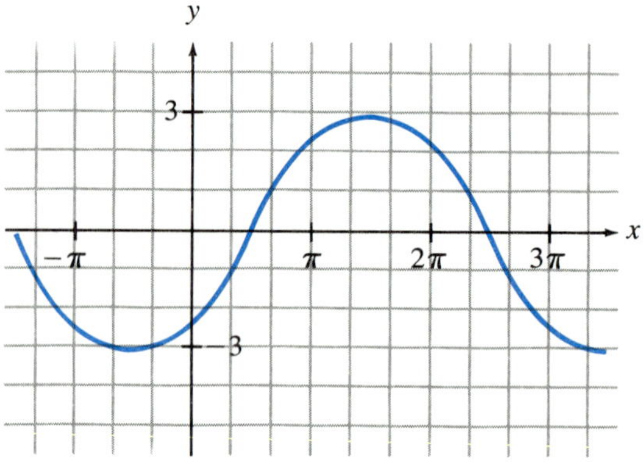

4.

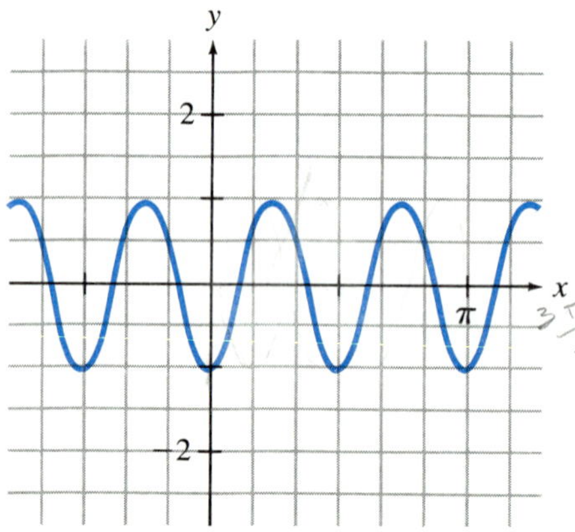

In Problems 5–8, determine the period and amplitude.

5. $y = -5 \sin 3x$

6. $y = \dfrac{1}{2} \cos 2x$

7. $y = 2 \cos \dfrac{x}{3}$

8. $y = \dfrac{-1}{3} \sin \dfrac{x}{2}$

In Problems 9 and 10, choose the equation that describes the given graph.

9. (a) $y = 2 \sin 3x$ (b) $y = 3 \sin 2x$

(c) $y = -3 \sin \dfrac{x}{2}$ (d) $y = 3 \cos 2x$

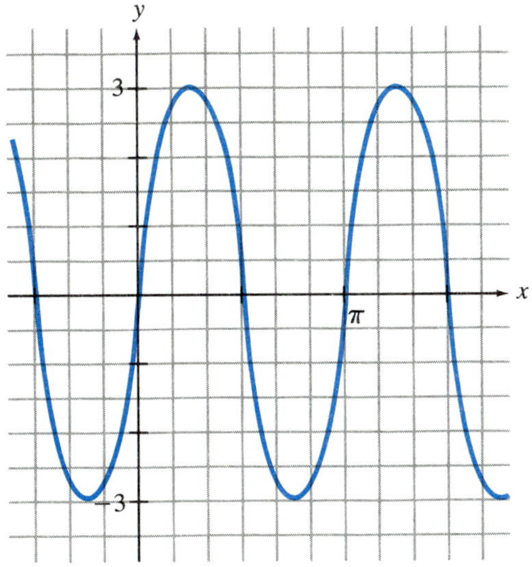

10. (a) $y = \cos 2x$ (b) $y = -2 \sin x$

(c) $y = \cos \dfrac{x}{2}$ (d) $y = 2 \cos x$

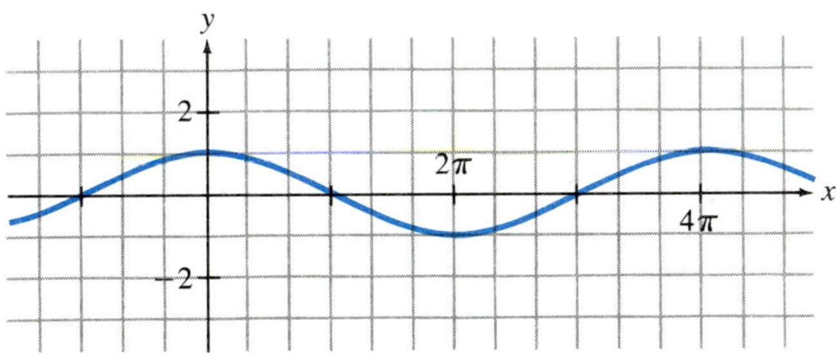

In Problems 11–26, sketch the graph.

11. $y = 4 \sin x$

12. $y = 4 \cos x$

13. $y = -\cos x$

14. $y = -\sin x$

15. $y = \sin 3x$

16. $y = -\cos 3x$

17. $y = -\cos \dfrac{x}{2}$

18. $y = \sin \dfrac{x}{2}$

19. $y = \sin 2\pi x$

20. $y = \cos 4\pi x$

21. $y = \sin (-x)$

22. $y = \cos (-x)$

23. $y = 3 \sin 2x$

24. $y = 3 \cos 2x$

25. $y = -2 \cos \dfrac{x}{2}$

26. $y = -2 \sin \dfrac{x}{2}$

27. Suppose a tuning fork vibrates 440 cycles per second with a maximum displacement of 0.0025 inch. What is the displacement (to two significant digits) when $t = 2.357$ seconds?

28. Suppose the displacement equation of a vibrating tuning fork is $y = 0.006 \sin 600\pi t$, where y is in inches and t is in seconds. What is the maximum displacement, and what is the frequency?

5.2 SINE AND COSINE GRAPHS CONTINUED

The graph of

$$y = a \sin (bx + c)$$

is a horizontal shift of $y = a \sin bx$. Thus, the period of such a function is still $2\pi/|b|$, and the amplitude is still $|a|$. Similarly, for

$$y = a \cos (bx + c),$$

the period is $2\pi/|b|$, and the amplitude is $|a|$. The graph is a horizontal shift of $y = a \cos bx$.

The directed amount of the shift is called the **phase shift** and equals $-c/b$. (If positive, the shift is to the right. If negative, it's to the left.) Other than memorizing the formula, we may find the phase shift by setting the argument to zero and solving for x. For $y = a \sin (bx + c)$ we have

$$bx + c = 0$$

$$x = \frac{-c}{b},$$

which is what we previously noted.

When the phase shift is a multiple of one-fourth of the period, then Section 5.1's technique of selecting x-values will still give us maximum, minimum, and crossing points. See Example 1.

EXAMPLE 1 Sketch the graph of $y = -2 \cos \left(\frac{x}{2} + \pi\right)$.

SOLUTION The period is $2\pi \div \frac{1}{2} = 4\pi$. We'll find the phase shift by setting the argument to zero and solving for x. We get

$$\frac{x}{2} + \pi = 0$$

$$\frac{x}{2} = -\pi$$

$$x = -2\pi.$$

The phase shift is a multiple of one-fourth of the period. So we may divide $[0, 4\pi]$ into equal fourths to get out x-values. We get the following table:

x	0	π	2π	3π	4π
$\frac{x}{2} + \pi$	π	$\frac{3\pi}{2}$	2π	$\frac{5\pi}{2}$	3π
y	2	0	-2	0	2

Plotting points and using the periodic property, we have the graph shown in Figure 5.14.

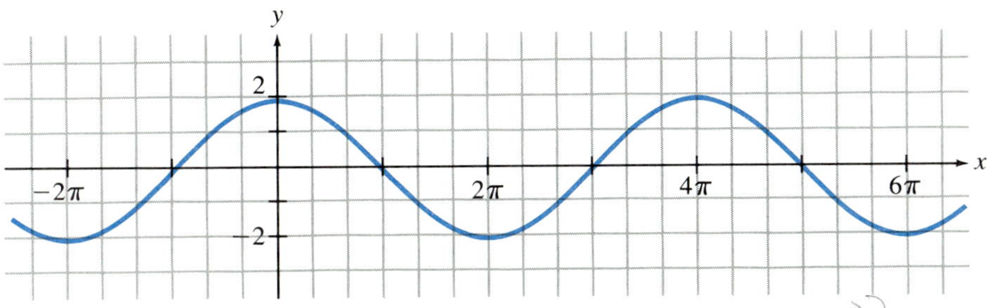

Figure 5.14

$$y = -2\cos\left(\frac{x}{2} + \pi\right)$$

▲

▲ *Exercise* Sketch the graph of $y = -2\sin\left(\frac{x}{2} - \pi\right)$.

Answer See Figure 5.15.

Figure 5.15

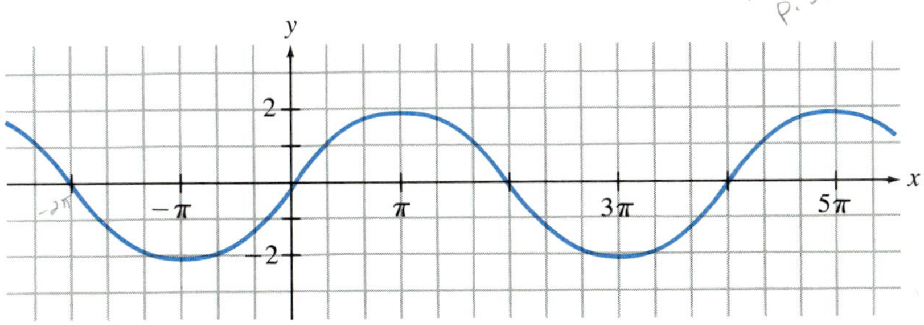

When the phase shift is not a multiple of one-fourth of the period, we must take it into account to get x-values that yield maximum, minimum, and crossing points. See Example 2.

EXAMPLE 2 Sketch the graph.

(a) $y = \sin\left(2x + \frac{\pi}{3}\right)$ (b) $y = 2\cos\left(\frac{-x}{2} + \frac{\pi}{4}\right)$

SOLUTION (a) To determine the phase shift we set the argument to zero and solve for x. We get

$$2x + \frac{\pi}{3} = 0$$

$$x = \frac{-\pi}{6}.$$

The period is $2\pi \div 2 = \pi$. So the "usual" x-values (dividing into equal fourths) would be 0, $\pi/4$, $\pi/2$, $3\pi/4$, and π. Taking the phase shift into account, however, we'll select

$$0 - \frac{\pi}{6},\quad \frac{\pi}{4} - \frac{\pi}{6},\quad \frac{\pi}{2} - \frac{\pi}{6},\quad \frac{3\pi}{4} - \frac{\pi}{6},\quad \text{and} \quad \pi - \frac{\pi}{6}.$$

Upon calculating the corresponding y-values, we get the following table,

x	$0 - \frac{\pi}{6}$	$\frac{\pi}{4} - \frac{\pi}{6}$	$\frac{\pi}{2} - \frac{\pi}{6}$	$\frac{3\pi}{4} - \frac{\pi}{6}$	$\pi - \frac{\pi}{6}$
$2x + \frac{\pi}{3}$	0	$\frac{\pi}{2}$	π	$\frac{3\pi}{2}$	2π
y	0	1	0	-1	0

which leads to the graph shown in Figure 5.16.

Figure 5.16
$y = \sin\left(2x + \frac{\pi}{3}\right)$

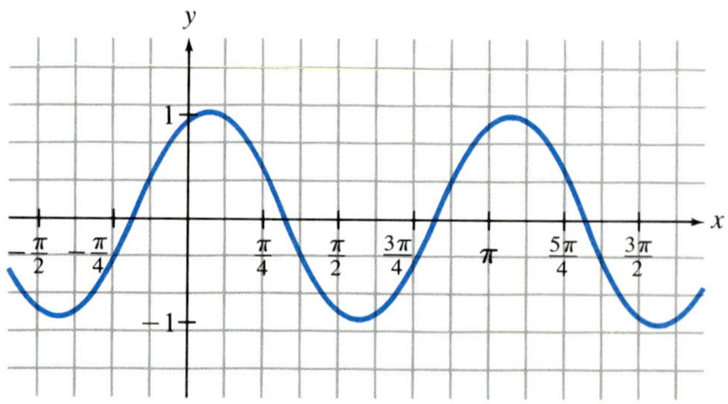

(b) We set the argument to zero to determine the phase shift. We get

$$\frac{-x}{2} + \frac{\pi}{4} = 0$$

$$-2x + \pi = 0$$

$$x = \frac{\pi}{2}.$$

The period is $2\pi \div \frac{1}{2} = 4\pi$. So the "usual" x-values would be $0, \pi, 2\pi, 3\pi$, and 4π. Adding the phase shift, however, gives

$$0 + \frac{\pi}{2}, \quad \pi + \frac{\pi}{2}, \quad 2\pi + \frac{\pi}{2}, \quad 3\pi + \frac{\pi}{2}, \quad \text{and} \quad 4\pi + \frac{\pi}{2}.$$

Our table then is

x	$0 + \frac{\pi}{2}$	$\pi + \frac{\pi}{2}$	$2\pi + \frac{\pi}{2}$	$3\pi + \frac{\pi}{2}$	$4\pi + \frac{\pi}{2}$
$\frac{-x}{2} + \frac{\pi}{4}$	0	$\frac{-\pi}{2}$	$-\pi$	$\frac{-3\pi}{2}$	-2π
y	2	0	-2	0	2

This leads to the graph in Figure 5.17.

Figure 5.17
$y = 2\cos\left(\frac{-x}{2} + \frac{\pi}{4}\right)$

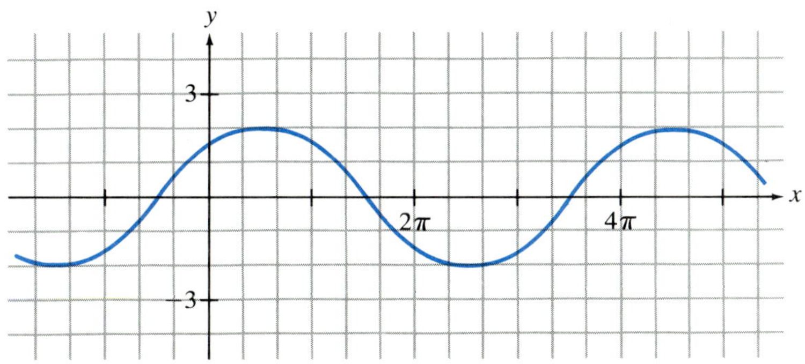

▲

▲ ***Exercise*** Sketch the graph.

(a) $y = \cos\left(2x - \frac{\pi}{3}\right)$ (b) $y = -2\sin\left(\frac{-x}{2} - \frac{\pi}{4}\right)$

Answer See Figure 5.18.

Figure 5.18 (a)

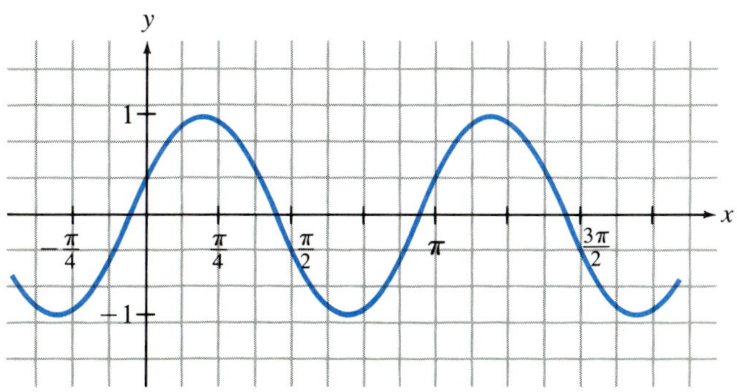

Figure 5.18 (*Continued*) (b)

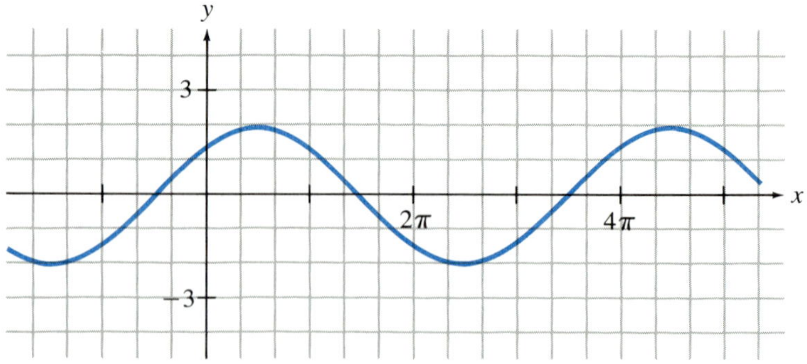

Many objects in the physical world exhibit a regular, repetitive pattern of behavior that may be described by

$$y = a \sin(bt + c),$$

where a and b are positive and t represents time. Such behavior is called **simple harmonic**. Examples include waves in an ocean, displacement of an engine's piston, sound waves, and electric current.

EXAMPLE 3 The sunrise time, as a function of the day of the year, is approximately simple harmonic. Write an equation that relates the sunrise time and the day of the year if the sunrise is at 7:40 on the shortest day of the year and is at 5:20 on the longest day of the year.

SOLUTION Because the behavior is simple harmonic, the graph will be something like that of Figure 5.19.

Figure 5.19

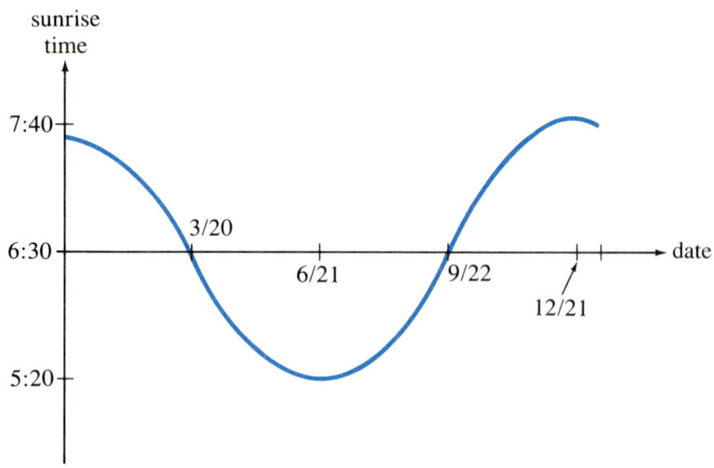

We must now define two variables. The horizontal axis variable is straightforward. Let

$$x = \text{day of year}.$$

The vertical axis variable is a bit more difficult. Noting that the horizontal axis intersects the vertical axis at 6:30, we let

$$y = \text{sunrise time } - 6{:}30 \text{ (in minutes)}.$$

So if $y = 10$, we know that the sunrise time is 6:40. If $y = -15$, the sunrise time is 6:15, etc. With these variables we have the graph in Figure 5.20.

Figure 5.20

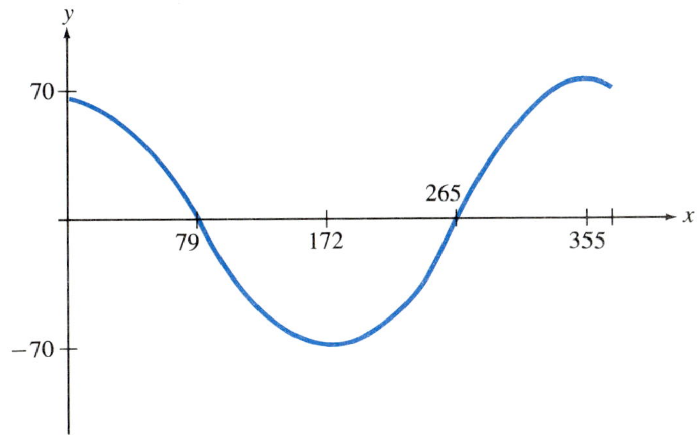

The general equation is

$$y = a \sin(bx + c).$$

We must determine a, b, and c. First we recognize that the period is 365 since the cycle repeats itself after one year. Hence

$$\frac{2\pi}{b} = 365$$

$$365b = 2\pi$$

$$b = \frac{2\pi}{365}.$$

Next we note that the amplitude is 70, which is $\frac{1}{2}$ the difference (in minutes) between the longest and shortest days. Of course, this is the same as the maximum value of the function since $y = 0$ corresponds to the average of the two extremes. Thus,

$$a = 70.$$

So we have

$$y = 70 \sin\left(\frac{2\pi}{365} x + c\right).$$

It remains to find c. We'll do this by examining the phase shift. Setting the argument to zero and solving for x gives

$$\frac{2\pi}{365}x + c = 0$$

$$\frac{2\pi}{365}x = -c$$

$$x = \frac{-365c}{2\pi}.$$

Now consider the graph of $y = \sin x$. At $x = 0$, this graph crosses the x-axis from negative y-values to positive y-values. However, for the graph of Figure 5.20, this occurs when $x = 265$. Hence the phase shift is 265. Therefore,

$$\frac{-365c}{2\pi} = 265.$$

Solving for c, we get

$$-365c = 530\pi$$

$$c = \frac{-530\pi}{365}.$$

Therefore, the equation is

$$y = 70\sin\left(\frac{2\pi}{365}x - \frac{530\pi}{365}\right)$$

$$= 70\sin\left(\frac{2\pi x - 530\pi}{365}\right).$$

▲

We can now use the formula of Example 3 to compute the sunrise time on any particular day. For instance, the sunrise time on February 15 should be computed as follows: when $x = 46$,

$$y = 70\sin\left(\frac{2\pi(46) - 530\pi}{365}\right) = 70\sin\left(\frac{-438\pi}{365}\right) \approx 41.$$

So the sunrise time occurs 41 minutes *after* 6:30—that is, at 7:11 (to the nearest minute).

▲ ***Exercise*** Write an equation that relates the sunrise time and the day of the year if the sunrise time is at 7:55 on the shortest day of the year and is at 5:05 on the longest day of the year.

Answer $y = 85\sin\left(\frac{2\pi x - 530\pi}{365}\right)$, where $x =$ day of year and $y =$ sunrise time $-$ 6:30 (in minutes).

Problem Set 5.2

In Problems 1–4, determine the period, amplitude, and phase shift.

1. $y = \frac{1}{2}\cos\left(\frac{x}{2} - \pi\right)$

2. $y = -4\sin(3x + \pi)$

3. $y = \pi\sin\left(2x - \frac{\pi}{3}\right)$

4. $y = \frac{-3}{4}\cos(\pi x + 2)$

In Problems 5–16, sketch the graph.

5. $y = \sin(x - \pi)$

6. $y = \cos(x + \pi)$

7. $y = \cos\left(x + \frac{\pi}{2}\right)$

8. $y = \sin\left(x - \frac{\pi}{2}\right)$

9. $y = 2\sin(-2x + \pi)$

10. $y = -2\cos(2x + \pi)$

11. $y = \sin\left(\frac{x}{2} + \frac{\pi}{6}\right)$

12. $y = \cos\left(\frac{-x}{2} + \frac{\pi}{6}\right)$

13. $y = 3\cos\left(3x - \frac{\pi}{2}\right)$

14. $y = -3\sin\left(4x + \frac{2\pi}{3}\right)$

15. $y = 1 + \cos\left(2x + \frac{\pi}{2}\right)$

16. $y = \sin\left(2x - \frac{\pi}{2}\right) - 1$

In Problems 17–20, the motion described is simple harmonic.

17. Suppose that water waves are rolling past a pier. Write the water's vertical displacement from equilibrium at the pier as a function of time if the distance between crest and trough is 10 feet, and consecutive crests pass every 4 seconds.

18. Suppose that a spacecraft has an orbit that takes it alternately north and south of the equator. Suppose that 10 minutes after liftoff the spacecraft reaches its maximum distance north of the equator: 4000 kilometers. Fifteen minutes later it reaches its maximum distance south of the equator: 4000 kilometers. Write its directed distance from the equator as a function of elapsed time since liftoff.

19. A body is suspended by a spring. It is then displaced downward 0.05 meter and released. Write its displacement from equilibrium as a function of elapsed time if its period is 0.6 second.

20. A pendulum is pulled to the right through an arc of 0.07 meter and released. Write its displacement from equilibrium, as measured along the arc, as a function of elapsed time if its period is 2.2 seconds.

5.3

THE TANGENT AND RECIPROCAL FUNCTIONS

Let's begin with the graph of the tangent function. Note that in the interval $[0, 2\pi]$, the tangent function isn't defined when $x = \pi/2$ or $x = 3\pi/2$. Consequently, the domain of the tangent function is $R - \{(\pi/2) + n\pi \mid n \in Z\}$. To plot some points we've constructed the following table:

x	0	$\frac{\pi}{6}$	$\frac{\pi}{3}$	$\frac{\pi}{2}$	$\frac{2\pi}{3}$	$\frac{5\pi}{6}$	π	$\frac{7\pi}{6}$	$\frac{4\pi}{3}$	$\frac{3\pi}{2}$	$\frac{5\pi}{3}$	$\frac{11\pi}{6}$	2π
y	0	$\frac{1}{\sqrt{3}}$	$\sqrt{3}$	not def.	$-\sqrt{3}$	$\frac{-1}{\sqrt{3}}$	0	$\frac{1}{\sqrt{3}}$	$\sqrt{3}$	not def.	$-\sqrt{3}$	$\frac{-1}{\sqrt{3}}$	0
$\approx y$	0	0.6	1.7	not def.	-1.7	-0.6	0	0.6	1.7	not def.	-1.7	-0.6	0

Although the tangent isn't defined at $\pi/2$, we may choose values of x as close to $\pi/2$ as we please. As x gets closer to $\pi/2$ from the negative side, the tangent increases without bound. (That is, the tangent is eventually greater than each positive number.) And as x gets closer to $\pi/2$ from the positive side, the tangent decreases without bound. (That is, the tangent is eventually less than each negative number.) See the following table:

x	**tan x (*nearest one*)**	x	**tan x (*nearest one*)**
1	2	2	-2
1.4	6	1.7	-8
1.5	14	1.6	-34
1.56	93	1.58	-109
$\pi/2$	not defined	$\pi/2$	not defined

Hence the graph approaches, but does not intersect, the vertical line passing through $x = \pi/2$. A line with the property that the distance from a point P to the line approaches zero as P moves away from the origin along some part of the graph is called an **asymptote**. Thus the vertical line passing through $x = \pi/2$ is an asymptote. Similarly, the vertical line passing through $x = 3\pi/2$ is an asymptote. Plotting points and indicating the asymptotes with dashed lines, we get the graph of $y = \tan x$ on the interval $[0, 2\pi]$. See Figure 5.21. The "complete" graph is shown in Figure 5.22.

Note that this graph repeats itself every π units. Hence the tangent function has period π. Also observe that the tangent graph is symmetric about the origin, so the tangent function is odd. Further note that the range is R and that there is a vertical asymptote at each value of x for which the function is not defined.

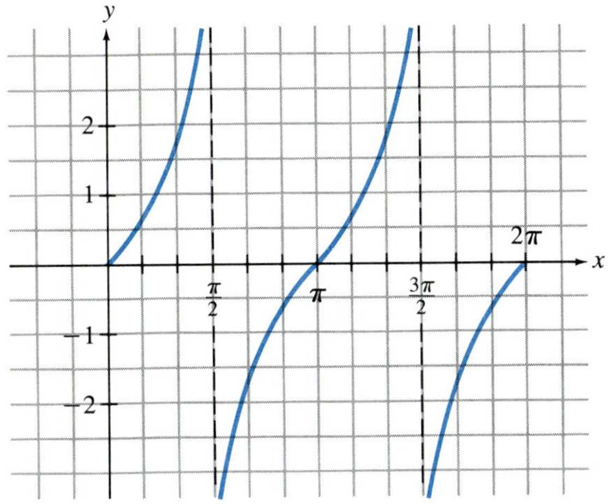

Figure 5.21
$y = \tan x$ *on* $[0, 2\pi]$

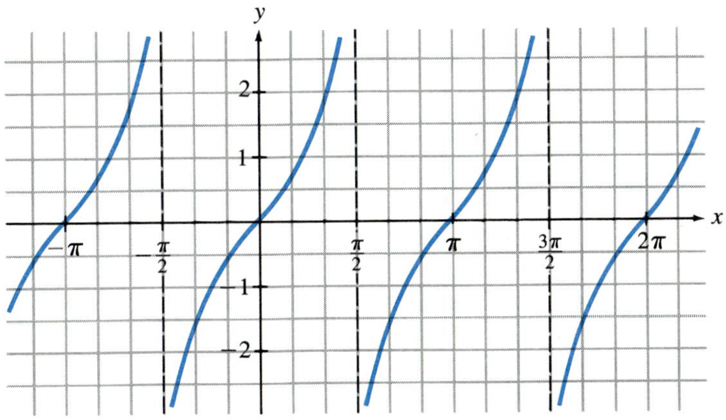

Figure 5.22
$y = \tan x$

The graph of the cotangent function is similar to the graph of the tangent function. Its period is π, the graph is symmetric about the origin, and it has vertical asymptotes. The asymptotes, however, occur for different values of x. The domain is $R - \{n\pi \mid n \in Z\}$, and the range is R. See Figure 5.23.

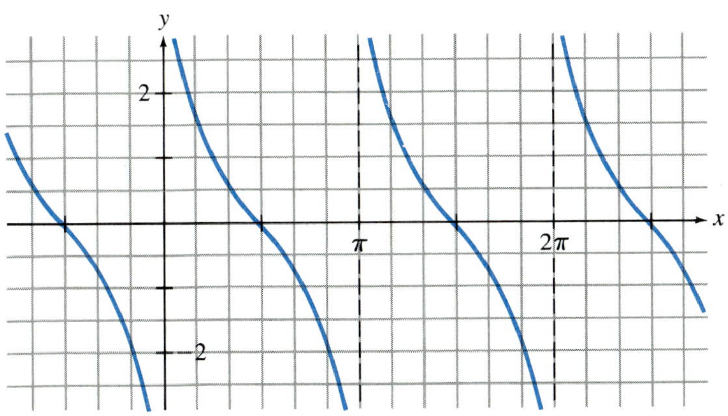

Figure 5.23
$y = \cot x$

EXAMPLE 1 Sketch the graph of $y = \tan\left(\frac{x}{3} + \frac{\pi}{6}\right)$.

SOLUTION The phase shift is $-\pi/2$ because we have

$$\frac{x}{3} + \frac{\pi}{6} = 0$$

$$x = \frac{-\pi}{2}.$$

The period is 3π (the usual period π divided by $\frac{1}{3}$). Dividing $[0, 3\pi]$ into equal fourths and using the phase shift, we get the following table:

x	$0 - \frac{\pi}{2}$	$\frac{3\pi}{4} - \frac{\pi}{2}$	$\frac{3\pi}{2} - \frac{\pi}{2}$	$\frac{9\pi}{4} - \frac{\pi}{2}$	$3\pi - \frac{\pi}{2}$
$\frac{x}{3} + \frac{\pi}{6}$	0	$\frac{\pi}{4}$	$\frac{\pi}{2}$	$\frac{3\pi}{4}$	π
y	0	1	not def.	-1	0

The vertical asymptote will occur at $x = 3\pi/2 - \pi/2 = \pi$, where the given function isn't defined. Sketching this asymptote, plotting points, and using our knowledge of the general shape of the graph of the tangent function, we get the graph on $[-\pi/2, 5\pi/2]$ shown in Figure 5.24. Using the periodic property we have the graph shown in Figure 5.25.

Figure 5.24 $y = \tan\left(\frac{x}{3} + \frac{\pi}{6}\right)$ *on* $\left[\frac{-\pi}{2}, \frac{5\pi}{2}\right]$

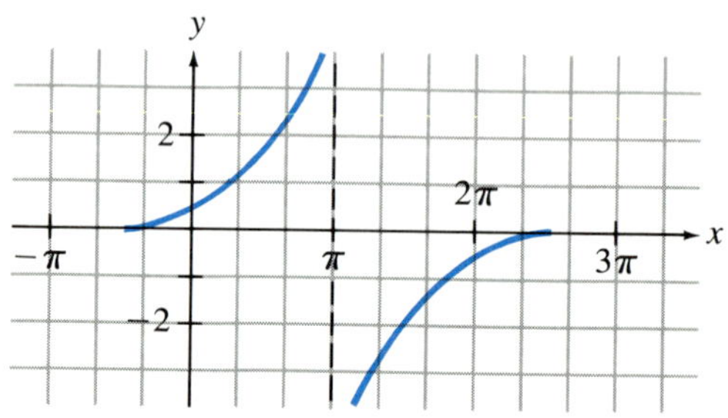

Figure 5.25 $y = \tan\left(\frac{x}{3} + \frac{\pi}{6}\right)$

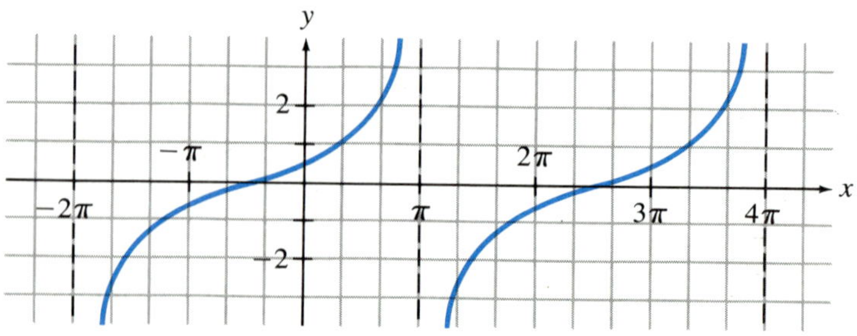

▲

▲ *Exercise* Sketch the graph of $y = \cot\left(\frac{x}{3} - \frac{\pi}{6}\right)$.

Answer See Figure 5.26.

Figure 5.26

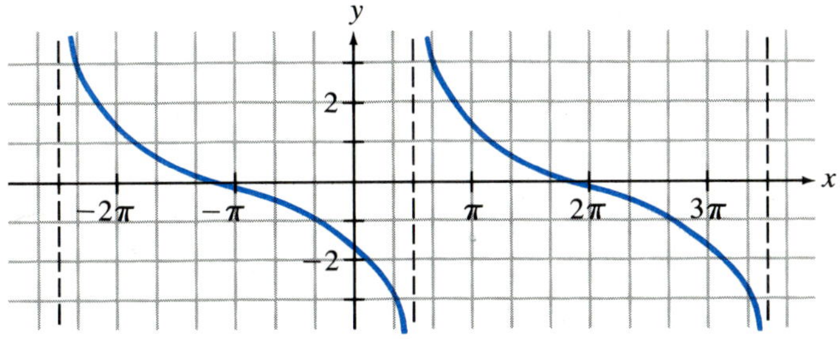

To graph the secant function, let's note that in the interval $[0, 2\pi]$, the secant function isn't defined when $x = \pi/2$ or $x = 3\pi/2$. Thus its domain is the same as that of the tangent function: $R - \{(\pi/2) + n\pi \mid n \in Z\}$. To plot some points, we've constructed the following table.

x	0	$\frac{\pi}{6}$	$\frac{\pi}{3}$	$\frac{\pi}{2}$	$\frac{2\pi}{3}$	$\frac{5\pi}{6}$	π	$\frac{7\pi}{6}$	$\frac{4\pi}{3}$	$\frac{3\pi}{2}$	$\frac{5\pi}{3}$	$\frac{11\pi}{6}$	2π
y	1	$\frac{2}{\sqrt{3}}$	2	not def.	-2	$\frac{-2}{\sqrt{3}}$	-1	$\frac{-2}{\sqrt{3}}$	-2	not def.	2	$\frac{2}{\sqrt{3}}$	1
$\approx y$	1	1.2	2	not def.	-2	-1.2	-1	-1.2	-2	not def.	2	1.2	1

Like the graph of the tangent function, the graph of the secant function has vertical asymptotes at $x = \pi/2$ and at $x = 3\pi/2$. Using this and plotting points, we get the graph of $y = \sec x$ on the interval $[0, 2\pi]$. See Figure 5.27. The "complete" graph is shown in Figure 5.28.

Figure 5.27
$y = \sec x$ on $[0, 2\pi]$

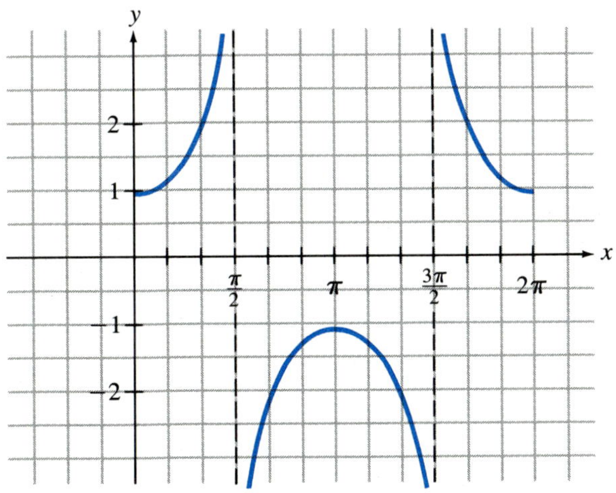

Figure 5.28
$y = \sec x$

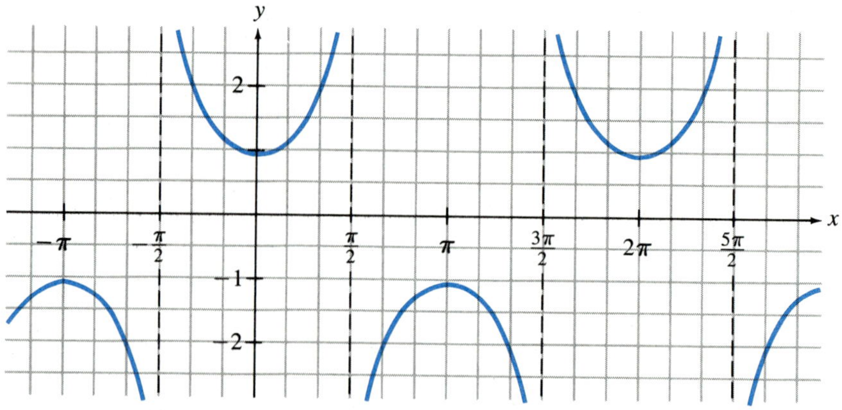

Note that the period of the secant function is 2π. Also observe that the secant function is even (the graph is symmetric about the y-axis), and the range is $(-\infty, -1] \cup [1, \infty)$. Further note that there is a vertical asymptote at each value of x for which the function is not defined.

The graph of the cosecant function is similar to the graph of the secant function. Its period is 2π, and the range is $(-\infty, -1] \cup [1, \infty)$. The cosecant graph, however, is symmetric about the origin. It has vertical asymptotes, but they occur for different values of x. The domain is $R - \{n\pi \mid n \in Z\}$. See Figure 5.29.

Figure 5.29
$y = \csc x$

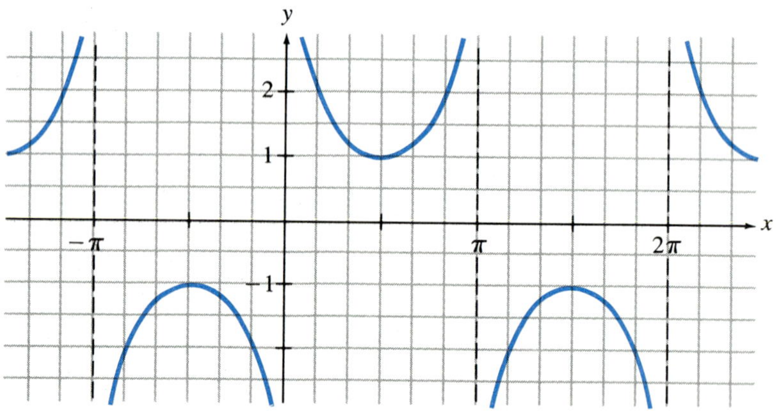

In fact, the graphs of the secant and cosecant functions have the same shape. They are simply "out of phase". If the cosecant graph were shifted to the left $\pi/2$ units, it would coincide with the secant graph.

EXAMPLE 2 Sketch the graph of $y = \csc(2x - \pi)$.

SOLUTION The phase shift is $\pi/2$ and the period is π (the usual period 2π divided by 2). Dividing $[0, \pi]$ into equal fourths and using the phase shift, we get the following table:

x	$0 + \frac{\pi}{2}$	$\frac{\pi}{4} + \frac{\pi}{2}$	$\frac{\pi}{2} + \frac{\pi}{2}$	$\frac{3\pi}{4} + \frac{\pi}{2}$	$\pi + \frac{\pi}{2}$
$2x - \pi$	0	$\frac{\pi}{2}$	π	$\frac{3\pi}{2}$	2π
y	not def.	1	not def.	-1	not def.

The vertical asymptotes will occur at $x = \pi/2$, π, and $3\pi/2$, where $\csc(2x - \pi)$ isn't defined. Sketching these asymptotes, plotting points, and using our knowledge of the shape of the graph of the cosecant function, we get the graph on the interval $[\pi/2, 3\pi/2]$ shown in Figure 5.30. Using the periodic property, we have the graph shown in Figure 5.31.

Figure 5.30
$y = \csc(2x - \pi)$ on $\left[\frac{\pi}{2}, \frac{3\pi}{2}\right]$

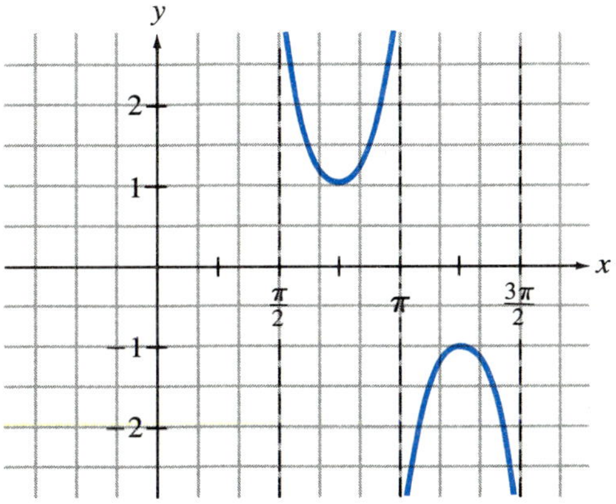

Figure 5.31
$y = \csc(2x - \pi)$

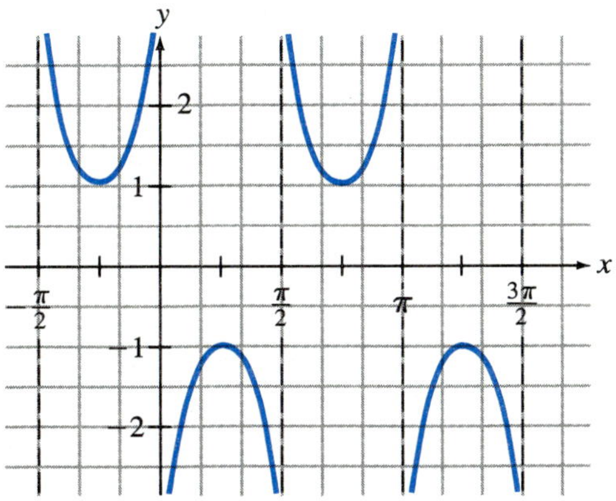

▲

▲ ***Exercise*** Sketch the graph of $y = \sec(2x + \pi)$.

Answer See Figure 5.32.

Figure 5.32

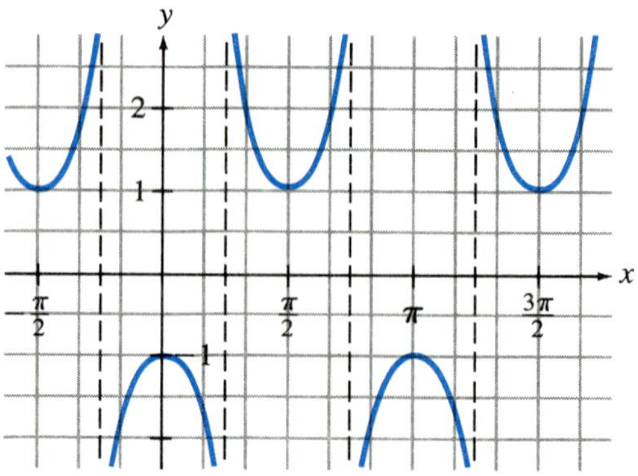

Problem Set 5.3

In Problems 1 and 2, determine the period.

1.

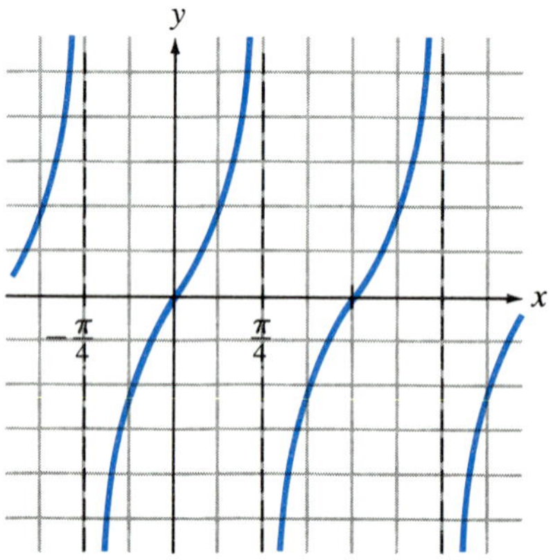

2.

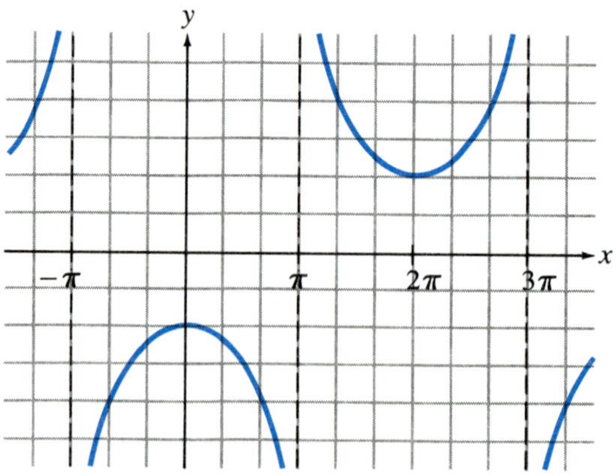

In Problems 3 and 4, choose the equation that describes the given graph.

3. (a) $y = 2\tan x$ (b) $y = \tan\dfrac{x}{2}$

(c) $y = \tan 2x$ (d) $y = 2 + \tan x$

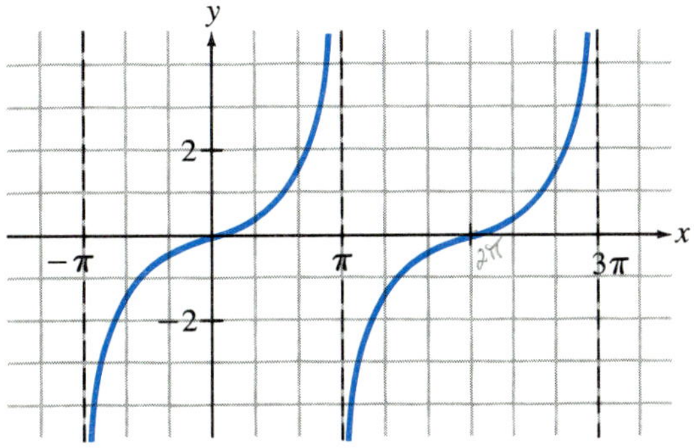

4. (a) $y = \sec\left(x - \dfrac{\pi}{2}\right)$ (b) $y = \csc x$

(c) $y = \sec\left(x + \dfrac{3\pi}{2}\right)$ (d) $y = \csc(x - 12\pi)$

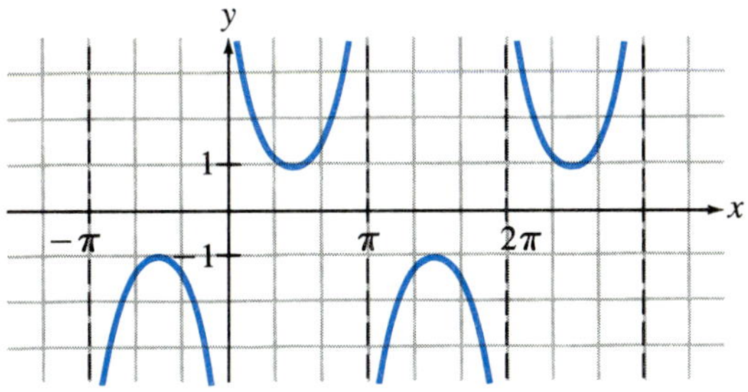

In Problems 5–10, determine the period and phase shift.

5. $y = \dfrac{1}{2}\tan\dfrac{x}{3}$

6. $y = -2\sec 2x$

7. $y = -\csc(3x - \pi)$

8. $y = \dfrac{1}{3}\cot\left(\dfrac{x}{3} + \pi\right)$

9. $y = \sec\left(\dfrac{x}{6} - 1\right)$

10. $y = \tan(5x + 2)$

In Problems 11–36, sketch the graph.

11. $y = 2\tan x$

12. $y = -\cot x$

13. $y = -2\sec x$

14. $y = 3\csc x$

15. $y = \tan 3x$

16. $y = \cot\dfrac{x}{2}$

17. $y = \sec \frac{x}{2}$

18. $y = \csc 3x$

19. $y = \tan(-x)$

20. $y = \csc(-x)$

21. $y = 2 \sec 2x$

22. $y = \frac{1}{2} \csc 2x$

23. $y = \frac{1}{2} \cot \frac{x}{2}$

24. $y = -2 \tan \frac{x}{2}$

25. $y = \cot(x + \pi)$

26. $y = \tan\left(x - \frac{\pi}{2}\right)$

27. $y = \csc(x - \pi)$

28. $y = \sec\left(x + \frac{\pi}{2}\right)$

29. $y = -2 \sec(-2x + \pi)$

30. $y = -2 \tan\left(2x + \frac{\pi}{2}\right)$

31. $y = \tan\left(\frac{x}{2} + \frac{\pi}{6}\right)$

32. $y = \cot\left(-3x + \frac{\pi}{2}\right)$

33. $y = 3 \sec\left(-4x - \frac{2\pi}{3}\right)$

34. $y = -3 \csc\left(\frac{x}{2} - \frac{\pi}{6}\right)$

35. $y = 1 + \cot(2x + \pi)$

36. $y = \csc\left(2x - \frac{\pi}{2}\right) - 1$

5.4

THE INVERSE TRIGONOMETRIC FUNCTIONS

In this section we study three new functions that are related to the sine, cosine, and tangent functions. We begin with the sine function,

$$y = \sin x,$$

whose graph is in Figure 5.33.

Figure 5.33
$y = \sin x$

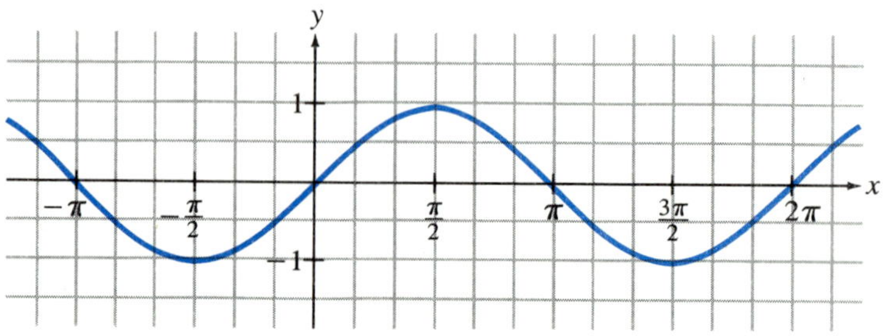

We see that the inverse doesn't define a function because the sine function is not one-to-one. (There is a horizontal line that intersects the graph in more than one point.) How can we restrict the sine function to get a one-to-one function? Suppose we don't want to alter the range $[-1, 1]$. Examining the graph in Figure 5.34, we see that one choice is to require that $-\pi/2 \leq x \leq \pi/2$.

Figure 5.34

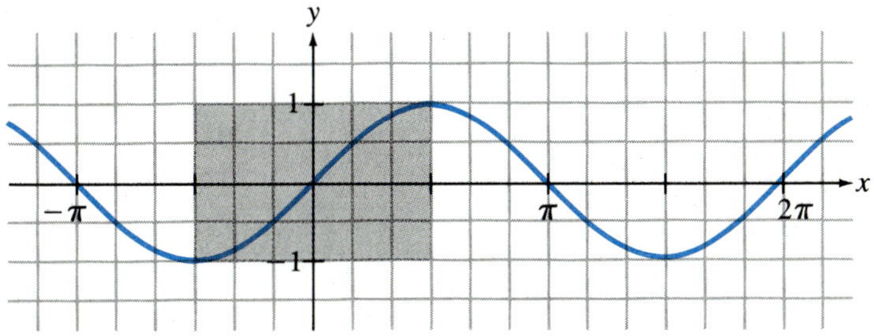

In this way, the inverse of

$$y = \sin x, \qquad \frac{-\pi}{2} \leq x \leq \frac{\pi}{2}$$

defines a function.

The function defined by

$$x = \sin y, \qquad \frac{-\pi}{2} \leq y \leq \frac{\pi}{2}$$

is called the **inverse sine** and is denoted **Arcsin**. Its graph is shown in Figure 5.35.

Figure 5.35
$y = \text{Arcsin } x$

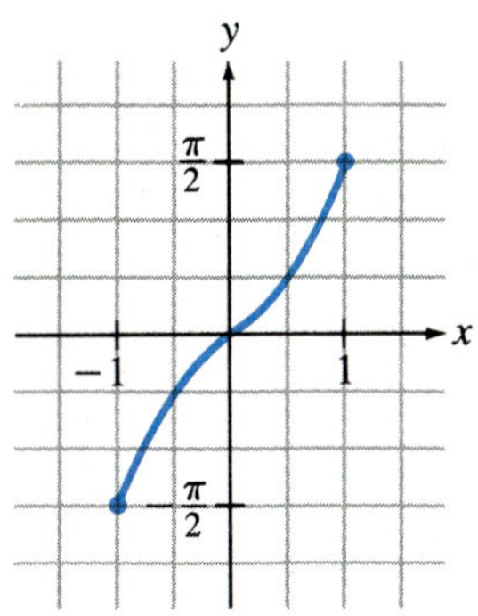

Its domain is $[-1, 1]$, and its range is $[-\pi/2, \pi/2]$. When we write

$$y = \text{Arcsin } x,$$

y is the number in the interval $[-\pi/2, \pi/2]$ whose sine is x. The notation is descriptive because, considering the wrapping function and the unit circle, y is the arc whose sine is x. Of course, y is also the radian measure in the interval

$[-\pi/2, \pi/2]$ of the angle whose sine is x. Consequently we frequently say that y is the angle (in $[-\pi/2, \pi/2]$) whose sine is x.

We have already been using the inverse sine function. In Chapter 4 we sometimes found the angle θ that had a specified sine. For example, let's find θ (in radians) if $\sin\theta = 0.63$. We enter .63 and push [inv] [sin] or [arc] [sin] or its equivalent to get 0.6815532. So we have taken the inverse sine of 0.63. Thus

$$\text{Arcsin } 0.63 = 0.6815532.$$

EXAMPLE 1 Find the value, exactly.

(a) $\text{Arcsin}\,\dfrac{1}{\sqrt{2}}$ (b) $\text{Arcsin}\left(\dfrac{-1}{2}\right)$

SOLUTION (a) We are looking for a first-quadrant angle whose sine is $1/\sqrt{2}$. See Figure 5.36.

Figure 5.36

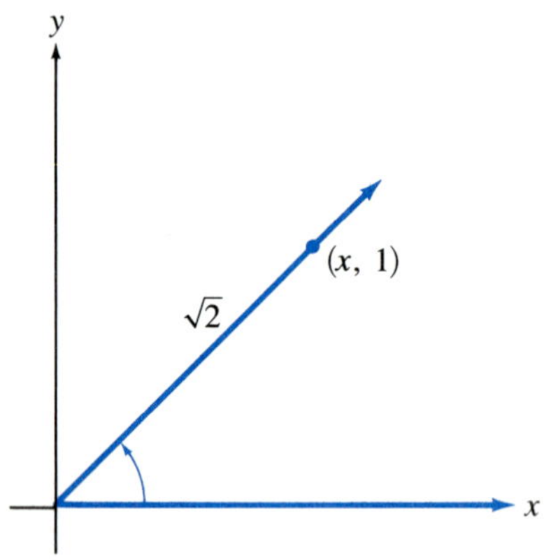

This is the special angle $\pi/4$. Thus

$$\text{Arcsin}\,\frac{1}{\sqrt{2}} = \frac{\pi}{4}.$$

(b) Because we are looking for an angle whose sine is negative, the result will be a fourth-quadrant angle. See Figure 5.37.

Figure 5.37

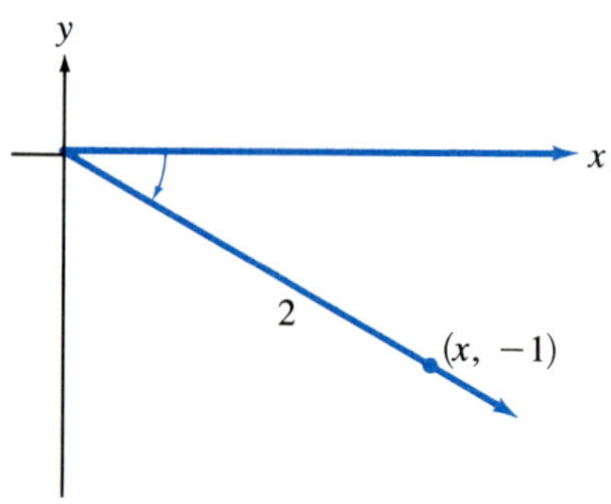

The reference angle that has a sine of $\frac{1}{2}$ is $\pi/6$. Hence

$$\text{Arcsin}\left(\frac{-1}{2}\right) = \frac{-\pi}{6}.$$

▲

Technically, Arcsin x is a real number and not the measure of an angle. But the real number equals the *radian* measure of an angle. Hence, if we think in terms of angles and their measures, we must think in *radians*. For example, *it is incorrect to write*

$$\text{Arcsin}\,\frac{1}{\sqrt{2}} = 45^\circ\,!$$

Arcsin $(1/\sqrt{2})$ is the *real number* $\pi/4$, which is the *radian* measure of an angle whose sine is $1/\sqrt{2}$. So if you think in terms of angle measure when using the inverse functions, *think radians.*

▲ ***Exercise*** Find the value, exactly.

(a) $\text{Arcsin}\,\dfrac{\sqrt{3}}{2}$ (b) $\text{Arcsin}\left(\dfrac{-1}{\sqrt{2}}\right)$

Answer (a) $\dfrac{\pi}{3}$ (b) $\dfrac{-\pi}{4}$

When approximating values of the inverse trigonometric functions with the calculator, we don't have to concern ourselves with a reference angle. Such considerations are "built in."

EXAMPLE 2 Find the value, to the nearest thousandth.

(a) Arcsin 0.75 (b) Arcsin (-0.315).

SOLUTION (a) We put the calculator in radian mode, enter .75, and then push [inv] [sin] or its equivalent. An eight-digit display reads 0.8480621. Thus,

$$\text{Arcsin}\,0.75 \approx 0.848\,.$$

(b) Entering $-.315$ and pushing [inv] [sin] gives

$$\text{Arcsin}\,(-0.315) \approx -0.320\,.$$

▲

▲ *Exercise* Find the value, to the nearest thousandth.

(a) Arcsin 0.512 (b) Arcsin (-0.25)

Answer (a) 0.538 (b) -0.253

To formulate the inverse of the cosine function, we must restrict the domain of the cosine to get a one-to-one function whose range is $[-1, 1]$. Examining the graph in Figure 5.38, we see that one choice is to have $0 \le x \le \pi$.

Figure 5.38

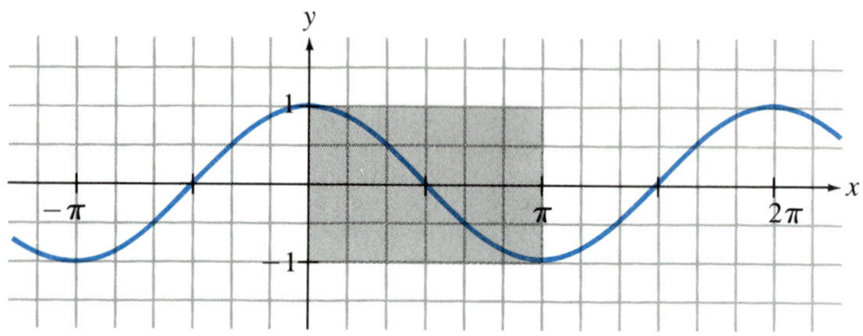

In this way the inverse of

$$y = \cos x, \qquad 0 \le x \le \pi$$

defines a function.

The function defined by

$$x = \cos y, \qquad 0 \le y \le \pi$$

is called the **inverse cosine** and is denoted **Arccos**. Its graph is in Figure 5.39.

Figure 5.39
$y = \text{Arccos } x$

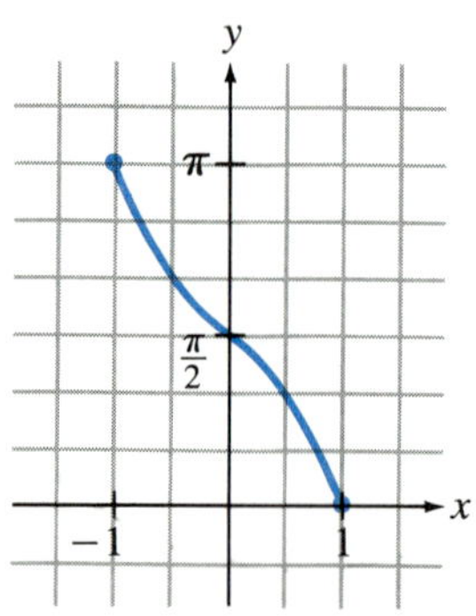

Its domain is $[-1, 1]$, and its range is $[0, \pi]$. When we write

$$y = \text{Arccos } x,$$

y is the number in the interval $[0, \pi]$ whose cosine is x. It is the radian measure in $[0, \pi]$ of the angle whose cosine is x.

EXAMPLE 3 Find the value, as specified.

(a) $\text{Arccos}\left(\dfrac{-\sqrt{3}}{2}\right)$, exactly (b) Arccos 0.8055, to the nearest thousandth

SOLUTION (a) Because we are looking for an angle whose cosine is negative, we know that the result will be a second-quadrant angle. See Figure 5.40.

Figure 5.40

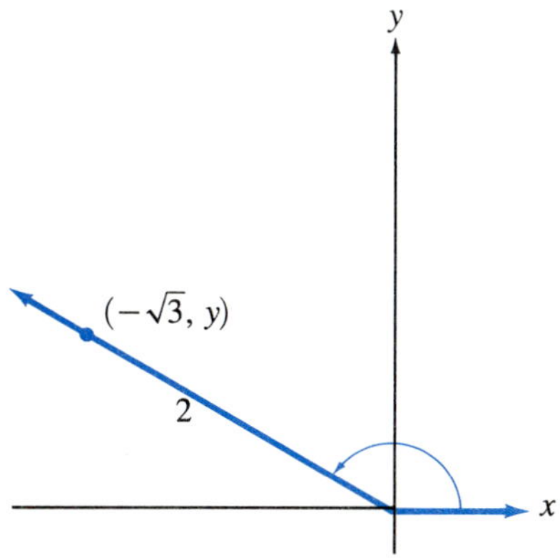

The reference angle that has a cosine of $\sqrt{3}/2$ is $\pi/6$. Hence

$$\text{Arccos}\left(\frac{-\sqrt{3}}{2}\right) = \pi - \frac{\pi}{6} = \frac{5\pi}{6}.$$

(b) In the radian mode we enter .8055. Then we push [inv] [cos] or its equivalent and round to the nearest thousandth. We get

$$\text{Arccos } 0.8055 \approx 0.634.$$

▲

▲ ***Exercise*** Find the value, as specified.

(a) $\text{Arccos}\left(\dfrac{-1}{2}\right)$, exactly (b) Arccos 0.155, to the nearest thousandth

Answer (a) $\dfrac{2\pi}{3}$ (b) 1.415

The last function to be considered is the inverse of a restricted tangent function. Examining the graph of the tangent function, Figure 5.41, we see that if we choose $-\pi/2 < x < \pi/2$, we'll have a one-to-one function whose range is R.

With this restriction, the inverse of

$$y = \tan x, \qquad \frac{-\pi}{2} < x < \frac{\pi}{2}$$

defines a function.

Figure 5.41

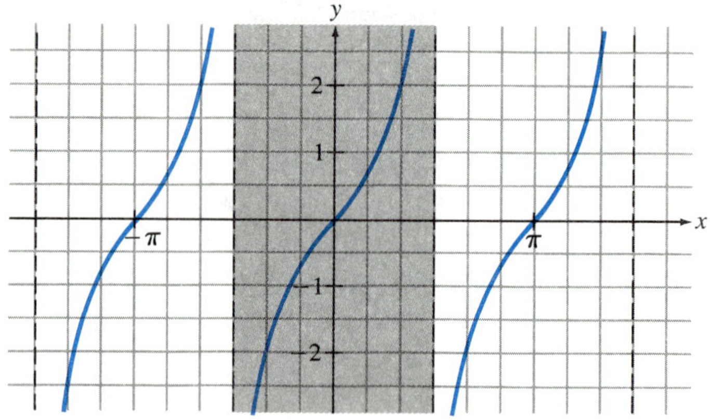

The function defined by

$$x = \tan y, \qquad \frac{-\pi}{2} < y < \frac{\pi}{2}$$

is called the **inverse tangent** and is denoted **Arctan**. Its graph is shown in Figure 5.42.

Figure 5.42
$y = \text{Arctan } x$

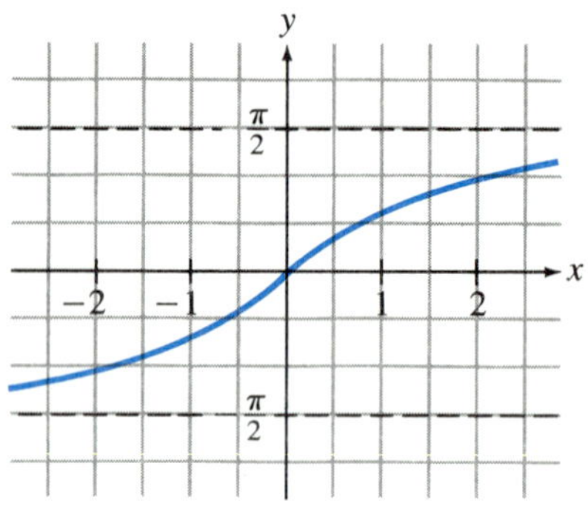

Its domain is R, and its range is $(-\pi/2, \pi/2)$. The horizontal lines passing through $y = \pi/2$ and $y = -\pi/2$ are asymptotes. When we write

$$y = \text{Arctan } x,$$

y is the number in the interval $(-\pi/2, \pi/2)$ whose tangent is x. It is the radian measure in $(-\pi/2, \pi/2)$ of the angle whose tangent is x.

EXAMPLE 4 Find the value, as specified.

(a) Arctan $\sqrt{3}$, exactly (b) Arctan (-35), to the nearest thousandth

SOLUTION (a) The angle whose tangent is $\sqrt{3}$ is the special angle $\pi/3$. So

$$\text{Arctan } \sqrt{3} = \frac{\pi}{3}.$$

(b) In radian mode we enter -35 and then push [inv] [tan]. Rounding off gives

$$\text{Arctan}\,(-35) \approx -1.542\,.$$

▲

▲ ***Exercise*** Find the value, as specified.

(a) Arctan 1, exactly (b) Arctan (-0.65), to the nearest thousandth

Answer (a) $\frac{\pi}{4}$ (b) -0.576

In this section we formulated inverses of the sine, cosine, and tangent functions. Although inverses of the remaining trigonometric functions won't be discussed, Arcsec, Arccsc, and Arccot can be developed in a similar fashion.

The following table summarizes the three inverse trigonometric functions discussed in this section.

Name	***Notation***	***Domain***	***Range***
Inverse sine	Arcsin	$[-1, 1]$	$\left[\frac{-\pi}{2}, \frac{\pi}{2}\right]$
Inverse cosine	Arccos	$[-1, 1]$	$[0, \pi]$
Inverse tangent	Arctan	R	$\left(\frac{-\pi}{2}, \frac{\pi}{2}\right)$

Some texts and calculators use an alternative notation for the inverse trigonometric functions:

$$\text{Sin}^{-1}\,x = \text{Arcsin}\,x\,,$$

$$\text{Cos}^{-1}\,x = \text{Arccos}\,x\,,$$

$$\text{Tan}^{-1}\,x = \text{Arctan}\,x, \text{ etc}\,.$$

In this context the -1 is not an exponent. $\text{Sin}^{-1}\,x$ is the inverse sine of x, and not the reciprocal. That is,

$$\text{Sin}^{-1}\,x \neq \frac{1}{\text{Sin}\,x}\,.$$

We'll continue to use the notation Arcsin x, Arccos x, and Arctan x.

EXAMPLE 5 Find the value, as specified.

(a) $\tan\left(\text{Arccos}\left(\frac{-2}{3}\right)\right)$, exactly

(b) cos(Arcsin 0.58), to four significant digits

SOLUTION (a) Letting $\theta = \text{Arccos}\left(\frac{-2}{3}\right)$, we choose a point on the terminal side such that $r = 3$ and $x = -2$. We have the situation shown in Figure 5.43.

Figure 5.43

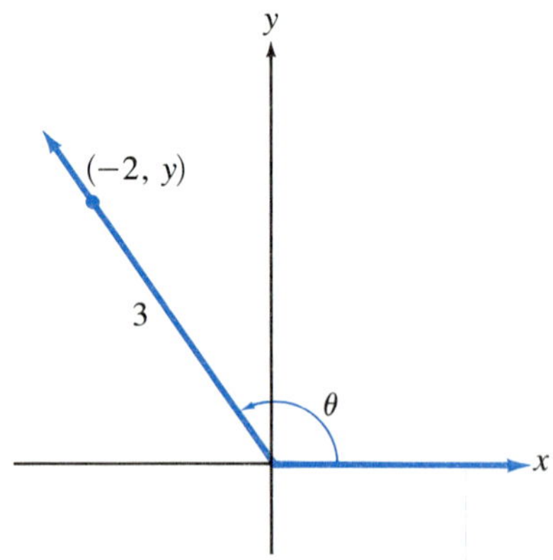

Hence

$$y = \pm\sqrt{r^2 - x^2} = \pm\sqrt{9-4} = \pm\sqrt{5}.$$

Since θ is a second-quadrant angle, we know that $y = \sqrt{5}$. From the definition of tangent we have $\tan\theta = y/x = \sqrt{5}/(-2)$. So

$$\tan\left(\text{Arccos}\left(\frac{-2}{3}\right)\right) = \frac{-\sqrt{5}}{2}.$$

(b) In radian mode we enter .58 and push [inv] [sin], which gives us Arcsin 0.58, and then we determine its cosine by pushing [cos]. We get 0.8146165. To four significant digits we have

$$\cos(\text{Arcsin } 0.58) \approx 0.8146.$$

▲

▲ ***Exercise*** Find the value, as specified.

(a) $\cos\left(\text{Arcsin}\left(\frac{-2}{5}\right)\right)$, exactly

(b) sin(Arccos 0.42), to four significant digits

Answer (a) $\frac{\sqrt{21}}{5}$ (b) 0.9075

EXAMPLE 6 Find the exact solutions of $\sin x = -0.65$, where $0 \leq x < 2\pi$.

SOLUTION Because x is not a special angle, we must use the inverse sine to write the exact solutions. It may appear that the reference angle we are seeking is

Arcsin(-0.65). But recall that the range of the inverse sine is $[-\pi/2, \pi/2]$. Hence Arcsin(-0.65) is a negative fourth-quadrant angle. The reference angle is Arcsin 0.65. So the third-quadrant solution is

$$x = \pi + \text{Arcsin } 0.65 .$$

And the fourth-quadrant solution is

$$x = 2\pi - \text{Arcsin } 0.65 .$$

Hence the exact solutions are π + Arcsin 0.65 and 2π − Arcsin 0.65.

▲

▲ ***Exercise*** Find the exact solutions of $\tan x = -3$, where $0 \le x < 2\pi$.

Answer π − Arctan 3, 2π − Arctan 3

EXAMPLE 7 Determine sin(Arctan t).

SOLUTION Let θ = Arctan t. Choose a point on the terminal side with a y-coordinate of t and an x-coordinate of 1. We have one of the situations shown in Figure 5.44.

Figure 5.44

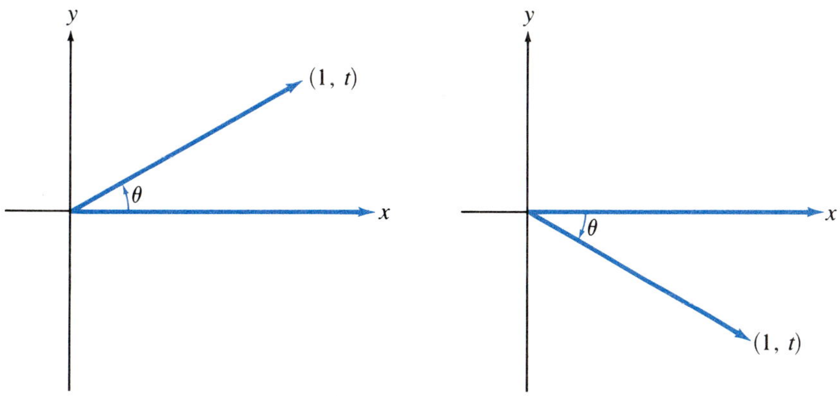

(i) $t > 0$ (ii) $t < 0$

In either case we get $r = \sqrt{1 + t^2}$. Thus $\sin \theta = t/\sqrt{1 + t^2}$. That is,

$$\sin(\text{Arctan } t) = \frac{t}{\sqrt{1 + t^2}} \quad \left(\text{or } \frac{t\sqrt{1 + t^2}}{1 + t^2}\right).$$

▲

In Example 7 note that even though there were two possibilities for θ (one positive and one negative), the points on the terminal sides were identically labeled. Hence the trigonometric ratios for each θ were the same.

Similarly, although there are two cases each for θ = Arcsin t and θ = Arccos t, the point on the terminal side will have the same label in each case. Consequently, in determining the value of a trigonometric function of

Arctan t, Arcsin t, or Arccos t, we may do so by considering one case only: $t > 0$. Thus, if it is in the first quadrant, the angle may be taken as an angle in a right triangle, and we use the definitions involving opposite, adjacent, and hypotenuse.

Knowing this, let's return to Example 7 where we let $\theta = \text{Arctan } t$. If t represents the side opposite and 1 the side adjacent, we have Figure 5.45.

Figure 5.45

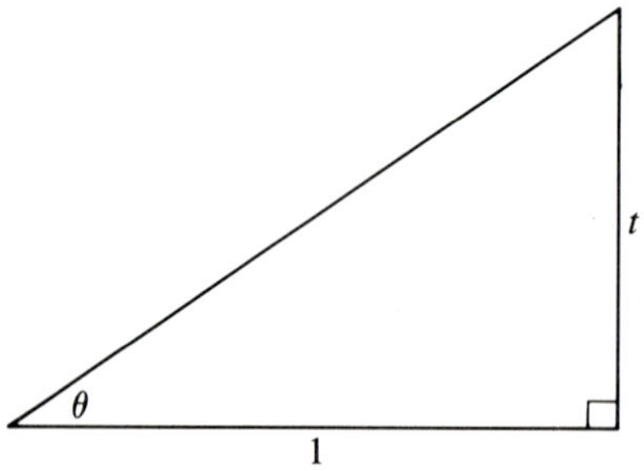

Hence the hypotenuse is $\sqrt{1 + t^2}$, and

$$\sin\theta = \frac{\text{opposite}}{\text{hypotenuse}} = \frac{t}{\sqrt{1+t^2}} \quad \left(\text{or } \frac{t\sqrt{1+t^2}}{1+t^2}\right),$$

the previous result.

▲ *Exercise* Determine tan (Arcsin t).

Answer $\dfrac{t}{\sqrt{1-t^2}} \quad \left(\text{or } \dfrac{t\sqrt{1-t^2}}{1-t^2}\right)$

Problem Set 5.4

In Problems 1–10, find the exact value.

1. $\text{Arcsin } \dfrac{1}{2}$
2. $\text{Arcsin}\left(\dfrac{-\sqrt{3}}{2}\right)$
3. $\text{Arccos}\left(\dfrac{-1}{2}\right)$
4. $\text{Arccos } \dfrac{1}{\sqrt{2}}$
5. $\text{Arctan } \dfrac{1}{\sqrt{3}}$
6. Arctan (-1)
7. Arctan 0
8. Arcsin (-1)
9. Arcsin (sin 1.2)
10. Arcsin (sin 2.2)

In Problems 11–18, find the value to the nearest thousandth.

11. Arcsin 0.58

12. Arcsin (-0.727)

13. Arccos (-0.124)

14. Arccos 0.95

15. Arctan 4

16. Arctan (-3.2)

17. Arctan (tan 1.2)

18. Arctan (tan 2.2)

In Problems 19–24, find the exact value.

19. $\tan\left(\text{Arcsin}\,\frac{3}{7}\right)$

20. $\sin\left(\text{Arccos}\,\frac{2}{7}\right)$

21. $\cos\left(\text{Arcsin}\left(\frac{-5}{8}\right)\right)$

22. $\cos\left(\text{Arctan}\left(\frac{-5}{2}\right)\right)$

23. $\sin\left(\text{Arcsin}\,\frac{\sqrt{17}}{5}\right)$

24. $\cot\left(\text{Arctan}\left(\frac{-1}{4}\right)\right)$

In Problems 25–28, find the value to four significant digits.

25. tan (Arcsin 0.52)

26. cos (Arcsin (-0.2771))

27. tan (Arccos (-0.51))

28. sin (Arctan 1.236)

29. Find cos (Arctan x).

30. Find cot (Arcsin x).

31. Find tan (Arccos x).

32. Find cos (Arcsin x).

In Problems 33–44, find the exact solutions, where $0 \le x < 2\pi$.

33. $\sin x = 0.8$

34. $\cos x = 0.42$

35. $\tan x = -5$

36. $\sin x = -0.65$

37. $\cos x = -0.25$

38. $\sec x = -10$

39. $\sin 2x = 0.68$

40. $\cos 2x = -0.84$

41. $\cos \frac{x}{2} = 0.81$

42. $\tan \frac{x}{2} = -0.21$

43. $\sin\left(x - \frac{\pi}{2}\right) = 0.7$

44. $\cos\left(x + \frac{\pi}{2}\right) = -0.4$

In Problems 45–50, sketch the graph.

45. $y = 2\,\text{Arcsin}\,x$

46. $y = -3\,\text{Arctan}\,x$

47. $y = \text{Arcsin}\,(x - 1)$

48. $y = \text{Arccos}\,(x + 1)$

49. $y = \text{Arctan}\,2x$

50. $y = 2\,\text{Arcsin}\,\frac{x}{2}$

CHAPTER 5 REVIEW

Key Terms and Concepts

amplitude, page 151

asymptote, page 170

cosecant graph, page 174

cosine graph, page 152

cotangent graph, page 171

inverse cosine function (Arccos), page 182

inverse sine function (Arcsin), page 179

inverse tangent function (Arctan), page 184

period, page 150

phase shift, page 162

secant graph, page 174

simple harmonic motion, page 166

sine graph, page 151

tangent graph, page 171

Problem Set

1. "Specify "true" or "false."

(a) The graph of $y = 2 \sin x$ is a horizontal stretching of the graph of $y = \sin x$.

(b) The graph of $y = \cos 2x$ is a horizontal compressing of the graph of $y = \cos x$.

(c) The amplitude of $y = -3 \sin x$ is -3.

(d) The range of the inverse sine function is $\left(\frac{-\pi}{2}, \frac{\pi}{2}\right)$.

(e) The domain of the inverse cosine function is $[0, \pi]$.

(f) $\text{Arctan}(-x) = -\text{Arctan}\, x$ for all $x \in R$.

(g) $\text{Arcsin}(-x) = -\text{Arcsin}\, x$ for all $x \in [-1, 1]$.

(h) $\text{Arccos}\,(-x) = \text{Arccos}\, x$ for all $x \in [-1, 1]$.

(i) $\sin(\text{Arcsin}\, x) = x$ for all $x \in [-1, 1]$.

(j) $\text{Arcsin}(\sin x) = x$ for all x.

(k) $\tan(-x) = -\tan x$ for all x in the domain.

(l) The cosecant function is even.

(m) $\text{Arccos}(-0.7) = \pi - \text{Arccos}\, 0.7$.

In Problems 2–4, choose the equation that best describes the given graph.

2. (a) $y = 2 \cos 2x$ (b) $y = 2 \cos x$

(c) $y = \sin 2x$ (d) $y = 2 \sin x$

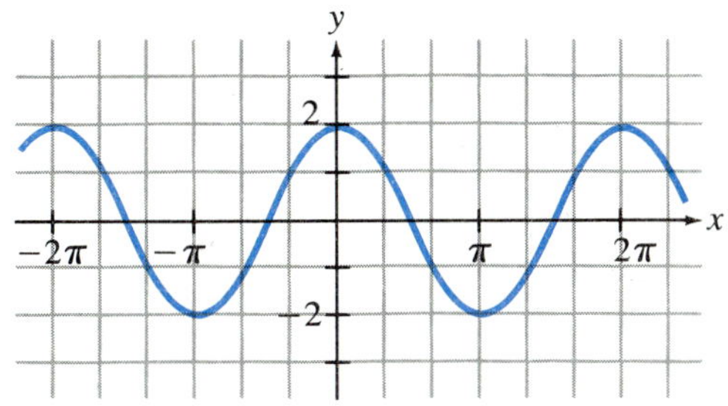

3. (a) $y = \tan\left(x + \dfrac{\pi}{2}\right)$ (b) $y = \tan\left(x - \dfrac{\pi}{2}\right)$

(c) $y = \cot\left(x - \dfrac{\pi}{2}\right)$ (d) $y = \cot\left(x + \dfrac{\pi}{2}\right)$

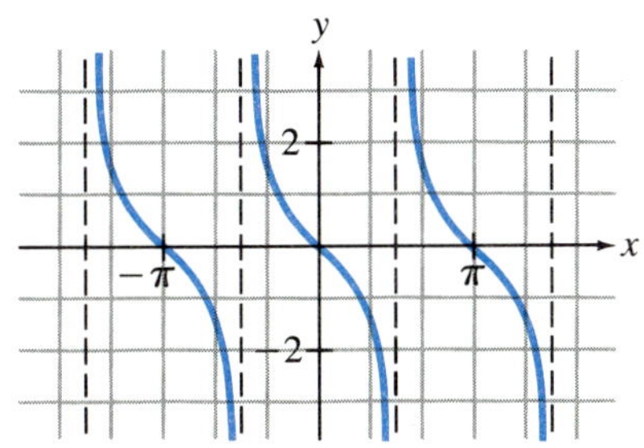

4. (a) $y = \sec \dfrac{x}{2}$ (b) $y = \sec 2x$

(c) $y = \csc \dfrac{x}{2}$ (d) $y = \csc 2x$

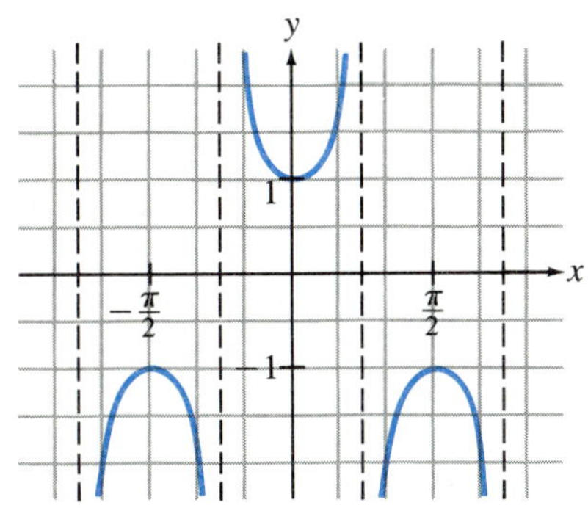

In Problems 5–20, sketch the graph.

5. $y = 3 \sin x$

6. $y = -4 \cos x$

7. $y = \sin 4x$

8. $y = \sin \dfrac{3x}{2}$

9. $y = \cos\left(x + \dfrac{3\pi}{4}\right)$

10. $y = \sin\left(-x + \dfrac{\pi}{2}\right)$

11. $y = \frac{1}{2}\tan 2x$

12. $y = 2 + \cos 2x$

13. $y = -2\sin(2x - \pi)$

14. $y = -2\tan\left(x - \frac{\pi}{6}\right)$

15. $y = \sec\left(2x + \frac{3\pi}{2}\right)$

16. $y = \cot(3x - \pi)$

17. $y = 3\sec x$

18. $y = \csc\left(2x - \frac{\pi}{3}\right)$

19. $y = -2\tan\left(\frac{x}{2} - \frac{\pi}{4}\right)$

20. $y = \tan\left(\frac{x}{3} + \frac{\pi}{2}\right)$

In Problems 21–28, find the exact value.

21. $\text{Arccos}\left(\frac{-\sqrt{3}}{2}\right)$

22. $\text{Arccos}\left(\frac{-1}{\sqrt{2}}\right)$

23. $\text{Arctan}(-\sqrt{3})$

24. $\text{Arcsin}\,\frac{\sqrt{2}}{2}$

25. Arcsin 1

26. Arctan (tan 2)

27 $\tan\left(\text{Arccos}\,\frac{8}{\sqrt{73}}\right)$

28. $\sin\left(\text{Arctan}\left(\frac{-1}{4}\right)\right)$

In Problems 29–32, find the value to the nearest thousandth.

29. Arcsin 0.247

30. Arccos (−0.18)

31. Arctan 8.25

32. Arcsin (−0.525)

In Problems 33 and 34, find the value to four significant digits.

33. sin (Arccos 0.9611)

34. cos (Arctan (−93))

In Problems 35 and 36, find the exact solutions, where $0 \leq x < 2\pi$.

35. $\cos\frac{x}{2} = 0.65$

36. $\tan 2x = -4$

37. Find sec (Arcsin x).

38. Sketch the graph of $y = -2\,\text{Arccos}(x - 1)$.

39. The displacement of the piston of a car engine is approximately simple harmonic. If the stroke is 10 centimeters, if the angular velocity of the flywheel is 3600 revolutions per minute, and if the piston is at the top of its stroke at $t = 0$, what is the equation of motion for the piston?

40. A pendulum is pulled to the right through an arc of 20°. Write its displacement from equilibrium, as measured along the arc, as a function of elapsed time if its period is 2.4 seconds and the length of the pendulum is 3 feet.

6
IDENTITIES

CHAPTER OUTLINE

An **identity** is an equation in which every element of the universe (sometimes called replacement set) of the variable is a solution. For example, $x(x + 2) = x^2 + 2x$ is an identity because the equation is true when x is replaced by any real number. On the other hand, $x + 2 = 7$ is not an identity because 5 is the only element from the universe R for which the equation is true. This chapter deals with identities that involve the trigonometric functions. These identities will permit us to rewrite complicated expressions in simplified form.

6.1 BASIC IDENTITIES

In this section we will be interested in some fundamental identities. We have already noted that the cosecant is the reciprocal of the sine. Hence

$$\csc\theta = \frac{1}{\sin\theta} \tag{6.1}$$

is an identity. The universe of θ is $R - \{n\pi \mid n \in Z\}$, and equation (6.1) is true for all such θ. (Recall that when the universe isn't specified, we assume it to be all real numbers for which all terms are defined and real. The cosecant isn't defined when $\theta = n\pi$, and neither is $1/\sin\theta$.)

Note than an equation may be true for infinitely many values and still not be an identity. For example,

$$\sin\theta + \cos\theta = 1 \tag{6.2}$$

is true for all $\theta = 2n\pi$ and $\theta = 2n\pi + \pi/2$ such that n is an integer. However, equation (6.2) is not an identity because there exists at least one element of the universe that is not a solution, $\pi/6$, for instance.

Although the identities in this chapter will be stated for angles and ratios, they will, of course, be true for corresponding real numbers and functions. There are eight basic identities:

Reciprocal:

$$\begin{cases} (1) \quad \csc\theta = \dfrac{1}{\sin\theta} \\ (2) \quad \sec\theta = \dfrac{1}{\cos\theta} \\ (3) \quad \cot\theta = \dfrac{1}{\tan\theta} \end{cases}$$

Quotient:

$$\begin{cases} (4) \quad \tan\theta = \dfrac{\sin\theta}{\cos\theta} \\ (5) \quad \cot\theta = \dfrac{\cos\theta}{\sin\theta} \end{cases}$$

Pythagorean:

$$\begin{cases} (6) \quad \sin^2\theta + \cos^2\theta = 1 \\ (7) \quad \sec^2\theta = 1 + \tan^2\theta \\ (8) \quad \csc^2\theta = 1 + \cot^2\theta \end{cases}$$

[Note: The notation $\sin^2\theta$ means $(\sin\theta)^2$, etc.]

Each of the basic identities may be proven from the definitions of the trigonometric ratios. To illustrate, let's prove that

$$\tan\theta = \frac{\sin\theta}{\cos\theta}.$$

Letting (x, y) be a point on the terminal side and $r = \sqrt{x^2 + y^2}$, we have

$$\tan\theta = \frac{y}{x}$$

by definition. Dividing both numerator and denominator of the right-hand side by $r (r \neq 0)$, we get

$$\tan\theta = \frac{y/r}{x/r},$$

but by definition, $\sin\theta = y/r$ and $\cos\theta = x/r$. Hence

$$\tan\theta = \frac{\sin\theta}{\cos\theta}.$$

To prove some identities, we may also use previously established identities. Once identities 2, 4, and 6 are proven, for example, they may be used to prove identity 7:

$$\sec^2\theta = 1 + \tan^2\theta.$$

We have

$$\begin{aligned} 1 + \tan^2\theta &= 1 + \frac{\sin^2\theta}{\cos^2\theta} && \text{(by identity 4)} \\ &= \frac{\cos^2\theta + \sin^2\theta}{\cos^2\theta} \\ &= \frac{1}{\cos^2\theta} && \text{(by identity 6)} \\ &= \sec^2\theta. && \text{(by identity 2)} \end{aligned}$$

The eight basic identities sometimes appear in alternative forms. For example, identity 1 is frequently written as

$$\sin\theta = \frac{1}{\csc\theta} \quad \text{or} \quad \sin\theta\csc\theta = 1.$$

Identities 2 and 3 have similar alternative forms. Identity 6 may be written as

$$\sin^2\theta = 1 - \cos^2\theta \quad \text{or} \quad \cos^2\theta = 1 - \sin^2\theta.$$

Identities 7 and 8 have similar alternative forms. You should readily recognize such forms as algebraic manipulations of the basic identities.

One use of the identities is to determine the values of the other trigonometric ratios given one value and a quadrant. The first example is Example 1 of Section 3.2, which was worked by sketching the angle and selecting a point on the terminal side. Here we'll use only identities.

EXAMPLE 1 Find the exact values of the other trigonometric ratios if $\sec\theta = -3$ and θ is in the third quadrant.

SOLUTION An alternative form of identity 2 will give us the cosine. We get

$$\cos\theta = \frac{1}{\sec\theta} = \frac{1}{-3} = \frac{-1}{3}.$$

We may now use identity 6 to get the sine. We have

$$\begin{aligned}\sin^2\theta + \cos^2\theta &= 1\\ \sin^2\theta &= 1 - \cos^2\theta\\ \sin\theta &= \pm\sqrt{1-\cos^2\theta}.\end{aligned}$$

Because θ is in the third quadrant, the sine is negative. Thus

$$\begin{aligned}\sin\theta &= -\sqrt{1-\cos^2\theta}\\ &= -\sqrt{1-\frac{1}{9}} = -\sqrt{\frac{8}{9}} = \frac{-2\sqrt{2}}{3}.\end{aligned}$$

Its reciprocal is

$$\begin{aligned}\csc\theta &= \frac{1}{\sin\theta}\\ &= \frac{-3}{2\sqrt{2}} \quad \left(\text{or } \frac{-3\sqrt{2}}{4}\right).\end{aligned}$$

We'll use identity 4 to determine the tangent. We have

$$\begin{aligned}\tan\theta &= \frac{\sin\theta}{\cos\theta}\\ &= \frac{\frac{-2\sqrt{2}}{3}}{\frac{-1}{3}} = 2\sqrt{2}.\end{aligned}$$

Its reciprocal is

$$\begin{aligned}\cot\theta &= \frac{1}{\tan\theta}\\ &= \frac{1}{2\sqrt{2}}\left(\text{or } \frac{\sqrt{2}}{4}\right).\end{aligned}$$

▲

▲ ***Exercise*** Use basic identities to find the exact values of the other trigonometric ratios if $\csc\theta = -4$ and θ is in the fourth quadrant.

Answer $\sin\theta = \dfrac{-1}{4}$, $\cos\theta = \dfrac{\sqrt{15}}{4}$, $\sec\theta = \dfrac{4}{\sqrt{15}}\left(\text{or } \dfrac{4\sqrt{15}}{15}\right)$, $\tan\theta = \dfrac{-1}{\sqrt{15}}\left(\text{or } \dfrac{-\sqrt{15}}{15}\right)$, $\cot\theta = -\sqrt{15}$

We shall also use the eight basic identities, together with algebraic manipulation, to verify new identities. Given an identity to be established, the general technique is to convert one side of the equation to the other by using algebraic manipulation and the basic identities. There are no set rules, and there is usually a variety of different proofs, but it is generally easier to convert the more complicated side to the simpler one.

EXAMPLE 2 Verify that $\dfrac{\sec\theta}{\csc\theta} = \tan\theta$.

SOLUTION Beginning with the more complicated side, the left-hand side, we have

$$\frac{\sec\theta}{\csc\theta} = \frac{\dfrac{1}{\cos\theta}}{\dfrac{1}{\sin\theta}} \qquad \text{(by identities 1 and 2, reciprocals)}$$

$$= \frac{1}{\cos\theta}\cdot\frac{\sin\theta}{1}$$

$$= \frac{\sin\theta}{\cos\theta}$$

$$= \tan\theta. \qquad \text{(by identity 4, quotient)}$$

▲

EXAMPLE 3 Verify that $\cos\theta(\sec\theta - \cos\theta) = \sin^2\theta$.

SOLUTION Beginning with the left-hand side, we have

$$\cos\theta(\sec\theta - \cos\theta) = \cos\theta\sec\theta - \cos^2\theta$$

$$= \cos\theta\cdot\frac{1}{\cos\theta} - \cos^2\theta \qquad \text{(by identity 2, reciprocal)}$$

$$= 1 - \cos^2\theta$$

$$= \sin^2\theta. \qquad \text{(by identity 6, Pythagorean)}$$

▲

EXAMPLE 4 Verify that $\tan^2 t = \dfrac{\sec^2 t}{1 + \cot^2 t}$.

SOLUTION Starting with the right-hand side, we get

$$\frac{\sec^2 t}{1+\cot^2 t} = \frac{\sec^2 t}{1 + \dfrac{1}{\tan^2 t}} \qquad \text{(by identity 3, reciprocal)}$$

$$= \frac{\tan^2 t \sec^2 t}{\tan^2 t + 1} \qquad \text{(by multiplying numerator and denominator by } \tan^2 t\text{)}$$

$$= \frac{\tan^2 t \sec^2 t}{\sec^2 t} \qquad \text{(by identity 7, Pythagorean)}$$

$$= \tan^2 t.$$

▲

For the identity of Example 4 we could have proceeded as follows:

$$\frac{\sec^2 t}{1+\cot^2 t} = \frac{\sec^2 t}{\csc^2 t} \qquad \text{(by identity 8, Pythagorean)}$$

$$= \frac{\dfrac{1}{\cos^2 t}}{\dfrac{1}{\sin^2 t}} \qquad \text{(by identities 1 and 2, reciprocal)}$$

$$= \frac{1}{\cos^2 t} \cdot \frac{\sin^2 t}{1}$$

$$= \frac{\sin^2 t}{\cos^2 t}$$

$$= \tan^2 t. \qquad \text{(by identity 4, quotient)}$$

Thus, as noted earlier, there may be a variety of different proofs. Sometimes writing all of the functions in terms of sines and cosines is the key. This technique is used in Examples 5 and 6.

EXAMPLE 5 Verify that $\dfrac{\tan \theta}{\sec \theta - \cos \theta} = \csc \theta$.

SOLUTION Starting with the left-hand side, we get

$$\frac{\tan \theta}{\sec \theta - \cos \theta} = \frac{\dfrac{\sin \theta}{\cos \theta}}{\dfrac{1}{\cos \theta} - \cos \theta} \qquad \text{(by identities 2 and 4, reciprocal and quotient)}$$

$$= \frac{\sin\theta}{1 - \cos^2\theta} \qquad \text{(by multiplying numerator and denominator by } \cos\theta\text{)}$$

$$= \frac{\sin\theta}{\sin^2\theta} \qquad \text{(by identity 6, Pythagorean)}$$

$$= \frac{1}{\sin\theta}$$

$$= \csc\theta. \qquad \text{(by identity 1, reciprocal)}$$

▲

EXAMPLE 6 Verify that $1 - 2\sin^2 x = \dfrac{1 - \tan^2 x}{1 + \tan^2 x}$.

SOLUTION Beginning with the right-hand side, we get

$$\frac{1 - \tan^2 x}{1 + \tan^2 x} = \frac{1 - \dfrac{\sin^2 x}{\cos^2 x}}{1 + \dfrac{\sin^2 x}{\cos^2 x}} \qquad \text{(by identity 4, quotient)}$$

$$= \frac{\cos^2 x - \sin^2 x}{\cos^2 x + \sin^2 x} \qquad \text{(by multiplying numerator and denominator by } \cos^2 x\text{)}$$

$$= \cos^2 x - \sin^2 x \qquad \text{(by identity 6, Pythagorean)}$$

$$= (1 - \sin^2 x) - \sin^2 x \qquad \text{(by identity 6, Pythagorean)}$$

$$= 1 - 2\sin^2 x.$$

▲

EXAMPLE 7 Verify that $1 - 2\sin^2 x = 2\cos^2 x - 1$.

SOLUTION We have

$$1 - 2\sin^2 x = 1 - 2(1 - \cos^2 x) \qquad \text{(by identity 6, Pythagorean)}$$

$$= 1 - 2 + 2\cos^2 x$$

$$= 2\cos^2 x - 1.$$

▲

The basic identities may also be used in solving conditional equations, as will be seen in Example 8.

EXAMPLE 8 Find the approximate solutions to the nearest thousandth, where $0 \le x < 2\pi$.

(a) $4\sin^2 x + 13\cos x - 7 = 0$ (b) $2\cos x + \sin x = 0$

SOLUTION (a) We must change this equation to one containing powers of $\sin x$ only or $\cos x$ only, so that the left side may be factored. Since $\sin^2 x = 1 - \cos^2 x$, we have

$$\begin{aligned} 4(1 - \cos^2 x) + 13 \cos x - 7 &= 0 \\ -4 \cos^2 x + 13 \cos x - 3 &= 0 \\ 4 \cos^2 x - 13 \cos x + 3 &= 0 \\ (\cos x - 3)(4 \cos x - 1) &= 0 \\ \cos x = 3 \quad \text{or} \quad \cos x &= \frac{1}{4}. \end{aligned}$$

Since $-1 \le \cos x \le 1$ for all x, there are no solutions to the equation $\cos x = 3$. The solutions to $\cos x = \frac{1}{4}$ are

$$\text{Arccos}\,\frac{1}{4} \approx 1.318 \quad \text{and} \quad 2\pi - \text{Arccos}\,\frac{1}{4} \approx 4.965.$$

(b) We need to change this equation to one containing a single trigonometric function. When $\cos x \ne 0$, the given equation is equivalent to

$$\frac{2 \cos x}{\cos x} + \frac{\sin x}{\cos x} = 0 \tag{6.3}$$

$$2 + \tan x = 0$$

$$\tan x = -2.$$

The solutions of the last equation are

$$\pi - \text{Arctan}\, 2 \approx 2.034 \quad \text{and} \quad 2\pi - \text{Arctan}\, 2 \approx 5.176.$$

However, equation (6.3) is equivalent to the given equation when $\cos x \ne 0$. What happens when $\cos x = 0$, that is, when $x = \pi/2 (\approx 1.571)$ or $x = 3\pi/2 (\approx 4.712)$? By direct substitution into the original equation, we see that neither is a solution. Hence the solutions are 2.034 and 5.176.

▲

▲ ***Exercise*** Find the approximate solutions to the nearest thousandth, where $0 \le x < 2\pi$.

(a) $3 \cos^2 x - 5 \sin x - 1 = 0$ (b) $3 \cos x - \sin x = 0$

Answer (a) 0.340, 2.802 (b) 1.249, 4.391

Problem Set 6.1

In Problems 1–6, use basic identities to find the exact values of the other trigonometric ratios for the given conditions.

1. $\cos\theta = \dfrac{-2}{5}$, θ in the second quadrant

2. $\sec\theta = -4$, θ in the second quadrant

3. $\sin\theta = \dfrac{-1}{6}$, θ in the fourth quadrant

4. $\csc\theta = 5$, θ in the first quadrant

5. $\tan\theta = \dfrac{3}{2}$, θ in the third quadrant

6. $\cot\theta = -4$, θ in the fourth quadrant

7. Express $\sin\theta$ in terms of $\sec\theta$.

8. Express $\cos\theta$ in terms of $\csc\theta$.

9. Express $\sin\theta$ in terms of $\tan\theta$.

10. Express $\cos\theta$ in terms of $\tan\theta$.

11. Express all trigonometric functions in terms of $\sin x$.

12. Express all trigonometric functions in terms of $\cos x$.

In Problems 13–16, use basic identities to reduce the expression to 1.

13. $\dfrac{\sin\theta}{\cos\theta\tan\theta}$

14. $\dfrac{\sin\theta}{\csc\theta} + \dfrac{\cos\theta}{\sec\theta}$

15. $\cos\theta\csc\theta\tan\theta$

16. $2\csc^2\theta - \dfrac{1+\cos^2\theta}{\sin^2\theta}$

In Problems 17–28, use basic identities to reduce the expression to a single trigonometric ratio.

17. $\tan\theta\cos\theta$

18. $\sin\theta\cot\theta$

19. $\cos\theta\csc\theta$

20. $\sec\theta\sin\theta$

21. $\dfrac{\tan\theta}{\sin\theta}$

22. $\dfrac{\cot\theta}{\cos\theta}$

23. $\cos\theta + \tan\theta\sin\theta$

24. $\cos\theta\cot\theta + \sin\theta$

25. $(1 - \sin x)(\sec x + \tan x)$

26. $(\sec x - \tan x)(1 + \csc x)$

27. $\dfrac{\tan x\sin x}{\sec^2 x - 1}$

28. $\dfrac{1 + \sec x}{\tan x + \sin x}$

In Problems 29–32, verify that the equation is *not* an identity.

29. $\sin(x + y) = \sin x + \sin y$

30. $\cos(x - y) = \cos x - \cos y$

31. $\cos 2x = 2\cos x$

32. $\sin\frac{x}{2} = \frac{\sin x}{2}$

In Problems 33–62, all but two of the equations are identities. Verify the identities and show that two of the equations are not identities.

33. $\cos^2\theta - \sin^2\theta = 1 - 2\sin^2\theta$

34. $\cos^2\theta - \sin^2\theta = 2\cos^2\theta - 1$

35. $\frac{\sin^2\theta}{\cos^2\theta} = \sec^2\theta - 1$

36. $\frac{\cos^2\theta}{\sin^2\theta} = \csc^2\theta - 1$

37. $1 + 2\tan^2\theta + \tan^4\theta = \sec^4\theta$

38. $\sec^4 x - \tan^4 x = \sec^2 x + \tan^2 x$

39. $\sec\theta - \cos\theta = \tan\theta\sin\theta$

40. $\tan\theta + \cot\theta = \sec\theta\csc\theta$

41. $\sec x + \csc x = (\tan x + \cot x)(\cos x + \sin x)$

42. $\sin^2 x + \tan^2 x = \sec^2 x - \cos^2 x$

43. $\frac{1}{1 - \sin\theta} + \frac{1}{1 + \sin\theta} = 2\sec^2\theta$

44. $\frac{1}{\csc\theta - \cot\theta} - \frac{1}{\csc\theta + \cot\theta} = 2\cot\theta$

45. $\frac{1 + \cot^2 x}{1 + \tan^2 x} = \sec^2 x$

46. $\frac{\cot^2\theta}{\csc\theta + 1} + 1 - \csc\theta = 0$

47. $\frac{\csc\theta + \cot\theta}{\tan\theta + \sin\theta} = \cot\theta\csc\theta$

48. $\frac{\sec x + \csc x}{\sec x - \csc x} = \frac{1 + \tan x}{\tan x - 1}$

49. $\sec^2 x\tan^2 x = \sin^2 x\sec^4 x$

50. $\cos^4 x - \sin^4 x = 1 - 2\sin^2 x$

51. $\tan^2 x - \sin^2 x = \tan^2 x\sin^2 x$

52. $\frac{\tan\theta}{\csc\theta - \cot\theta} - \frac{\sin\theta}{\csc\theta + \cot\theta} = \sec\theta + \cos\theta$

53. $\dfrac{\cos\theta}{1+\sin\theta}+\dfrac{1+\sin\theta}{\cos\theta}=2\sec\theta$

54. $\dfrac{1+\sin\theta}{\cos\theta}=\dfrac{\cos\theta}{\sin\theta-1}$

55. $\sin^2 x(2+\tan^2 x)=\sin^2 x+\tan^2 x$

56. $(\sec x-\tan x)^2=\dfrac{1-\sin x}{1+\sin x}$

57. $\dfrac{2\sin^2 x+3\cos x-3}{\sin^2 x}=\dfrac{2\cos x-1}{1+\cos x}$

58. $\dfrac{\tan x-\sin x}{\sin^3 x}=\dfrac{\sec x}{1+\cos x}$

59. $\dfrac{\tan x+\cot x}{1-\sin^2 x}-\sin x\sec^3 x=\sec x\csc x$

60. $\csc^2\theta\sec\theta+\csc\theta\sec\theta-\cot\theta\csc\theta-\cot\theta=\sec\theta+\tan\theta$

61. $\dfrac{\sin x\tan x-\cos x\cot x}{\sec x-\csc x}=1+\sin x\cos x$

62. $\sin^6 x+\cos^6 x=1-3\sin^2 x\cos^2 x$

In Problems 63–70, find the exact solutions, where $0\le x<2\pi$.

63. $\cos^2 x-3\sin^2 x=0$

64. $\tan^2 x+2\sec^2 x=3$

65. $\dfrac{1}{1+\cos x}+\dfrac{1}{1-\cos x}=4$

66. $\dfrac{4\tan^2 x}{1+\tan^2 x}=1$

67. $\dfrac{\cos^4 x+\cos^4 x\cot^2 x}{1-\sin^2 x}=3$

68. $\dfrac{\tan x}{1+\sec x}-\dfrac{\tan x}{1-\sec x}=4$

69. $\cos^2 x+\cot^2 x+\sin^2 x=9$

70. $\sin^2 x\tan^2 x+\sin^2 x=9$

In Problems 71–78, find the approximate solutions to the nearest thousandth, where $0\le x<2\pi$.

71. $2\sec x-3\cos x+5=0$

72. $3\sin x=4+4\csc x$

73. $5\sin x+1=3\cos^2 x$

74. $2\tan^2 x+9\sec x-3=0$

75. $\sin x=-2\cos x$

76. $\cos x=2\sin x$

77. $\sin^2 x(1-\cot^2 x)=1$

78. $(1+\sin x)(\sec x-\tan x)=\dfrac{1}{4}$

6.2 IDENTITIES FOR THE COSINE OF SUMS AND DIFFERENCES

Unlike the last section, this one will deal primarily with identities that involve two angles. The first formula is for the cosine of the difference of two angles. For all angles α and β,

$$\cos(\alpha - \beta) = \cos\alpha\cos\beta + \sin\alpha\sin\beta.$$

To establish this formula let α and β be in standard position (Figure 6.1).

Figure 6.1

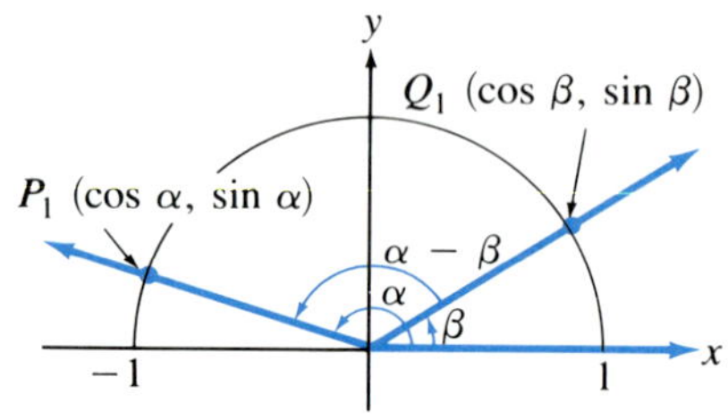

Although α is drawn in the second quadrant and β in the first, the following remarks are true for any α and β. Let P_1 be the point on the terminal side of α that is 1 unit from the origin, and let Q_1 be the point on the terminal side of β that is 1 unit from the origin. (Thus $r = 1$ for each angle.) Therefore, if $P_1 = (x, y)$, then

$$\cos\alpha = \frac{x}{1} \quad \text{and} \quad \sin\alpha = \frac{y}{1}.$$

Hence $P_1 = (x, y) = (\cos\alpha, \sin\alpha)$. Similarly, $Q_1 = (\cos\beta, \sin\beta)$.

The distance from P_1 to Q_1 is given by

$$P_1Q_1 = \sqrt{(\cos\alpha - \cos\beta)^2 + (\sin\alpha - \sin\beta)^2}.$$

Squaring both sides and simplifying, we get

$$\begin{aligned}(P_1Q_1)^2 &= (\cos\alpha - \cos\beta)^2 + (\sin\alpha - \sin\beta)^2\\ &= \cos^2\alpha - 2\cos\alpha\cos\beta + \cos^2\beta + \sin^2\alpha\\ &\quad -2\sin\alpha\sin\beta + \sin^2\beta\\ &= (\cos^2\alpha + \sin^2\alpha) + (\cos^2\beta + \sin^2\beta) - 2\cos\alpha\cos\beta\\ &\quad -2\sin\alpha\sin\beta\\ &= 2 - 2\cos\alpha\cos\beta - 2\sin\alpha\sin\beta. \qquad \textbf{(6.4)}\end{aligned}$$

Now suppose that we move the angle $\alpha - \beta$ into standard position, letting P_2 be the point on the terminal side of $\alpha - \beta$ that is 1 unit from the origin, and letting Q_2 be the point on the initial side of $\alpha - \beta$ that is 1 unit from the origin. Thus $P_2 = (\cos(\alpha - \beta), \sin(\alpha - \beta))$, and $Q_2 = (1, 0)$. See Figure 6.2.

Figure 6.2

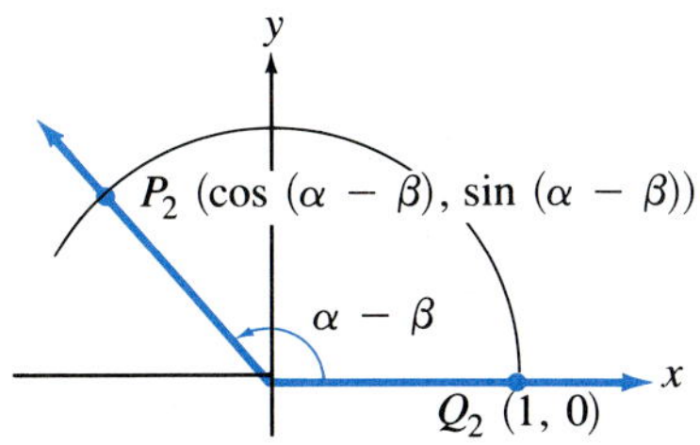

The distance between P_2 and Q_2 is

$$P_2Q_2 = \sqrt{(\cos(\alpha - \beta) - 1)^2 + \sin^2(\alpha - \beta)}.$$

Squaring both sides, we get

$$\begin{aligned}(P_2Q_2)^2 &= (\cos(\alpha - \beta) - 1)^2 + \sin^2(\alpha - \beta)\\ &= \cos^2(\alpha - \beta) - 2\cos(\alpha - \beta) + 1 + \sin^2(\alpha - \beta)\\ &= \cos^2(\alpha - \beta) + \sin^2(\alpha - \beta) - 2\cos(\alpha - \beta) + 1\\ &= 1 - 2\cos(\alpha - \beta) + 1\\ &= 2 - 2\cos(\alpha - \beta). \qquad \textbf{(6.5)}\end{aligned}$$

The central angle $\alpha - \beta$ intercepts equal chords in the unit circles in Figures 6.1 and 6.2. Thus

$$\begin{aligned}P_2Q_2 &= P_1Q_1\\ (P_2Q_2)^2 &= (P_1Q_1)^2. \qquad \textbf{(6.6)}\end{aligned}$$

Hence, using equations (6.4) and (6.5) to substitute into equation (6.6), we get

$$\begin{aligned}2 - 2\cos(\alpha - \beta) &= 2 - 2\cos\alpha\cos\beta - 2\sin\alpha\sin\beta\\ -2\cos(\alpha - \beta) &= -2\cos\alpha\cos\beta - 2\sin\alpha\sin\beta\\ \cos(\alpha - \beta) &= \cos\alpha\cos\beta + \sin\alpha\sin\beta.\end{aligned}$$

EXAMPLE 1 Find the exact value of $\cos 15°$.

SOLUTION In order to find the exact value, we cant't use the calculator. It would give an approximation. Instead we'll use the formula for the cosine of the difference. We must write $15°$ as the difference of two angles whose cosines and sines can be calculated exactly, that is, as the difference of two special angles. Since

$$15° = 60° - 45°,$$

we have

$$\begin{aligned}\cos 15^\circ &= \cos(60^\circ - 45^\circ)\\ &= \cos 60^\circ \cos 45^\circ + \sin 60^\circ \sin 45^\circ\\ &= \frac{1}{2}\cdot\frac{1}{\sqrt{2}} + \frac{\sqrt{3}}{2}\cdot\frac{1}{\sqrt{2}} = \frac{1+\sqrt{3}}{2\sqrt{2}}\left(\text{or } \frac{\sqrt{2}+\sqrt{6}}{4}\right).\end{aligned}$$

▲

Note that

$$15^\circ = 45^\circ - 30^\circ.$$

So we could have used two other special angles to work Example 1. (Try it.)

▲ ***Exercise*** Find the exact value of $\cos 75^\circ$.

Answer $\dfrac{\sqrt{3}-1}{2\sqrt{2}}\left(\text{or } \dfrac{\sqrt{6}-\sqrt{2}}{4}\right)$

EXAMPLE 2 Find the exact value of $\cos(\alpha - \beta)$ if $\cos \alpha = \frac{1}{3}$, where α is in quadrant I, and $\cos \beta = \frac{-3}{4}$, where β is in quadrant III.

SOLUTION We will use

$$\cos(\alpha - \beta) = \cos \alpha \cos \beta + \sin \alpha \sin \beta,$$

but we must first determine $\sin \alpha$ and $\sin \beta$. Using the Pythagorean identity for sine and cosine, we have

$$\begin{aligned}\sin \alpha &= +\sqrt{1 - \cos^2 \alpha}\\ &= \sqrt{1 - \frac{1}{9}} = \sqrt{\frac{8}{9}} = \frac{2\sqrt{2}}{3}.\end{aligned}$$

Note that a plus sign precedes the radical because the sine is positive in the first quadrant. In the third quadrant, however, the sine is negative. Thus,

$$\begin{aligned}\sin \beta &= -\sqrt{1 - \cos^2 \beta}\\ &= -\sqrt{1 - \frac{9}{16}} = -\sqrt{\frac{7}{16}} = \frac{-\sqrt{7}}{4}.\end{aligned}$$

Therefore,

$$\begin{aligned}\cos(\alpha - \beta) &= \cos \alpha \cos \beta + \sin \alpha \sin \beta\\ &= \frac{1}{3}\cdot\frac{-3}{4} + \frac{2\sqrt{2}}{3}\cdot\frac{-\sqrt{7}}{4} = \frac{-3 - 2\sqrt{14}}{12}.\end{aligned}$$

▲

▲ *Exercise* Find the exact value of $\cos(\alpha - \beta)$ if $\cos\alpha = \frac{2}{3}$, where α is in quadrant IV, and $\cos\beta = \frac{-1}{4}$, where β is in quadrant II.

Answer $\dfrac{-2 - 5\sqrt{3}}{12}$

In Section 2.1 we noted that the sine of an angle equals the cosine of its complement, and vice versa. Although this statement was made for acute angles only, we now see that this is true for any angles whose sum is $\pi/2$ (or $90°$). That is, for any angle θ,

$$\cos\left(\frac{\pi}{2} - \theta\right) = \sin\theta \quad \text{and} \quad \sin\left(\frac{\pi}{2} - \theta\right) = \cos\theta.$$

These identities are established by using the formula for the cosine of the difference.

$$\begin{aligned}\cos\left(\frac{\pi}{2} - \theta\right) &= \cos\frac{\pi}{2}\cos\theta + \sin\frac{\pi}{2}\sin\theta \\ &= 0 \cdot \cos\theta + 1 \cdot \sin\theta \\ &= \sin\theta.\end{aligned}$$

This identity may be used to establish the other. We have

$$\begin{aligned}\sin\left(\frac{\pi}{2} - \theta\right) &= \cos\left(\frac{\pi}{2} - \left(\frac{\pi}{2} - \theta\right)\right) \\ &= \cos\theta.\end{aligned}$$

Thus, if the sum of any two angles is $\pi/2$ (or $90°$), the sine of each equals the cosine of the other. This property is called the **cofunction identity**.

Three other important identities involve negative angles. For any angle θ,

$$\begin{aligned}\cos(-\theta) &= \cos\theta, \\ \sin(-\theta) &= -\sin\theta, \\ \tan(-\theta) &= -\tan\theta.\end{aligned}$$

These identities establish some facts noted in Chapter 5: the cosine function is even, and the sine and tangent functions are odd.

The first relationship comes directly from the formula for the cosine of the difference.

$$\begin{aligned}\cos(-\theta) &= \cos(0-\theta)\\ &= \cos 0 \cos\theta + \sin 0 \sin\theta\\ &= 1\cdot\cos\theta + 0\cdot\sin\theta\\ &= \cos\theta.\end{aligned}$$

To establish the second identity we must use more ingenuity.

$$\begin{aligned}\sin(-\theta) &= \cos\left(\frac{\pi}{2}-(-\theta)\right)\\ &= \cos\left(\frac{\pi}{2}+\theta\right)\\ &= \cos\left(\theta-\left(\frac{-\pi}{2}\right)\right)\\ &= \cos\theta\cos\left(\frac{-\pi}{2}\right)+\sin\theta\sin\left(\frac{-\pi}{2}\right)\\ &= (\cos\theta)\cdot 0+\sin\theta\cdot(-1)\\ &= -\sin\theta.\end{aligned}$$

The final identity is established by using the other two along with a quotient identity.

$$\begin{aligned}\tan(-\theta) &= \frac{\sin(-\theta)}{\cos(-\theta)}\\ &= \frac{-\sin\theta}{\cos\theta}\\ &= -\tan\theta.\end{aligned}$$

The formula for the cosine of the sum of two angles may be derived from the formula for the difference, together with the formulas for sine and cosine of a negative angle. We have

$$\begin{aligned}\cos(\alpha+\beta) &= \cos(\alpha-(-\beta))\\ &= \cos\alpha\cos(-\beta)+\sin\alpha\sin(-\beta)\\ &= \cos\alpha\cos\beta+\sin\alpha(-\sin\beta)\\ &= \cos\alpha\cos\beta-\sin\alpha\sin\beta.\end{aligned}$$

Thus, for all angles α and β,

$$\cos(\alpha+\beta) = \cos\alpha\cos\beta - \sin\alpha\sin\beta.$$

EXAMPLE 3 Simplify $\cos(\theta + 180°)$.

SOLUTION Applying the formula for the cosine of a sum,

$$\begin{aligned}\cos(\theta + 180°) &= \cos\theta\cos 180° - \sin\theta\sin 180° \\ &= \cos\theta\cdot(-1) - (\sin\theta)\cdot 0 \\ &= -\cos\theta.\end{aligned}$$

▲

▲ ***Exercise*** Simplify $\cos(\theta + 90°)$.

Answer $-\sin\theta$

Problem Set 6.2

In Problems 1–6, simplify the expression.

1. $\cos(\theta - 90°)$

2. $\cos(180° - \theta)$

3. $\cos(\theta + \pi)$

4. $\cos\left(\theta + \frac{3\pi}{2}\right)$

5. $\sec\left(\theta + \frac{\pi}{2}\right)$

6. $\sec(180° + \theta)$

In Problems 7–12, find the *exact* value.

7. $\cos 165°$

8. $\cos 105°$

9. $\sec 75°$

10. $\sec 15°$

11. $\cos\left(\text{Arcsin}\frac{1}{4} - \text{Arctan}(-1)\right)$

12. $\cos\left(\text{Arcsin}\frac{1}{3} + \text{Arctan}(-1)\right)$

In Problems 13–16, write the expression as a single term.

13. $\cos 2x\cos x + \sin 2x\sin x$

14. $\cos 4x\cos x - \sin 4x\sin x$

15. $\cos\frac{\pi}{5}\cos\frac{\pi}{4} - \sin\frac{\pi}{5}\sin\frac{\pi}{4}$

16. $\cos 70°\cos 20° + \sin 70°\sin 20°$

In Problems 17–20, find the exact value of $\cos(\alpha - \beta)$ for the given conditions,

17. $\cos\alpha = \frac{1}{6}$, where α is in quadrant I, and $\cos\beta = \frac{-3}{4}$, where β is in quadrant III

18. $\cos\alpha = \frac{-4}{5}$, where α is in quadrant II, and $\cos\beta = \frac{1}{3}$, where β is in quadrant IV

19. $\sin\alpha = \frac{2}{3}$, where α is in quadrant II, and $\sin\beta = \frac{1}{4}$, where β is in quadrant I

20. $\sin\alpha = \frac{-1}{3}$, where α is in quadrant III, and $\sin\beta = \frac{-2}{5}$, where β is in quadrant IV

In Problems 21–24, find the exact value of $\cos(\alpha + \beta)$ for the given conditions.

21. $\cos\alpha = \frac{2}{3}$, where α is in quadrant I, and $\sin\beta = \frac{3}{4}$, where β is in quadrant II

22. $\cos\alpha = \frac{-1}{3}$, where α is in quadrant II, and $\sin\beta = \frac{-1}{4}$, where β is in quadrant III

23. $\tan\alpha = \frac{3}{4}$, where α is in quadrant I, and $\tan\beta = \frac{-1}{4}$, where β is in quadrant II

24. $\tan\alpha = \frac{-1}{3}$, where α is in quadrant II, and $\tan\beta = \frac{4}{5}$, where β is in quadrant III

In Problems 25–27, verify the identity.

25. $\cos(\alpha - \beta) - \cos(\alpha + \beta) = 2\sin\alpha\sin\beta$

26. $\cos(x + y) + \cos(x - y) = 2\cos x\cos y$

27. $\cos(\pi + x - y) = -\cos x\cos y - \sin x\sin y$

In Problems 28–30, find the exact solutions, where $0 \leq x < 2\pi$.

28. $\cos 3x\cos x - \sin 3x\sin x = 1$

29. $\cos 2x\cos x + \sin 2x\sin x = \frac{-1}{2}$

30. $\cos 3x\cos x + \sin 3x\sin x = -0.7$

6.3 IDENTITIES FOR THE SINE AND TANGENT OF SUMS AND DIFFERENCES

In this section we develop the formulas for the sine of sums and differences and the tangent of sums and differences.

By the cofunction identity for sine and cosine,

$$\sin(\alpha + \beta) = \cos\left(\frac{\pi}{2} - (\alpha + \beta)\right).$$

Rewriting the right side and using the formula for the cosine of a difference gives

$$\begin{aligned}\cos\left(\frac{\pi}{2} - (\alpha + \beta)\right) &= \cos\left(\left(\frac{\pi}{2} - \alpha\right) - \beta\right)\\ &= \cos\left(\frac{\pi}{2} - \alpha\right)\cos\beta + \sin\left(\frac{\pi}{2} - \alpha\right)\sin\beta\\ &= \sin\alpha\cos\beta + \cos\alpha\sin\beta,\end{aligned}$$

where the last step uses the cofunction identities again. Thus, for all angles α and β,

$$\sin(\alpha + \beta) = \sin\alpha\cos\beta + \sin\beta\cos\alpha.$$

The formula for the difference follows directly from this.

$$\begin{aligned}\sin(\alpha - \beta) &= \sin(\alpha + (-\beta))\\ &= \sin\alpha\cos(-\beta) + \sin(-\beta)\cos\alpha\\ &= \sin\alpha\cos\beta - \sin\beta\cos\alpha.\end{aligned}$$

Therefore, for all angles α and β,

$$\sin(\alpha - \beta) = \sin\alpha\cos\beta - \sin\beta\cos\alpha.$$

EXAMPLE 1 Find the exact value of $\sin\dfrac{5\pi}{12}$.

SOLUTION We must write $5\pi/12$ as the sum or difference of special angles and apply the appropriate formula. We may use the fact that

$$\frac{5\pi}{12}=\frac{2\pi}{12}+\frac{3\pi}{12}=\frac{\pi}{6}+\frac{\pi}{4} \quad \text{or} \quad \frac{5\pi}{12}=\frac{8\pi}{12}-\frac{3\pi}{12}=\frac{2\pi}{3}-\frac{\pi}{4}.$$

With the former we have

$$\begin{aligned}\sin\frac{5\pi}{12}&=\sin\left(\frac{\pi}{6}+\frac{\pi}{4}\right)\\&=\sin\frac{\pi}{6}\cos\frac{\pi}{4}+\sin\frac{\pi}{4}\cos\frac{\pi}{6}\\&=\frac{1}{2}\cdot\frac{1}{\sqrt{2}}+\frac{1}{\sqrt{2}}\cdot\frac{\sqrt{3}}{2}=\frac{1+\sqrt{3}}{2\sqrt{2}}\quad\left(\text{or }\frac{\sqrt{2}+\sqrt{6}}{4}\right).\end{aligned}$$

▲

▲ *Exercise* Find the exact value of $\sin\frac{11\pi}{12}$.

Answer $\frac{\sqrt{3}-1}{2\sqrt{2}}\quad\left(\text{or }\frac{\sqrt{6}-\sqrt{2}}{4}\right)$

EXAMPLE 2 Find the exact value of $\sin(\alpha-\beta)$ if $\sin\alpha=\frac{2}{5}$, where α is in quadrant I, and $\sin\beta=\frac{-1}{3}$, where β is in quadrant III.

SOLUTION We will use

$$\sin(\alpha-\beta)=\sin\alpha\cos\beta-\sin\beta\cos\alpha,$$

but we must first determine $\cos\alpha$ and $\cos\beta$. Using the Pythagorean identity for sine and cosine, we have

$$\begin{aligned}\cos\alpha&=+\sqrt{1-\sin^2\alpha}\\&=\sqrt{1-\frac{4}{25}}=\sqrt{\frac{21}{25}}=\frac{\sqrt{21}}{5}.\end{aligned}$$

Note that a plus sign precedes the radical because the cosine is positive in the first quadrant. In the third quadrant, however, the cosine is negative. Thus,

$$\begin{aligned}\cos\beta&=-\sqrt{1-\sin^2\beta}\\&=-\sqrt{1-\frac{1}{9}}\\&=-\sqrt{\frac{8}{9}}=\frac{-2\sqrt{2}}{3}.\end{aligned}$$

Therefore,

$$\begin{aligned}\sin(\alpha-\beta) &= \sin\alpha\cos\beta - \sin\beta\cos\alpha \\ &= \frac{2}{5}\left(\frac{-2\sqrt{2}}{3}\right) - \left(\frac{-1}{3}\right)\frac{\sqrt{21}}{5} \\ &= \frac{-4\sqrt{2}}{15} + \frac{\sqrt{21}}{15} = \frac{-4\sqrt{2}+\sqrt{21}}{15}.\end{aligned}$$

▲

▲ *Exercise* Find the exact value of $\sin(\alpha+\beta)$ if $\sin\alpha = \frac{1}{5}$, where α is in quadrant I, and $\sin\beta = \frac{-1}{4}$, where β is in quadrant III.

Answer $\dfrac{-\sqrt{15}-2\sqrt{6}}{20}$

The formula for the tangent of a sum is

$$\tan(\alpha+\beta) = \frac{\tan\alpha + \tan\beta}{1 - \tan\alpha\tan\beta}.$$

We may verify this by first noting that

$$\begin{aligned}\tan(\alpha+\beta) &= \frac{\sin(\alpha+\beta)}{\cos(\alpha+\beta)} \\ &= \frac{\sin\alpha\cos\beta + \sin\beta\cos\alpha}{\cos\alpha\cos\beta - \sin\alpha\sin\beta}.\end{aligned}$$

Dividing both the numerator and denominator by $\cos\alpha\cos\beta$ gives

$$\begin{aligned}\tan(\alpha+\beta) &= \frac{\dfrac{\sin\alpha\cos\beta}{\cos\alpha\cos\beta} + \dfrac{\sin\beta\cos\alpha}{\cos\alpha\cos\beta}}{\dfrac{\cos\alpha\cos\beta}{\cos\alpha\cos\beta} - \dfrac{\sin\alpha\sin\beta}{\cos\alpha\cos\beta}} \\ &= \frac{\dfrac{\sin\alpha}{\cos\alpha} + \dfrac{\sin\beta}{\cos\beta}}{1 - \dfrac{\sin\alpha}{\cos\alpha}\cdot\dfrac{\sin\beta}{\cos\beta}} \\ &= \frac{\tan\alpha + \tan\beta}{1 - \tan\alpha\tan\beta}.\end{aligned}$$

The formula for the tangent of a difference follows.

$$\tan(\alpha - \beta) = \frac{\tan\alpha - \tan\beta}{1 + \tan\alpha\tan\beta}$$

EXAMPLE 3 Write $\dfrac{\tan 4x + \tan x}{1 - \tan 4x \tan x}$ as a single term.

SOLUTION This expression is the right-hand side of the formula for the tangent of the sum of α and β with $\alpha = 4x$ and $\beta = x$. Thus

$$\frac{\tan 4x + \tan x}{1 - \tan 4x \tan x} = \tan(4x + x) = \tan 5x.$$

▲

▲ ***Exercise*** Write $\dfrac{\tan 5x - \tan 2x}{1 + \tan 5x \tan 2x}$ as a single term.

Answer $\tan 3x$

Problem Set 6.3

In Problems 1–10, simplify the expression.

1. $\sin\left(\theta - \dfrac{\pi}{2}\right)$

2. $\sin(\theta + \pi)$

3. $\sin(\theta - 180°)$

4. $\sin(\theta + 90°)$

5. $\tan(\theta - \pi)$

6. $\tan\left(\theta + \dfrac{3\pi}{4}\right)$

7. $\tan(\theta + 45°)$

8. $\tan(\theta - 135°)$

9. $\cot(\pi - \theta)$

10. $\csc(\theta - 180°)$

In Problems 11–18, find the *exact* value.

11. $\sin\dfrac{\pi}{12}$

12. $\sin 255°$

13. $\tan 105°$

14. $\tan\dfrac{\pi}{12}$

15. $\csc 105°$

16. $\cot\dfrac{5\pi}{12}$

17. $\sin\left(\text{Arctan}\sqrt{3} - \text{Arccos}\dfrac{1}{4}\right)$

18. $\tan\left(\text{Arctan } 2 + \text{Arcsin}\dfrac{1}{3}\right)$

In Problems 19–22, write the expression as a single term.

19. $\sin \frac{\pi}{5} \cos \frac{3\pi}{5} - \sin \frac{3\pi}{5} \cos \frac{\pi}{5}$

20. $\sin 3x \cos x + \sin x \cos 3x$

21. $\dfrac{\tan 4x + \tan 3x}{1 - \tan 4x \tan 3x}$

22. $\dfrac{\tan 2 - \tan 3}{1 + \tan 2 \tan 3}$

In Problems 23 and 24, find the exact value of $\sin(\alpha - \beta)$ for the given conditions.

23. $\sin \alpha = \frac{1}{6}$, where α is in quadrant I, and $\sin \beta = \frac{2}{5}$, where β is in quadrant II

24. $\sin \alpha = \frac{-4}{5}$, where α is in quadrant III, and $\sin \beta = \frac{-1}{3}$, where β is in quadrant IV

In Problems 25 and 26, find the exact value of $\sin(\alpha + \beta)$ for the given conditions.

25. $\sin \alpha = \frac{-2}{3}$, where α is in quadrant III, and $\cos \beta = \frac{1}{4}$, where β is in quadrant IV

26. $\cos \alpha = \frac{1}{3}$, where α is in quadrant I, and $\cos \beta = \frac{-1}{6}$, where β is in quadrant II

In Problems 27 and 28, find the exact values of $\tan(\alpha + \beta)$ and $\tan(\alpha - \beta)$ for the given conditions.

27. $\tan \alpha = 3$ and $\tan \beta = \frac{1}{4}$

28. $\tan \alpha = -4$ and $\tan \beta = \frac{2}{3}$

29. Find $\tan(\alpha + \beta)$, exactly, if $\sin \alpha = \frac{1}{3}$, where α is in quadrant II, $\cos \beta = \frac{1}{4}$, where β is in quadrant I.

30. Find $\tan(\alpha - \beta)$, exactly, if $\sin \alpha = \frac{-2}{3}$, where α is in quadrant III, and $\cos \beta = \frac{3}{4}$, where β is in quadrant IV.

In Problems 31–33, verify the identity.

31. $\sin(x + y) + \sin(x - y) = 2 \sin x \cos y$

32. $\dfrac{\sin(x + y) + \sin(x - y)}{\cos(x + y) + \cos(x - y)} = \tan x$

33. $\dfrac{\sin(x + y)}{\sin(x - y)} = \dfrac{\cot y + \cot x}{\cot y - \cot x}$

In Problems 34–36, find the exact solutions, where $0 \leq x < 2\pi$.

34. $\sin 2x \cos x + \sin x \cos 2x = -1$

35. $\sin x \cos 2x - \sin 2x \cos x = \dfrac{1}{\sqrt{2}}$

36. $\dfrac{\tan 3x - \tan x}{1 + \tan 3x \tan x} = 1$

6.4

DOUBLE-ANGLE AND HALF-ANGLE IDENTITIES

In this section we develop formulas for writing the trigonometric ratios of double angles and half angles in terms of the single angle. We begin with the formula for the cosine of a double angle. For every angle θ,

$$\cos 2\theta = \cos^2 \theta - \sin^2 \theta.$$

This formula follows immediately from the formula for the cosine of the sum of two angles. We have

$$\begin{aligned} \cos 2\theta &= \cos(\theta + \theta) \\ &= \cos \theta \cos \theta - \sin \theta \sin \theta \\ &= \cos^2 \theta - \sin^2 \theta. \end{aligned}$$

Since $\cos^2 \theta = 1 - \sin^2 \theta$ and $\sin^2 \theta = 1 - \cos^2 \theta$, two alternate forms of the formula for the cosine of a double angle are

$$\cos 2\theta = 1 - 2 \sin^2 \theta \quad \text{and} \quad \cos 2\theta = 2 \cos^2 \theta - 1.$$

The formula for the sine of double angles and the one for the tangent of double angles are based on the formulas for the sum of two angles.

$$\sin 2\theta = 2 \sin \theta \cos \theta,$$

$$\tan 2\theta = \frac{2 \tan \theta}{1 - \tan^2 \theta}$$

EXAMPLE 1 Find the exact values of $\sin 2\theta$, $\cos 2\theta$, and $\tan 2\theta$ if θ is in quadrant II and $\tan \theta = \frac{-3}{4}$

SOLUTION We use the double-angle formulas. Because $\sin 2\theta = 2 \sin \theta \cos \theta$, we first need to determine $\cos \theta$ and $\sin \theta$. Let's choose $(-4, 3)$ on the terminal side of θ as shown in Figure 6.3.

Figure 6.3

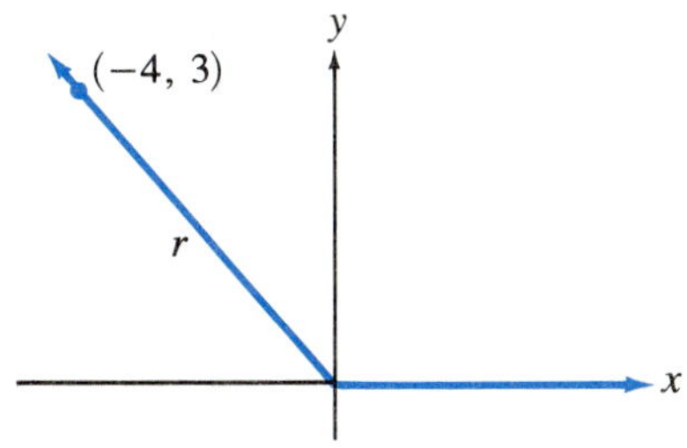

Thus $r = 5$. By definition then,

$$\sin \theta = \frac{3}{5} \quad \text{and} \quad \cos \theta = \frac{-4}{5}.$$

Now, using the double-angle formulas, we get

$$\begin{aligned} \sin 2\theta &= 2 \sin \theta \cos \theta \\ &= 2\left(\frac{3}{5}\right)\left(\frac{-4}{5}\right) = \frac{-24}{25}, \end{aligned}$$

$$\begin{aligned} \cos 2\theta &= \cos^2 \theta - \sin^2 \theta \\ &= \left(\frac{-4}{5}\right)^2 - \left(\frac{3}{5}\right)^2 = \frac{7}{25}, \end{aligned}$$

$$\begin{aligned} \tan 2\theta &= \frac{2 \tan \theta}{1 - \tan^2 \theta} \\ &= \frac{2\left(\frac{-3}{4}\right)}{1 - \left(\frac{-3}{4}\right)^2} = \frac{\frac{-3}{2}}{1 - \frac{9}{16}} \\ &= \frac{-3}{2} \cdot \frac{16}{7} = \frac{-24}{7}. \end{aligned}$$

▲

Of course, we could have used a basic identity to determine $\tan 2\theta$.

$$\begin{aligned} \tan 2\theta &= \frac{\sin 2\theta}{\cos 2\theta} \\ &= \frac{-24}{25} \div \frac{7}{25} = \frac{-24}{7}. \end{aligned}$$

▲ *Exercise* Find the exact values of $\sin 2\theta$, $\cos 2\theta$, and $\tan 2\theta$ if θ is in quadrant IV and $\tan \theta = \frac{-12}{5}$.

Answer $\sin 2\theta = \dfrac{-120}{169}$, $\cos 2\theta = \dfrac{-119}{169}$, $\tan 2\theta = \dfrac{120}{119}$

EXAMPLE 2 Use double-angle formulas to simplify the expression.

(a) $\cos^2 2x - \sin^2 2x$ (b) $2 \sin \dfrac{x}{4} \cos \dfrac{x}{4}$ (c) $\dfrac{2 \tan 1.5}{1 - \tan^2 1.5}$

SOLUTION We use the double-angle formula in reverse, doubling the given angles.

(a) $\cos^2 2x - \sin^2 2x = \cos(2(2x)) = \cos 4x$

(b) $2 \sin \dfrac{x}{4} \cos \dfrac{x}{4} = \sin\left(2\left(\dfrac{x}{4}\right)\right) = \sin \dfrac{x}{2}$

(c) $\dfrac{2 \tan 1.5}{1 - \tan^2 1.5} = \tan(2(1.5)) = \tan 3.0$

▲

▲ *Exercise* Use double-angle formulas to simplify the expression.

(a) $\cos^2 \dfrac{x}{2} - \sin^2 \dfrac{x}{2}$ (b) $2 \sin 3x \cos 3x$ (c) $\dfrac{2 \tan 2.1}{1 - \tan^2 2.1}$

Answer (a) $\cos x$ (b) $\sin 6x$ (c) $\tan 4.2$

EXAMPLE 3 Express $\sin 3\theta$ in terms of $\sin \theta$.

SOLUTION Writing 3θ as $2\theta + \theta$, we first use the formula for the sine of a sum, and then use the double-angle formulas. We have

$$\begin{aligned}
\sin 3\theta &= \sin(2\theta + \theta) \\
&= \sin 2\theta \cos \theta + \sin \theta \cos 2\theta \\
&= (2 \sin \theta \cos \theta) \cos \theta + \sin \theta (1 - 2 \sin^2 \theta) \\
&= 2 \sin \theta \cos^2 \theta + \sin \theta - 2 \sin^3 \theta \\
&= 2 \sin \theta (1 - \sin^2 \theta) + \sin \theta - 2 \sin^3 \theta \\
&= 3 \sin \theta - 4 \sin^3 \theta.
\end{aligned}$$

▲

▲ *Exercise* Express $\cos 3\theta$ in terms of $\cos \theta$.

Answer $4 \cos^3 \theta - 3 \cos \theta$

The next three identities establish formulas for half angles.

$$\cos\frac{\theta}{2} = \pm\sqrt{\frac{1+\cos\theta}{2}},$$

$$\sin\frac{\theta}{2} = \pm\sqrt{\frac{1-\cos\theta}{2}},$$

$$\tan\frac{\theta}{2} = \frac{\sin\theta}{1+\cos\theta}, \quad \text{also} \quad \tan\frac{\theta}{2} = \frac{1-\cos\theta}{\sin\theta}$$

Note the appearance of the $\pm$ in the formulas for the sine and cosine of a half angle. The particular sign, $+$ or $-$, depends on the quadrant in which $\theta/2$ lies.

The first identity may be established by using the double-angle formula. We have

$$\cos\theta = \cos\left(2\left(\frac{\theta}{2}\right)\right) = 2\cos^2\frac{\theta}{2} - 1.$$

Solving for $\cos^2(\theta/2)$, we get

$$\cos^2\frac{\theta}{2} = \frac{1+\cos\theta}{2}.$$

Thus,

$$\cos\frac{\theta}{2} = \pm\sqrt{\frac{1+\cos\theta}{2}}.$$

The second identity may be established in a similar fashion. We may prove the third identity as follows:

$$\tan\frac{\theta}{2} = \frac{\sin\frac{\theta}{2}}{\cos\frac{\theta}{2}} \cdot \frac{2\cos\frac{\theta}{2}}{2\cos\frac{\theta}{2}} = \frac{2\sin\frac{\theta}{2}\cos\frac{\theta}{2}}{2\cos^2\frac{\theta}{2}}$$

$$= \frac{\sin\left(2\left(\frac{\theta}{2}\right)\right)}{1-\sin^2\frac{\theta}{2}+\cos^2\frac{\theta}{2}}$$

$$= \frac{\sin\theta}{1+\cos\left(2\left(\frac{\theta}{2}\right)\right)}$$

$$= \frac{\sin}{1+\cos\theta}.$$

The alternative form for the tangent may be established from this:

$$\tan\frac{\theta}{2} = \frac{\sin\theta}{1+\cos\theta}\cdot\frac{1-\cos\theta}{1-\cos\theta}$$

$$= \frac{\sin\theta(1-\cos\theta)}{1-\cos^2\theta}$$

$$= \frac{\sin\theta(1-\cos\theta)}{\sin^2\theta} = \frac{1-\cos\theta}{\sin\theta}.$$

EXAMPLE 4 Find the exact values of $\sin\frac{\theta}{2}$, $\cos\frac{\theta}{2}$, and $\tan\frac{\theta}{2}$ if $\pi < \theta < \frac{3\pi}{2}$ and $\tan\theta = \frac{3}{4}$.

SOLUTION We'll use the half-angle formulas, but we will first determine $\cos\theta$ and $\sin\theta$. Choosing $(-4, -3)$ on the terminal side of θ, we have $r = 5$, as shown in Figure 6.4.

Figure 6.4

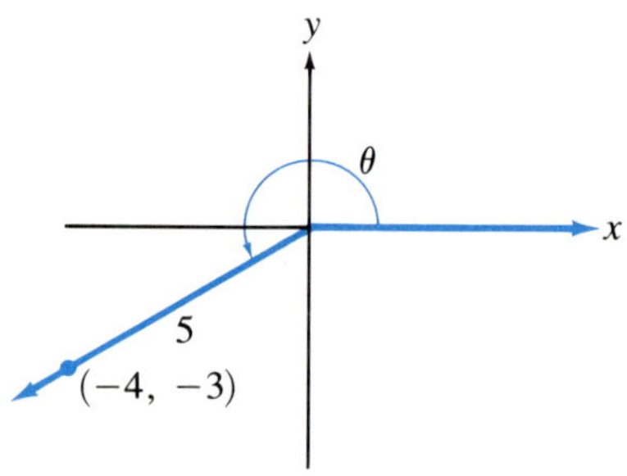

By definition, then,

$$\cos\theta = \frac{-4}{5} \quad\text{and}\quad \sin\theta = \frac{-3}{5}.$$

Since

$$\pi < \theta < \frac{3\pi}{2},$$

$$\frac{\pi}{2} < \frac{\theta}{2} < \frac{3\pi}{4}.$$

Thus $\theta/2$ is in the second quadrant. Now, using the half-angle formulas, we get

$$\sin\frac{\theta}{2} = \pm\sqrt{\frac{1-\cos\theta}{2}}$$

$$= +\sqrt{\frac{1+\frac{4}{5}}{2}} \qquad \text{(plus because } \theta/2 \text{ is in quadrant II)}$$

$$= \frac{3}{\sqrt{10}} \quad \left(\text{or } \frac{3\sqrt{10}}{10}\right),$$

$$\cos\frac{\theta}{2} = \pm\sqrt{\frac{1+\cos\theta}{2}}$$

$$= -\sqrt{\frac{1-\frac{4}{5}}{2}} \qquad \text{(minus because } \theta/2 \text{ is in quadrant II)}$$

$$= \frac{-1}{\sqrt{10}} \quad \left(\text{or } \frac{-\sqrt{10}}{10}\right),$$

$$\tan\frac{\theta}{2} = \frac{\sin\theta}{1+\cos\theta}$$

$$= \frac{\frac{-3}{5}}{1-\frac{4}{5}} = -3.$$

▲

Note in Example 4 that the tangent could have been calculated by dividing $\sin(\theta/2)$ by $\cos(\theta/2)$. This would have avoided the formula for the tangent of a half angle and would have made the determination of $\sin\theta$ unnecessary.

Also note that knowing in which quadrant θ lies does not, by itself, tell us in which quadrant $\theta/2$ lies. For example, suppose we know that θ is in qudarant III. We may have

$$\pi < \theta < \frac{3\pi}{2},$$

which would mean that $\theta/2$ is a second-quadrant angle since

$$\frac{\pi}{2} < \frac{\theta}{2} < \frac{3\pi}{4}.$$

On the other hand, we may have

$$-\pi < \theta < \frac{-\pi}{2},$$

which would mean that $\theta/2$ is a fourth-quadrant angle since

$$\frac{-\pi}{2} < \frac{\theta}{2} < \frac{-\pi}{4}.$$

Thus, if θ is in the third quadrant, we can conclude only that $\theta/2$ is in the second or fourth.

▲ ***Exercise*** Find the exact values of $\sin\frac{\theta}{2}$, $\cos\frac{\theta}{2}$, and $\tan\frac{\theta}{2}$ if $\pi < \theta < \frac{3\pi}{2}$ and $\tan\theta = \frac{5}{12}$.

Answer $\sin\frac{\theta}{2} = \frac{5}{\sqrt{26}}\left(\text{or } \frac{5\sqrt{26}}{26}\right)$, $\cos\frac{\theta}{2} = \frac{-1}{\sqrt{26}}\left(\text{or } \frac{-\sqrt{26}}{26}\right)$, $\tan\frac{\theta}{2} = -5$

EXAMPLE 5 Find the exact solutions of $\cos 2x + \sin x - 1 = 0$, where $0 \leq x < 2\pi$.

SOLUTION We first note that the arguments are different. There is $2x$ and x. We need identical arguments, so let's substitute for $\cos 2x$ by using the double-angle formula. There are three forms of this identity, but since we want the ratios to match as well, we use $\cos 2x = 1 - 2\sin^2 x$. This gives

$$1 - 2\sin^2 x + \sin x - 1 = 0$$

$$-2\sin^2 x + \sin x = 0$$

$$\sin x(-2\sin x + 1) = 0$$

$$\sin x = 0 \quad \text{or} \quad \sin x = \frac{1}{2}$$

Hence the solutions are 0, π, $\pi/6$, and $5\pi/6$.

▲

▲ ***Exercise*** Find the exact solutions of $\sin 2x + \cos x = 0$, where $0 \leq x < 2\pi$.

Answer $\frac{\pi}{2}, \frac{3\pi}{2}, \frac{7\pi}{6}, \frac{11\pi}{6}$

Problem Set 6.4

In Problems 1–4, find the exact values of $\sin 2\theta$, $\cos 2\theta$, and $\tan 2\theta$ for the given conditions.

1. $\sin\theta = \frac{-4}{5}$, θ in the third quadrant

2. $\tan\theta = \frac{-3}{4}$, θ in the fourth quadrant

3. $\cos\theta = \frac{-5}{13}$, θ in the second quadrant

4. $\sec\theta = \sqrt{10}$, θ in the fourth quadrant

5. Find $\tan 2\theta$, exactly, if $\tan\theta = \sqrt{5}$.

6. Find $\tan 2\theta$, exactly, if $\cot\theta = \frac{-1}{3}$.

In Problems 7–12, simplify the expression.

7. $\cos^2 \frac{x}{3} - \sin^2 \frac{x}{3}$

8. $\sin^2 x - \cos^2 x$

9. $2 \sin 1.6 \cos 1.6$

10. $\sin 2x \cos 2x$

11. $\dfrac{2 \tan 4x}{1 - \tan^2 4x}$

12. $\dfrac{2 \tan 40^\circ}{1 - \tan^2 40^\circ}$

13. Express $\cos 4x$ in terms of $\cos x$.

14. Express $\cos 4x$ in terms of $\sin x$.

In Problems 15–18, find the exact values of $\sin(\theta/2)$, $\cos(\theta/2)$, and $\tan(\theta/2)$ for the given conditions.

15. $\cot \theta = \dfrac{-12}{5}$ and $\dfrac{\pi}{2} < \theta < \pi$

16. $\sec \theta = \dfrac{-4}{3}$ and $\dfrac{-3\pi}{2} < \theta < -\pi$

17. $\sin \theta = \dfrac{-3}{5}$, θ in the third quadrant

18. $\cos \theta = \dfrac{1}{3}$ and $\pi < \theta < 2\pi$

19. Use the half-angle formulas to find the exact values of $\sin(\pi/8)$, $\cos(\pi/8)$, and $\tan(\pi/8)$.

20. Use the half-angle formulas to find the exact values of $\sin(\pi/12)$, $\cos(\pi/12)$, and $\tan(\pi/12)$.

21. Find the exact value of $\tan\left(\dfrac{\text{Arcsin } \frac{4}{5}}{2}\right)$.

22. Find the exact value of $\sin\left(2 \text{ Arccos } \frac{2}{5}\right)$.

In Problems 23–28, verify the identity.

23. $(1 + \tan x) \tan 2x = \dfrac{2 \tan x}{1 - \tan x}$

24. $\dfrac{2 \sin 2x}{\sin 4x} = \sec 2x$

25. $\sin^2 \theta = \frac{1}{2}(1 - \cos 2\theta)$

26. $\cos^2 \theta = \frac{1}{2}(1 + \cos 2\theta)$

27. $\left(\sin \frac{\theta}{2} + \cos \frac{\theta}{2}\right)^2 = 1 + \sin \theta$

28. $\dfrac{7 + \cos 2x}{1 + \cos 2x} - 1 = 3 \sec^2 x$

In Problems 29–38, find the exact solutions, where $0 \le x < 2\pi$.

29. $\sin 2x = 3 \cos x$

30. $\cos 2x + \cos x = 0$

31. $\cos 2x + \cos^2 x = \sin^2 x$

32. $\sin 4x = \sin 2x$

33. $\cos x = \sin \dfrac{x}{2}$

34. $\cos x + \cos \dfrac{x}{2} = 0$

35. $\dfrac{\sin x}{1 + \cos x} = \sqrt{3}$

36. $\dfrac{1 - \cos x}{\sin x} = -1$

37. $\dfrac{1}{\csc x + \cot x} = 1$

38. $\dfrac{1 - \cos 2x}{2 \sin x + \sin 2x} = -\sqrt{3}$

In Problems 39–46, find the approximate solutions to the nearest thousandth, where $0 \le x < 2\pi$.

39. $\dfrac{2 - 4\cos^2 x}{\sin 2x} + \cot x = 3$

40. $\dfrac{2 \cos 2x}{2\cos^2 x - \sin 2x} = -1$

41. $2 \sin x - \sin \dfrac{x}{2} = 0$

42. $2 \sin x \sin \dfrac{x}{2} = 1 - \cos x$

43. $2 \sin x + \cos x = 1$

44. $\sin x - 5 \cos x = 5$

45. $\cos x - \cos 3x = 0$

46. $\sin 2x - \sin 3x = 0$

(Refer to Figure 6.5.) Suppose a camera is focused for a specified distance u. If a point Q is closer to the camera, its projected image lies behind the film at Q'. Consequently, Q's image on the film is a circle rather than a point. If we let d be the diameter of this circle, v the distance from lens to film, and θ the measure of the indicated angle, then

$$\tan \frac{\theta}{2} = \frac{d/2}{v}. \tag{6.7}$$

Points not too far from the focus plane are roughly in focus. The distance between the nearest such points (to the camera) and the farthest such points is called the depth of field. Suppose Q is at the near limit of the depth of field. Then the distance from the focus plane to Q is given by

$$w = \frac{u^2 \tan \theta}{\dfrac{F}{f} + u \tan \theta}, \tag{6.8}$$

where F is the focal length of the lens and f is the f-stop. This information is needed in Problems 47 and 48.

Figure 6.5

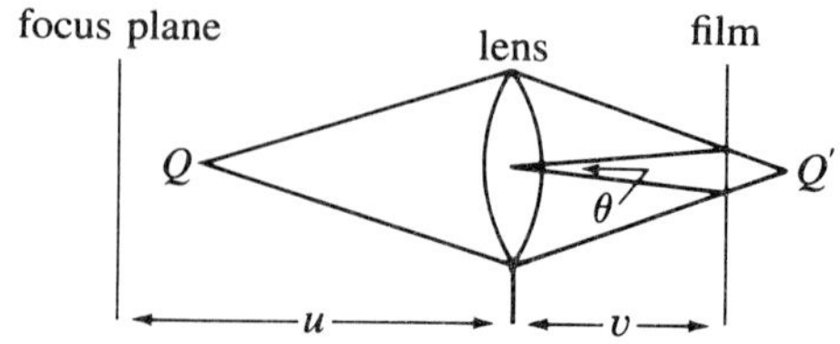

47. Use the double-angle formula for tangent and the fact that for small x, $\tan^2 x \approx 0$, to show that formula (6.7) may be approximated with

$$\tan\theta = \frac{d}{v}.$$

48. Use formula (6.8) to compute w (to three significant digits) if

$F = 50$ millimeters, f-stop $= 8$, $u = 5$ meters, and $\theta = (\frac{1}{30})^\circ$.

6.5

PRODUCT-TO-SUM AND SUM-TO-PRODUCT IDENTITIES

The identities studied in this section are needed for more advanced work in mathematics. The identities that enable us to convert products to sums are derived from the identities for the sine and cosine of sums and differences. For example, adding the two identities for $\sin(\alpha+\beta)$ and $\sin(\alpha-\beta)$ gives

$$\begin{aligned} \sin(\alpha+\beta) &= \sin\alpha\cos\beta + \sin\beta\cos\alpha \\ \sin(\alpha-\beta) &= \sin\alpha\cos\beta - \sin\beta\cos\alpha \\ \hline \sin(\alpha+\beta)+\sin(\alpha-\beta) &= 2\sin\alpha\cos\beta. \end{aligned}$$

Therefore,

$$\sin\alpha\cos\beta = \frac{1}{2}(\sin(\alpha+\beta)+\sin(\alpha-\beta)).$$

Using the identities for $\cos(\alpha+\beta)$ and $\cos(\alpha-\beta)$ in a similar fashion gives

$$\cos\alpha\cos\beta = \frac{1}{2}(\cos(\alpha+\beta)+\cos(\alpha-\beta)),$$

$$\sin\alpha\sin\beta = \frac{1}{2}(\cos(\alpha-\beta)-\cos(\alpha+\beta)).$$

EXAMPLE 1 Express the product as a sum or difference.

(a) $\sin 2\theta\cos\theta$ (b) $\cos x\cos 3x$ (c) $\sin^2 x$

SOLUTION (a) Using the identity for $\sin\alpha\cos\beta$ we get

$$\sin 2\theta \cos\theta = \frac{1}{2}(\sin(2\theta+\theta) + \sin(2\theta-\theta))$$

$$= \frac{1}{2}(\sin 3\theta + \sin\theta)$$

$$= \frac{\sin 3\theta}{2} + \frac{\sin\theta}{2}.$$

(b) Using the identity for $\cos\alpha\cos\beta$ gives

$$\cos x \cos 3x = \frac{1}{2}(\cos(x+3x) + \cos(x-3x))$$

$$= \frac{1}{2}(\cos 4x + \cos(-2x))$$

$$= \frac{1}{2}(\cos 4x + \cos 2x)$$

$$= \frac{\cos 4x}{2} + \frac{\cos 2x}{2}.$$

(c) We note that $\sin^2 x$ is $\sin x \sin x$ and use the identity for $\sin\alpha\sin\beta$. We have

$$\sin^2 x = \sin x \sin x$$

$$= \frac{1}{2}(\cos(x-x) - \cos(x+x))$$

$$= \frac{1}{2}(\cos 0 - \cos 2x)$$

$$= \frac{1}{2}(1 - \cos 2x)$$

$$= \frac{1}{2} - \frac{\cos 2x}{2}.$$

▲

▲ ***Exercise*** Express the product as a sum or difference.

(a) $\sin 3\theta \cos\theta$ (b) $\cos^2 x$ (c) $\sin x \sin 2x$

Answer (a) $\dfrac{\sin 4\theta}{2} + \dfrac{\sin 2\theta}{2}$ (b) $\dfrac{\cos 2x}{2} + \dfrac{1}{2}$ (c) $\dfrac{\cos x}{2} - \dfrac{\cos 3x}{2}$

EXAMPLE 2 Use the product-to-sum identities to evaluate the expression, exactly.

(a) $\sin\dfrac{\pi}{12}\cos\dfrac{\pi}{4}$ (b) $\sin 135^\circ \sin 75^\circ$

SOLUTION (a) We have

$$\sin\frac{\pi}{12}\cos\frac{\pi}{4} = \frac{1}{2}\left(\sin\left(\frac{\pi}{12}+\frac{\pi}{4}\right)+\sin\left(\frac{\pi}{12}-\frac{\pi}{4}\right)\right)$$
$$= \frac{1}{2}\left(\sin\frac{\pi}{3}+\sin\left(\frac{-\pi}{6}\right)\right)$$
$$= \frac{1}{2}\left(\sin\frac{\pi}{3}-\sin\frac{\pi}{6}\right)$$
$$= \frac{1}{2}\left(\frac{\sqrt{3}}{2}-\frac{1}{2}\right) = \frac{\sqrt{3}-1}{4}.$$

(b) We get

$$\sin 135^\circ \sin 75^\circ = \frac{1}{2}(\cos(135^\circ - 75^\circ) - \cos(135^\circ + 75^\circ))$$
$$= \frac{1}{2}(\cos 60^\circ - \cos 210^\circ)$$
$$= \frac{1}{2}(\cos 60^\circ + \cos 30^\circ)$$
$$= \frac{1}{2}\left(\frac{1}{2}+\frac{\sqrt{3}}{2}\right) = \frac{1+\sqrt{3}}{4}.$$

▲

▲ ***Exercise*** Use the product-to-sum identities to evaluate the expression, exactly.

(a) $\sin 135^\circ \cos 75^\circ$ (b) $\cos\frac{\pi}{12}\cos\frac{\pi}{4}$

Answer (a) $\frac{\sqrt{3}-1}{4}$ (b) $\frac{1+\sqrt{3}}{4}$

The preceding product-to-sum formulas may be used to develop formulas for the reverse process: writing a sum of sines or cosines as a product. We have

$$\sin\left(\frac{x+y}{2}\right)\cos\left(\frac{x-y}{2}\right) = \frac{1}{2}\left(\sin\frac{x+y+x-y}{2} + \sin\frac{x+y-x+y}{2}\right)$$
$$= \frac{1}{2}(\sin x + \sin y).$$

Therefore,

$$\sin x + \sin y = 2\sin\left(\frac{x+y}{2}\right)\cos\left(\frac{x-y}{2}\right).$$

Note that this identity is for the sum of sines, *not* the sine of a sum—that is, *not* $\sin(x + y)$.

In a similar fashion we obtain three more identities:

$$\sin x - \sin y = 2\cos\left(\frac{x+y}{2}\right)\sin\left(\frac{x-y}{2}\right),$$

$$\cos x + \cos y = 2\cos\left(\frac{x+y}{2}\right)\cos\left(\frac{x-y}{2}\right),$$

$$\cos x - \cos y = -2\sin\left(\frac{x+y}{2}\right)\sin\left(\frac{x-y}{2}\right).$$

These four identities are known as the sum-to-product identities and are used in Examples 3 and 4.

EXAMPLE 3 Express the sum or difference as a product.

(a) $\sin 2\theta - \sin 4\theta$ (b) $\cos 3\theta + \cos\theta$

SOLUTION (a) Using the identity for $\sin x - \sin y$ we get

$$\begin{aligned}\sin 2\theta - \sin 4\theta &= 2\cos\left(\frac{2\theta+4\theta}{2}\right)\sin\left(\frac{2\theta-4\theta}{2}\right)\\ &= 2\cos 3\theta\sin(-\theta)\\ &= -2\cos 3\theta\sin\theta.\end{aligned}$$

(b) Using the identity for $\cos x + \cos y$ gives

$$\begin{aligned}\cos 3\theta + \cos\theta &= 2\cos\left(\frac{3\theta+\theta}{2}\right)\cos\left(\frac{3\theta-\theta}{2}\right)\\ &= 2\cos 2\theta\cos\theta.\end{aligned}$$

▲

▲ *Exercise* Express the sum or difference as a product.

(a) $\sin\theta + \sin 3\theta$ (b) $\cos 4\theta - \cos 2\theta$

Answer (a) $2\sin 2\theta\cos\theta$ (b) $-2\sin 3\theta\sin\theta$

EXAMPLE 4 Use the sum-to-product identities to evaluate $\sin\frac{5\pi}{12} + \sin\frac{\pi}{12}$, exactly.

SOLUTION We have

$$\sin\frac{5\pi}{12} + \sin\frac{\pi}{12} = 2\sin\left(\frac{\frac{5\pi}{12}+\frac{\pi}{12}}{2}\right)\cos\left(\frac{\frac{5\pi}{12}-\frac{\pi}{12}}{2}\right)$$

$$= 2\sin\frac{\pi}{4}\cos\frac{\pi}{6}$$

$$= 2\cdot\frac{1}{\sqrt{2}}\cdot\frac{\sqrt{3}}{2} = \sqrt{\frac{3}{2}}\ \left(\text{or } \frac{\sqrt{6}}{2}\right).$$

▲

▲ ***Exercise*** Use the sum-to-product identities to evaluate $\cos\frac{\pi}{12} + \cos\frac{5\pi}{12}$, exactly.

Answer $\sqrt{\frac{3}{2}}\left(\text{or } \frac{\sqrt{6}}{2}\right)$

Problem Set 6.5

In Problems 1–8, express the product as a sum or difference.

1. $\sin 4\theta \cos 3\theta$
2. $\sin 2\theta \cos 3\theta$
3. $\cos 2x \sin 3x$
4. $\cos 5x \sin 3x$
5. $\cos 2t \cos t$
6. $\cos 3t \cos 4t$
7. $\sin x \sin 3x$
8. $\sin 3x \sin 2x$

In Problems 9–12, use the product-to-sum formulas to evaluate, exactly.

9. $\sin 105^\circ \cos 165^\circ$
10. $\sin 225^\circ \cos 105^\circ$
11. $\sin\frac{3\pi}{8}\sin\frac{\pi}{8}$
12. $\cos\frac{\pi}{8}\cos\frac{7\pi}{8}$

In Problems 13–20, express the sum or difference as a product.

13. $\sin\theta + \sin 2\theta$
14. $\sin 2\theta + \sin 4\theta$
15. $\sin x - \sin 3x$
16. $\sin 2x - \sin x$
17. $\cos x + \cos 2x$
18. $\cos 2x + \cos 4x$
19. $\cos 5t - \cos 3t$
20. $\cos t - \cos 2t$

In Problems 21–24, use the sum-to-product formulas to evaluate, exactly.

21. $\sin \frac{7\pi}{12} + \sin \frac{13\pi}{12}$

22. $\sin 105^\circ - \sin 165^\circ$

23. $\cos 105^\circ - \cos 15^\circ$

24. $\cos \frac{13\pi}{12} + \cos \frac{5\pi}{12}$

In Problems 25–28, verify the identity.

25. $\frac{\sin 3x + \sin x}{\cos x + \cos 3x} = \tan 2x$

26. $\frac{\sin 5\theta + \sin 3\theta}{\cos 3\theta - \cos 5\theta} = \cot \theta$

27. $\sin^2 4x - \sin^2 2x = \sin 2x \sin 6x$

28. $\frac{\sin 9x - \sin 5x}{\sin 4x} = \frac{\cos 7x}{\cos 2x}$

In Problems 29–32, find the exact solutions, where $0 \leq x < 2\pi$.

29. $\sin 3x - \sin x = 0$

30. $\cos x = \cos 5x$

31. $\sin x + \sin 3x = \cos x$

32. $\cos 2x - \cos 4x = 2 \sin x$

CHAPTER 6 REVIEW

Key Terms and Formulas

identity, page 193

basic identities, page 194

$$\csc \theta = \frac{1}{\sin \theta} \qquad \tan \theta = \frac{\sin \theta}{\cos \theta} \qquad \sin^2 \theta + \cos^2 \theta = 1$$

$$\sec \theta = \frac{1}{\cos \theta} \qquad \cot \theta = \frac{\cos \theta}{\sin \theta} \qquad \sec^2 \theta = 1 + \tan^2 \theta$$

$$\cot \theta = \frac{1}{\tan \theta} \qquad \csc^2 \theta = 1 + \cot^2 \theta$$

negative-angle identities, page 207

$$\sin(-\theta) = -\sin \theta \qquad \cos(-\theta) = \cos \theta \qquad \tan(-\theta) = -\tan \theta$$

sum and difference identities, pages 204, 208, 211, 213, 214

$$\sin(\alpha + \beta) = \sin \alpha \cos \beta + \sin \beta \cos \alpha$$

$$\sin(\alpha - \beta) = \sin \alpha \cos \beta - \sin \beta \cos \alpha$$

$$\cos(\alpha + \beta) = \cos \alpha \cos \beta - \sin \alpha \sin \beta$$

$$\cos(\alpha - \beta) = \cos \alpha \cos \beta + \sin \alpha \sin \beta$$

$$\tan(\alpha + \beta) = \frac{\tan \alpha + \tan \beta}{1 - \tan \alpha \tan \beta} \qquad \tan(\alpha - \beta) = \frac{\tan \alpha - \tan \beta}{1 + \tan \alpha \tan \beta}$$

double-angle and half-angle identities, pages 216, 219

$$\sin 2\theta = 2 \sin \theta \cos \theta \qquad \sin \frac{\theta}{2} = \pm\sqrt{\frac{1 - \cos \theta}{2}}$$

$$\cos 2\theta = \cos^2 \theta - \sin^2 \theta \qquad \cos \frac{\theta}{2} = \pm\sqrt{\frac{1 + \cos \theta}{2}}$$

$$\tan 2\theta = \frac{2 \tan \theta}{1 - \tan^2 \theta} \qquad \tan \frac{\theta}{2} = \frac{\sin \theta}{1 + \cos \theta} = \frac{1 - \cos \theta}{\sin \theta}$$

product-to-sum identities, page 225

$$\sin \alpha \cos \beta = \frac{1}{2}(\sin(\alpha + \beta) + \sin(\alpha - \beta))$$

$$\cos \alpha \cos \beta = \frac{1}{2}(\cos(\alpha + \beta) + \cos(\alpha - \beta))$$

$$\sin \alpha \sin \beta = \frac{1}{2}(\cos(\alpha - \beta) - \cos(\alpha + \beta))$$

sum-to-product identities, pages 227, 228

$$\sin x + \sin y = 2 \sin\left(\frac{x + y}{2}\right)\cos\left(\frac{x - y}{2}\right)$$

$$\sin x - \sin y = 2 \cos\left(\frac{x + y}{2}\right)\sin\left(\frac{x - y}{2}\right)$$

$$\cos x + \cos y = 2 \cos\left(\frac{x + y}{2}\right)\cos\left(\frac{x - y}{2}\right)$$

$$\cos x - \cos y = -2 \sin\left(\frac{x + y}{2}\right)\sin\left(\frac{x - y}{2}\right)$$

Problem Set

1. Specify "true" or "false."

(a) If an equation has infinitely many solutions, then it is an identity.

(b) If any two angles add up to 90°, then the sine of each equals the cosine of the other.

(c) $\csc(-x)\sec(-x) = \csc x \sec x$

(d) $\cot(-\theta) = -\cot \theta$

(e) $\sin(-x) = -\sin\left(\frac{\pi}{2} - x\right)$

In Problems 2–4, use basic identities to find the exact values of the other trigonometric ratios for the given conditions.

2. $\cos \theta = \frac{-5}{8}$, θ in the second quadrant

3. $\csc \theta = \frac{3}{2}$, θ in the first quadrant

4. $\tan\theta = \frac{3}{5}$, θ in the third quadrant

In Problems 5 and 6, use basic identities to reduce the expression to 1.

5. $\dfrac{\cot x \cos x}{1 - \sin x} - \csc x$

6. $\dfrac{\sin^3 x + \cos^3 x}{\sin x + \cos x} + \sin x \cos x$

In Problems 7 and 8, use basic identities to reduce the expression to a single trigonometric function.

7. $\dfrac{1 + \sec x}{\sin x + \tan x}$

8. $\dfrac{1}{\sec x - \tan x} - \sec x$

In Problems 9–14, verify the identity.

9. $\dfrac{\csc^4 x - 1}{\cot^2 x} = 2 + \cot^2 x$

10. $\sec^2\theta + \csc^2\theta = \sec^2\theta\csc^2\theta$

11. $\dfrac{1 + \tan x \sin^2 x - \tan x}{\sin x} = \csc x - \cos x$

12. $\dfrac{1 - \tan^4 x}{\sec^2 x(\sin x + \cos x)} = \sec x(1 - \tan x)$

13. $\dfrac{1}{\tan\theta + \cot\theta} = \sin\theta\cos\theta$

14. $\sec^2 x - \csc^2 x = \dfrac{\tan x - \cot x}{\sin x \cos x}$

In Problems 15–18, simplify the expression.

15. $\cos\left(\theta + \dfrac{3\pi}{2}\right)$

16. $\sin(\pi - \theta)$

17. $\tan(\theta + \pi)$

18. $\sec(\theta - 90^\circ)$

In Problems 19–22, find the *exact* value.

√*19.* $\tan\dfrac{5\pi}{12}$

20. $\cos 165^\circ$

21. $\sin\left(\dfrac{-\pi}{12}\right)$

22. $\cos\left(\text{Arcsin}\,\dfrac{1}{5} - \text{Arctan}\,(-1)\right)$

In Problems 23–25, write the expression as a single term.

23. $\sin x \cos 3x - \sin 3x \cos x$

24. $\cos 40^\circ \cos 10^\circ - \sin 40^\circ \sin 10^\circ$

25. $\dfrac{\tan 2x + \tan x}{1 - \tan 2x \tan x}$

26. Express $\tan 3\theta$ in terms of $\tan \theta$.

In Problems 27–32, answers are to be exact.

27. Find $\cos(\alpha - \beta)$ if $\cos \alpha = \dfrac{4}{5}$, where α is in quadrant I, and $\cos \beta = \dfrac{-1}{7}$, where β is in quadrant III.

28. Find $\cos(\alpha + \beta)$ if $\sin \alpha = \dfrac{2}{3}$, where α is in quadrant II, and $\cos \beta = \dfrac{-1}{6}$, where β is in quadrant II.

29. Find $\sin(\alpha - \beta)$ if $\sin \alpha = \dfrac{1}{6}$, where α is in quadrant I, and $\sin \beta = \dfrac{-1}{5}$, where β is in quadrant IV.

30. Find $\sin(\alpha + \beta)$ if $\cos \alpha = \dfrac{1}{3}$, where α is in quadrant I, and $\cos \beta = \dfrac{-1}{4}$, where β is in quadrant II.

31. Find $\tan(\alpha + \beta)$ if $\tan \alpha = 6$ and $\tan \beta = \dfrac{-1}{4}$.

32. Find $\tan(\alpha - \beta)$ if $\sin \alpha = \dfrac{-1}{3}$, where α is in quadrant III, and $\cos \beta = \dfrac{2}{3}$, where β is in quadrant IV.

In Problems 33 and 34, find the exact values of $\sin 2\theta$, $\cos 2\theta$, and $\tan 2\theta$ for the given conditions.

33. $\tan \theta = 2$, θ in the third quadrant

34. $\csc \theta = \dfrac{\sqrt{6}}{2}$, θ in the second quadrant

In Problems 35 and 36, find the exact values of $\sin(\theta/2)$, $\cos(\theta/2)$, and $\tan(\theta/2)$ for the given conditions.

35. $\cos \theta = \dfrac{-5}{13}$ and $-\pi < \theta < \dfrac{-\pi}{2}$

36. $\sec \theta = \dfrac{-7}{2}$ and $\dfrac{\pi}{2} < \theta < \pi$

In Problems 37 and 38, simplify the expression.

37. $\cos^2 3x - \sin^2 3x$

38. $2 \sin \dfrac{x}{5} \cos \dfrac{x}{5}$

In Problems 39–44, verify the identity.

39. $\cos^4\theta - \sin^4\theta = \cos 2\theta$

40. $\dfrac{\cos(x-y)}{\cos(x+y)} = \dfrac{1+\tan x\tan y}{1-\tan x\tan y}$

41. $\sin(x+y) - \sin(x-y) = 2\cos x\sin y$

42. $\cot 2\theta + \csc 2\theta = \cot\theta$

43. $\dfrac{2}{\tan\dfrac{x}{2} + \cot\dfrac{x}{2}} = \sin x$

44. $\sin^3\theta = \dfrac{1}{4}(3\sin\theta - \sin 3\theta)$

In Problems 45–48, express the product as a sum or difference.

45. $\sin 4x\cos 5x$

46. $\cos 4\theta\sin 3\theta$

47. $\sin x\sin 4x$

48. $\cos 3x\cos 2x$

In Problems 49–52, express the sum or difference as a product.

49. $\sin 2x + \sin 3x$

50. $\sin 3x - \sin 5x$

51. $\cos\theta - \cos 3\theta$

52. $\cos 5x + \cos x$

In Problems 53–55, use the product-to-sum and sum-to-product identities to evaluate the expression, exactly.

53. $\cos 75^\circ\cos 135^\circ$

54. $\sin\dfrac{\pi}{12}\sin\dfrac{\pi}{4}$

55. $\sin 75^\circ - \sin 15^\circ$

56. Verify the identity $\dfrac{\sin 2x + \sin 2y}{\cos 2x + \cos 2y} = \tan(x+y)$.

In Problems 57–64, find the exact solutions, where $0 \le x < 2\pi$.

57. $\cos 7x\cos 5x + \sin 7x\sin 5x = 0.81$

58. $\sin 3x\cos x - \sin x\cos 3x = 1$

59. $\cos x\cos 2x + \sin x\sin 2x = 1$

60. $\cos x + 2\sin x\tan x = 1$

61. $\cos x - 2\sin^2\dfrac{x}{2} = 0$

62. $3\tan^2 x + 4\sec x - 1 = 0$

63. $\sin 3x + \sin x = \sin 2x$

64. $\cos x - \cos 3x = \sin x$

In Problems 65–70, find the approximate solutions to the nearest thousandth, where $0 \le x < 2\pi$.

65. $\dfrac{\tan x + 3 \sin x}{\sec x + 3} = \dfrac{1}{5}$

66. $\dfrac{\cot x - 1}{1 - \tan x} = 3$

67. $\dfrac{\sin x}{1 + \cos x} + \dfrac{1 + \cos x}{\sin x} = 8$

68. $3 - 3(\sin x - \cos x)^2 = 1$

69. $\dfrac{\cot(-x) + \tan(-x)}{\csc(-x)} = 3$

70. $1 + \cos x = 2 \sin x$

7
OBLIQUE TRIANGLES AND VECTORS

CHAPTER OUTLINE

In Section 2.3 we studied the solution of right triangles. In the first two sections of this chapter we study two key formulas that enable us to solve the oblique triangle, a triangle that has no right angle. Two formulas for area are also introduced. In the last two sections we study quantities that have direction as well as magnitude.

7.1 THE LAW OF SINES

In this section and the next, we shall use a standard notation. Given triangle ABC, α, β, and γ (gamma) will represent the measures of angles A, B, and C, respectively, while a, b, and c will denote the lengths of the opposite sides. See Figure 7.1.

Figure 7.1

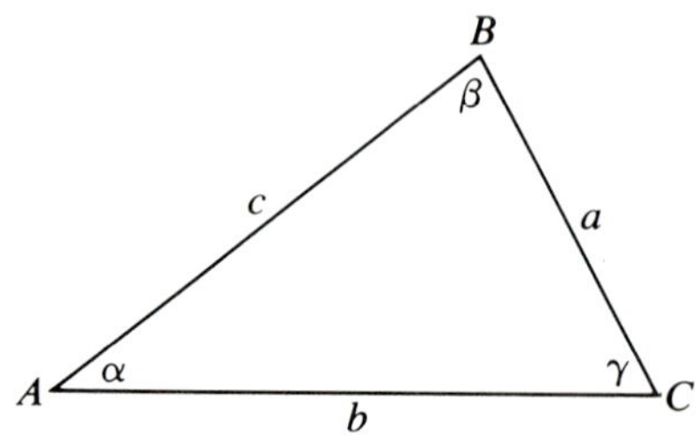

We begin by considering the area of a triangle. If K represents the area, then

$$K = \frac{1}{2} bh,$$

where b is the length of the base and h is the height. In practice, however, the height of the triangle is frequently unknown. The next formula addresses such circumstances.

For triangle ABC, the area K is given by

$$K = \frac{1}{2} bc \sin \alpha = \frac{1}{2} ac \sin \beta = \frac{1}{2} ab \sin \gamma .$$

To prove this, we must consider two possible situations. Either all angles are acute, or one angle is obtuse. See Figure 7.2.

Figure 7.2

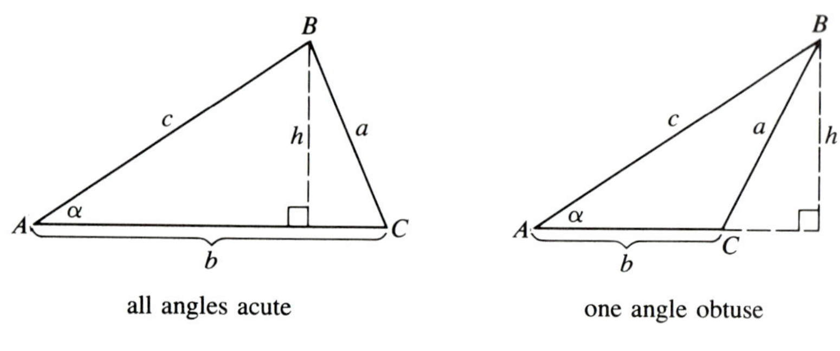

Regardless of which situation exists, we see that

$$\sin \alpha = \frac{h}{c}$$

$$h = c \sin \alpha .$$

Thus,

$$K = \frac{1}{2} bh = \frac{1}{2} bc \sin \alpha .$$

In a similar fashion we may show that

$$K = \frac{1}{2} ac \sin \beta \quad \text{and} \quad K = \frac{1}{2} ab \sin \gamma .$$

Therefore,

$$K = \frac{1}{2} bc \sin \alpha = \frac{1}{2} ac \sin \beta = \frac{1}{2} ab \sin \gamma .$$

This formula, then, says that *the area of a triangle is $\frac{1}{2}$ the product of any two sides times the sine of the included angle.*

EXAMPLE 1 Find the area (to four significant digits) of triangle ABC, given that $a = 5$, $c = 11$, and $\beta = 52°$.

Figure 7.3

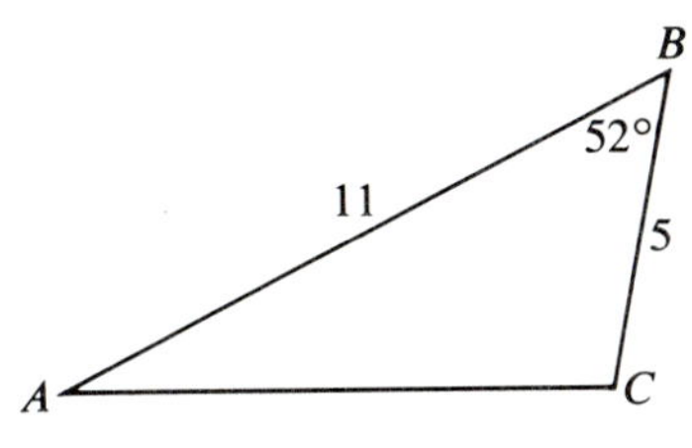

SOLUTION We have the situation shown in Figure 7.3. Because we have two sides and an included angle, we have

$$K = \frac{1}{2}(5)(11) \sin 52° \approx 21.67 .$$

▲

▲ ***Exercise*** Find the area (to four significant digits) of triangle ABC, given that $b = 12$, $c = 8$, and $\alpha = 37°$.

Answer 28.89

From this formula for the area of a triangle we know that

$$\frac{1}{2}bc\sin\alpha = \frac{1}{2}ac\sin\beta = \frac{1}{2}ab\sin\gamma.$$

Upon multiplying by 2 and dividing by abc, we get

$$\frac{\sin\alpha}{a} = \frac{\sin\beta}{b} = \frac{\sin\gamma}{c}.$$

This result is known as the **law of sines**. It says that, in any triangle, *the sines of the angles are proportional to the lengths of the opposite sides.* And, of course, it may be stated in the alternative form

$$\frac{a}{\sin\alpha} = \frac{b}{\sin\beta} = \frac{c}{\sin\gamma}.$$

EXAMPLE 2 In triangle ABC, $\alpha = 42°$, $\beta = 65°$, and $a = 20$. Find b.

SOLUTION We have the triangle shown in Figure 7.4.

Figure 7.4

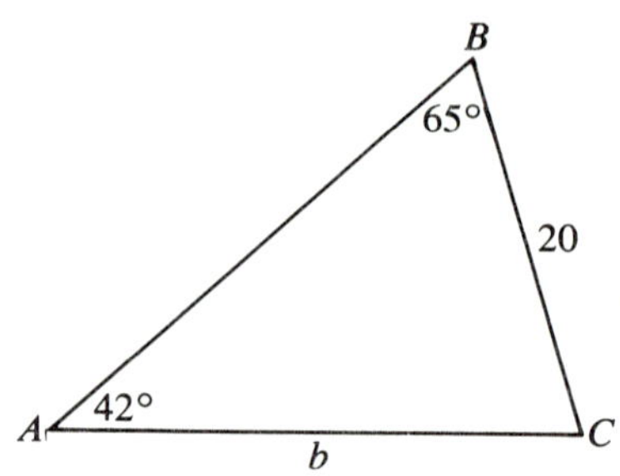

By the law of sines,

$$\frac{b}{\sin 65°} = \frac{20}{\sin 42°}.$$

Solving for b, we get

$$b = \frac{20\sin 65°}{\sin 42°} \approx 27.09.$$

▲

Note that we will continue to follow the convention, introduced in Section 2.3, for nonapplied triangles. The lengths and measures are taken as exact, but we will round off lengths to four significant digits and angle measures to the nearest hundredth of a degree.

EXAMPLE 3 In triangle ABC, $a = 12$, $c = 20$, and $\alpha = 25^\circ$. Find γ.

SOLUTION We have the triangle shown in Figure 7.5.

Figure 7.5

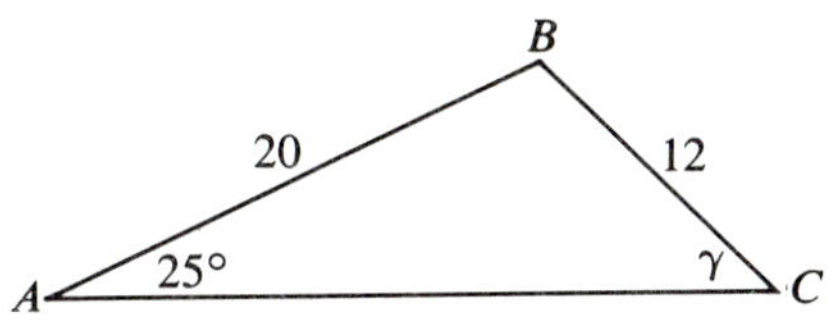

By the law of sines,

$$\frac{\sin \gamma}{20} = \frac{\sin 25^\circ}{12}.$$

Solving for $\sin \gamma$ gives

$$\sin \gamma = \frac{20 \sin 25^\circ}{12} = 0.7043638.$$

The acute angle with this sine is approximately 44.78°. However, $180^\circ - 44.78^\circ = 135.22^\circ$ has the same sine. Hence

$$\gamma \approx 44.78^\circ \quad \text{or} \quad \gamma \approx 135.22^\circ.$$

▲

Example 3 is an illustration of what is called the ambiguous case. This is because there are two different triangles that satisfy the given conditions, one where γ is acute and the other where γ is obtuse. See Figure 7.6.

Figure 7.6

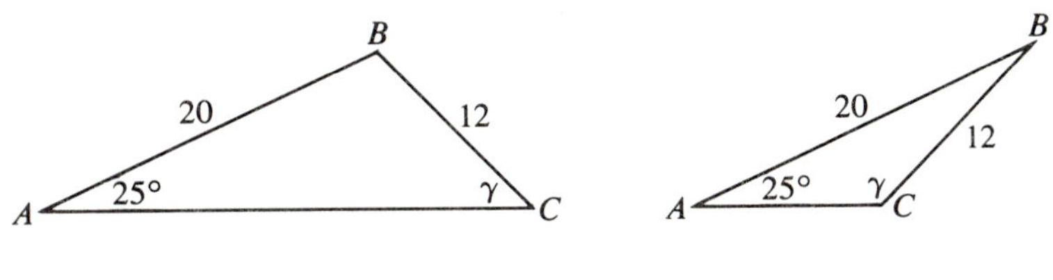

acute γ obtuse γ

If we are given two sides and an opposite angle and are asked to determine the angle opposite the longer side, then an acute solution will always give an obtuse solution as well.

EXAMPLE 4 In triangle ABC, $a = 10$, $b = 8$, and $\alpha = 65^\circ$. Find β.

SOLUTION We have the triangle shown in Figure 7.7.

By the law of sines,

$$\frac{\sin \beta}{8} = \frac{\sin 65^\circ}{10}.$$

Figure 7.7

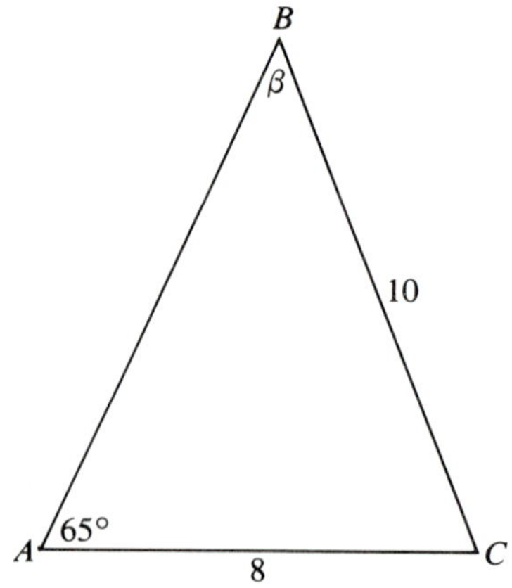

This gives

$$\sin \beta = \frac{8 \sin 65^\circ}{10} = 0.7250462.$$

The acute angle with this sine is approximately 46.47°. Thus it appears that $\beta \approx 46.47^\circ$, or $\beta \approx 180^\circ - 46.47^\circ = 133.53^\circ$. However, if $\beta \approx 133.53^\circ$, the sum of α (65°) and β exceeds 180°, which isn't possible. (The sum of the three angles in a triangle equals 180°). Hence there is but one answer: $\beta \approx 46.47^\circ$.

▲

If we are given two sides and an opposite angle and are asked to determine the angle opposite the shorter side, there is but one solution—acute. The reason is that, in a triangle, longer sides are opposite greater angles. Hence, the angle to be determined must be less than the given angle, and since no triangle can have two obtuse angles, the angle to be found must be acute. With this information we can immediately reject the obtuse possibility for β in Example 4.

▲ ***Exercise***

(a) In $\triangle ABC$, $\alpha = 110^\circ$, $\beta = 20^\circ$, and $a = 10$. Find b.
(b) In $\triangle ABC$, $a = 9$, $b = 5$, and $\beta = 7.5^\circ$. Find α.
(c) In $\triangle ABC$, $a = 15$, $c = 6$, and $\alpha = 35^\circ$. Find γ.

Answer (a) $b \approx 3.640$ (b) $\alpha \approx 13.59^\circ$ or $\alpha \approx 166.41^\circ$ (c) $\gamma \approx 13.26^\circ$

EXAMPLE 5 A telephone pole stands vertically on a hill that makes an angle of 12° with the horizontal. From a point down the hill 40 feet from the base of the pole, the angle of elevation to the top of the pole is 65°. What is the length (to two significant digits) of the pole?

SOLUTION We have the situation shown in Figure 7.8, where

$$x = \text{length of pole (in feet)}.$$

Figure 7.8

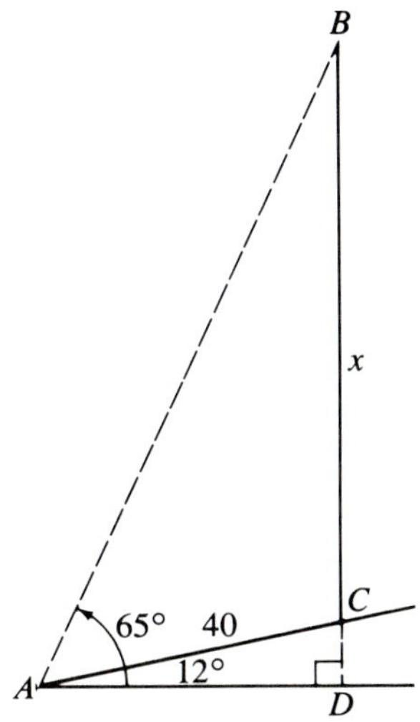

The measure of $\angle BAC$ is $65^\circ - 12^\circ = 53^\circ$ since the angle of elevation is measured with the horizontal and *not* the hill. From right triangle ADB, we see that the measure of $\angle B$ is $90^\circ - 65^\circ = 25^\circ$. Thus, by the law of sines,

$$\frac{x}{\sin 53^\circ} = \frac{40}{\sin 25^\circ}.$$

Solving for x, we get

$$x = \frac{40 \sin 53^\circ}{\sin 25^\circ} \approx 76.$$

To the nearest foot, the pole has a length of 76 feet.

▲

▲ *Exercise* A telephone pole stands vertically on a hill that makes an angle of 10° with the horizontal. From a point down the hill 30 feet from the base of the pole, the angle of elevation to the top of the pole is 55°. What is the length (to two significant digits) of the pole?

Answer 37 ft

Problem Set 7.1

1. In $\triangle ABC$, $\alpha = 22^\circ$, $\beta = 71^\circ$, and $a = 10$. Find b.
2. In $\triangle ABC$, $\alpha = 29^\circ$, $\gamma = 35^\circ$, and $c = 20$. Find a.
3. In $\triangle ABC$, $\beta = 31^\circ$, $\gamma = 42^\circ$, and $b = 15$. Find c.

4. In $\triangle ABC$, $\alpha = 47^\circ$, $\beta = 51^\circ$, and $b = 13$. Find a.

5. In $\triangle ABC$, $a = 8$, $b = 10$, and $\alpha = 22^\circ$. Find β.

6. In $\triangle ABC$, $b = 10$, $c = 15$, and $\beta = 37^\circ$. Find γ.

7. In $\triangle ABC$, $a = 12$, $b = 8$, and $\alpha = 70^\circ$. Find β.

8. In $\triangle ABC$, $a = 8$, $b = 10$, and $\beta = 84^\circ$. Find α.

9. In $\triangle ABC$, $a = 8$, $b = 20$, and $\alpha = 27^\circ$. Find β.

10. In $\triangle ABC$, $a = 25$, $c = 8$, and $\gamma = 35^\circ$. Find α.

In Problems 11–18, find the area (to four significant digits) of triangle *ABC* for the given conditions.

11. $a = 6$, $c = 8.2$, and $\beta = 25^\circ$

12. $a = 8.5$, $c = 9.1$, and $\beta = 59^\circ$

13. $b = 5.2$, $c = 8.9$, and $\beta = 28.3^\circ$

14. $a = 12.3$, $b = 11.2$, and $\alpha = 71.3^\circ$

15. $a = 12$, $\beta = 35^\circ$, and $\gamma = 52^\circ$

16. $c = 17$, $\alpha = 20^\circ$, and $\beta = 105^\circ$

17. $b = 16.1$, $\alpha = 110^\circ$, and $\beta = 31^\circ$

18. $c = 8.4$, $\beta = 85^\circ$, and $\gamma = 48^\circ$

19. A triangular lot has frontages of 125 feet and 185 feet on two streets that intersect at an angle of 75°. What is the area of the lot (to two significant digits)?

20. If each side of a roof is 24 feet long, and the angle formed by the two sides is 115°, what is the area of the gable (to two significant digits)?

21. A hill slopes at an angle of 12°. From the base of the hill, the angle of elevation to the top of a vertical tower on the hill is 41°. If the tower is 300 feet high, what is the distance (to two significant digits) from the base of the tower to the base of the hill?

22. The stone statues on Easter Island stand vertically on an incline of 5°. If each statue casts a 14-foot shadow up the slope when the angle of elevation of the sun is 70°, how tall (to two significant digits) are the statues?

23. At a point 500 feet from the base of the Tower of Pisa, opposite the direction of the lean, the angle of elevation to the top is 19.6°. If the tower leans at an angle of 5.45° from the vertical, what is the length (to three significant digits) of the tower?

24. A rock lies on the side of a hill 60 meters from the base of the hill. From a point level with the base of the hill, 75 meters from the base, the angle of elevation to the rock is 28°. What angle (to the nearest degree) does the hill make with the horizontal?

25. From the bottom of a building the angle of elevation to the top of the Eiffel Tower is 39°. From the top of the building the angle of elevation is 30°. If the building is 280 feet high, how tall (to two significant digits) is the Eiffel Tower?

26. The angle of elevation from an observer to the top of the Great Pyramid in Egypt is 17°. If the observer walks 460 feet toward the pyramid, the angle of elevation is 24°. If the observer and the pyramid are on level ground, how tall (to two significant digits) is the Great Pyramid?

27. Forest rangers at stations A and B, which are 12 miles apart, spot a fire at F. If $\angle FAB$ is 49° and $\angle FBA$ is 37°, how far (to two significant digits) is each station from the fire?

28. A weather balloon is hovering directly between two towns located 3 miles apart. If the angle of elevation is 42° from one city and 35° from the other, what is the altitude of the balloon (to two significant digits)?

29. A ship is sighted from a point on the coast, with the line of sight making an angle of 130° with due north at the observation point. Thirty minutes later, the line of sight makes an angle of 35° with due north. If the ship is known to be moving at 18 miles per hour along a straight line, how far (to two significant digits) is the ship from the observation point at the second sighting?

30. Use the fact that, in the following triangle, area $\triangle DEF =$ area $\triangle DEG +$ area $\triangle EGF$ to show that $\sin(\alpha + \beta) = \sin\alpha\cos\beta + \sin\beta\cos\alpha$.

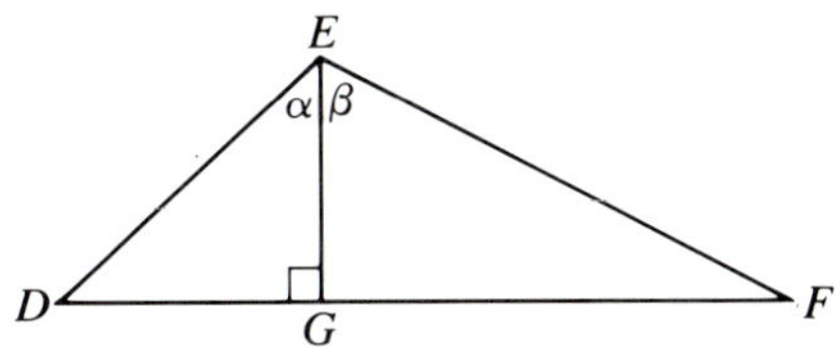

31. Use the law of sines to show that in $\triangle ABC$, $\alpha < \beta$ if and only if $a < b$.

7.2 THE LAW OF COSINES

In this section we encounter another important technique for finding unknown sides and angles. It is called the **law of cosines** and says that for any triangle ABC,

$$c^2 = a^2 + b^2 - 2ab\cos\gamma,$$
$$a^2 = b^2 + c^2 - 2bc\cos\alpha,$$
$$b^2 = a^2 + c^2 - 2ac\cos\beta.$$

To establish this, it suffices to prove one of the three equations, as the remaining two may be proven in a similar manner. We'll verify the first of the three equations. We begin by placing $\angle C$ in standard position (see Figure 7.9).

Figure 7.9

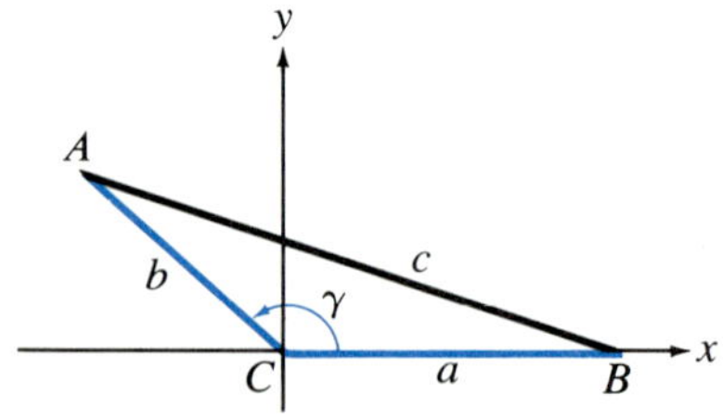

Let (x, y) represent A, whose distance from the origin is b. Thus,

$$\frac{x}{b} = \cos\gamma \quad \text{and} \quad \frac{y}{b} = \sin\gamma .$$

Therefore,

$$x = b\cos\gamma \quad \text{and} \quad y = b\sin\gamma .$$

So $A = (b\cos\gamma, b\sin\gamma)$. Since the coordinates of B are $(a, 0)$, we have, by the distance formula,

$$\begin{aligned} c^2 &= (b\cos\gamma - a)^2 + (b\sin\gamma)^2 \\ &= b^2\cos^2\gamma - 2ab\cos\gamma + a^2 + b^2\sin^2\gamma \\ &= a^2 + b^2(\cos^2\gamma + \sin^2\gamma) - 2ab\cos\gamma . \end{aligned}$$

Because $\cos^2\gamma + \sin^2\gamma = 1$, we get

$$c^2 = a^2 + b^2 - 2ab\cos\gamma .$$

Thus the law of cosines says that in any triangle, *the square of the length of a side is equal to the sum of the squares of the lengths of the other two sides minus twice the product of the other two sides and the cosine of the included angle.*

Note that if γ is a right angle, the law of cosines simplifies to

$$\begin{aligned} c^2 &= a^2 + b^2 - 2ab\cos\gamma \\ &= a^2 + b^2 - 2ab(0) \\ &= a^2 + b^2 , \end{aligned}$$

the Pythagorean Theorem. In some sense, then, the law of cosines may be thought of as a generalized Pythagorean Theorem.

EXAMPLE 1 In $\triangle ABC$, $a = 20$, $b = 10$, and $\gamma = 42°$. Find c.

SOLUTION We have the triangle shown in Figure 7.10.

Figure 7.10

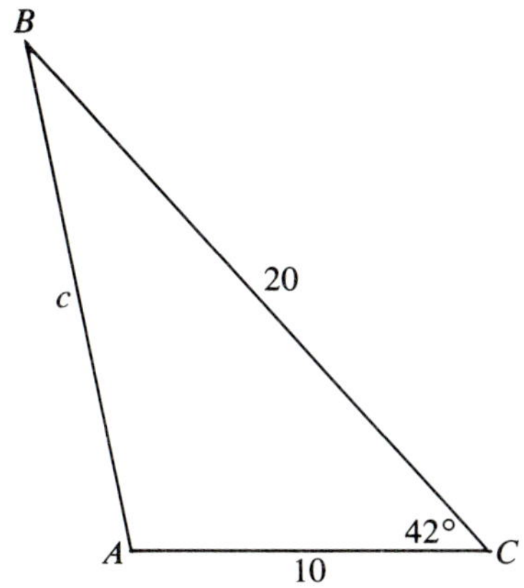

Using the law of cosines, we get

$$\begin{aligned} c^2 &= a^2 + b^2 - 2ab \cos \gamma \\ &= 20^2 + 10^2 - 2(20)(10) \cos 42^\circ \\ &= 202.74207. \end{aligned}$$

Thus

$$c = \sqrt{202.74207} \approx 14.24.$$

▲

EXAMPLE 2 In $\triangle ABC$, $a = 15$, $b = 10$, and $c = 20$. Find γ.

SOLUTION We have the triangle shown in Figure 7.11.

Figure 7.11

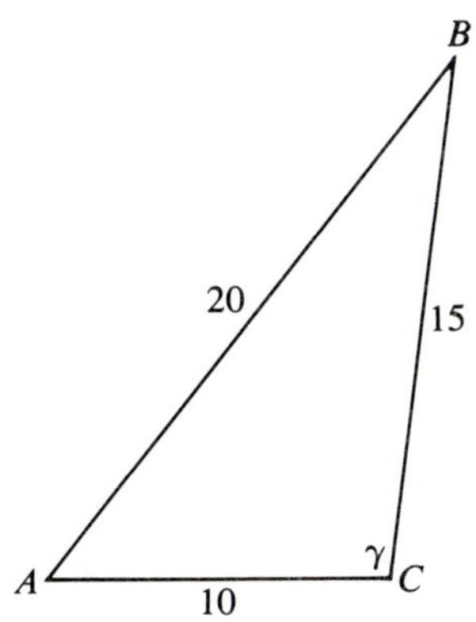

Using the law of cosines, we get

$$20^2 = 15^2 + 10^2 - 2(15)(10) \cos \gamma.$$

Solving for $\cos \gamma$, we have

$$\cos \gamma = \frac{20^2 - 15^2 - 10^2}{-2(15)(10)} = -0.25.$$

Since the cosine is negative, the angle must be obtuse. So we find the acute angle whose cosine is 0.25 and then subtract from 180°. We get $\gamma \approx 104.48°$.

▲

▲ *Exercise*

(a) In $\triangle ABC$, $a = 15$, $b = 12$, and $\gamma = 55°$. Find c.
(b) In $\triangle ABC$, $a = 12$, $b = 10$, and $c = 20$. Find γ.

Answer (a) $c \approx 12.75$ (b) $\gamma \approx 130.54°$

EXAMPLE 3 On a baseball diamond, the distance from the pitcher's mound to home plate is 60 feet, 6 inches, while the distance from home plate to first base is 90 feet. What is the distance (to three significant digits) from the pitcher's mound to first base?

SOLUTION We have the situation shown in Figure 7.12, where

$$x = \text{distance from the pitcher's mound to first base (in feet).}$$

Figure 7.12

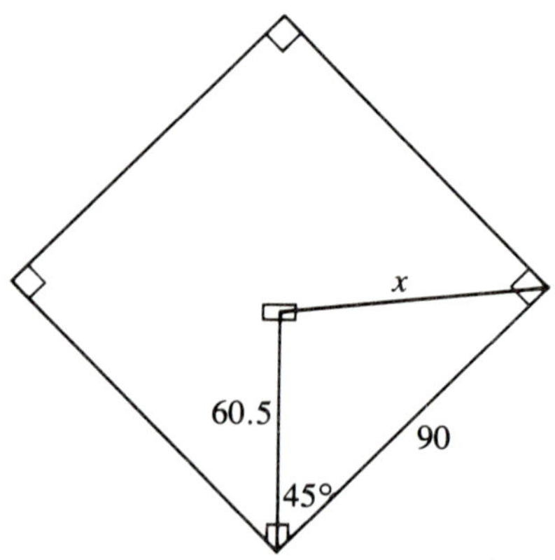

Using the law of cosines, we get

$$\begin{aligned} x^2 &= (60.5)^2 + 90^2 - 2(60.5)(90)\cos 45° \\ &= 4059.8572. \end{aligned}$$

Thus,

$$x \approx 63.7.$$

The distance is approximately 63.7 feet.

▲

▲ *Exercise* On a softball diamond the distance from the pitcher's mound to home plate is 46 feet, while the distance from home plate to first base is 60 feet. What is the distance (to three significant digits) from pitcher's mound to first base?

Answer 42.6 ft

Using the law of cosines and the area formula developed in the preceding section, we can express the area of a triangle in terms of its three sides. Known as **Heron's** (or **Hero's**) **formula**, it says that if a, b, and c are the lengths of the three sides and s is the semiperimeter—that is, $s = \frac{1}{2}(a + b + c)$, then the area K is

$$K = \sqrt{s(s - a)(s - b)(s - c)}\,.$$

(The proof of this formula is problem 32.)

EXAMPLE 4 Find the area (to four significant digits) of a triangle whose sides are 3.5, 6, and 7.5.

SOLUTION We first compute the semiperimeter s. We get

$$s = \frac{1}{2}(3.5 + 6 + 7.5) = 8.5\,.$$

Applying Heron's formula gives

$$\begin{aligned} K &= \sqrt{8.5(8.5 - 3.5)(8.5 - 6)(8.5 - 7.5)} \\ &= \sqrt{8.5(5)(2.5)(1)} \approx 10.31\,. \end{aligned}$$

▲

▲ *Exercise* Find the area (to four significant digits) of a triangle whose sides are 4.5, 8, and 5.5.

Answer 11.91

Recall that solving a triangle means determining all remaining parts from the given parts. If we know one side and two of the remaining parts, the triangle may, in general, be determined. There are three possibilities.

1. Three sides.
2. Two sides and one angle.
3. One side and two angles.

Let's consider case 1. If we have all three sides, we must use the law of cosines to determine one of the angles. To determine a second angle, we may use either the law of cosines or the law of sines (although the law of sines is generally less difficult). Of course, the third angle is easily determined since the sum of the angles in a triangle is 180°.

In case 2 we know two sides and an angle. If the angle is between the two sides, we use the law of cosines to get the third side. Then we use either law to find a second angle, and this gives the third angle as well.

If the angle of case 2 is opposite one of the known sides, we apply the law of sines to find a second angle (and thus the third). The third side is then obtained with either law. In this situation we may get two possible second angles (as in Example 3 of the preceding section) and thus two different triangles. Two sides and an opposite angle sometimes give us an ambiguous case.

In case 3 we obviously know all angles since we know two. We may then use the law of sines to determine the remaining two sides.

EXAMPLE 5 Solve triangle ABC if $a = 6.6$, $b = 9$, and $\gamma = 37.2°$.

SOLUTION We have the triangle shown in Figure 7.13.

Figure 7.13

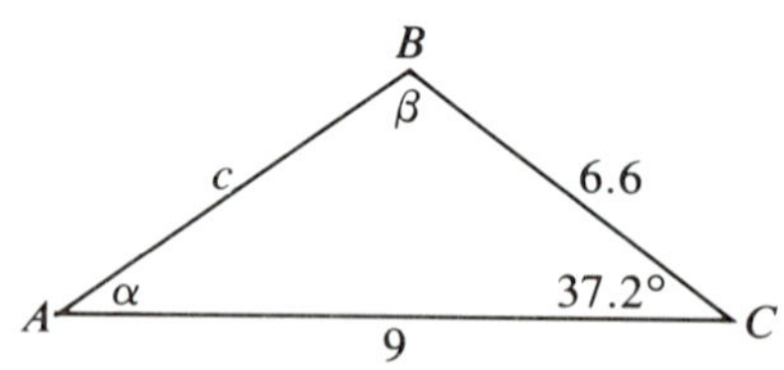

We must first calculate c using the law of cosines:

$$\begin{aligned} c^2 &= 9^2 + (6.6)^2 - 2(9)(6.6)\cos 37.2° \\ &= 29.932246 \\ c &= 5.471037 \approx 5.471\,. \end{aligned}$$

Now, using the law of sines, let's determine α.

$$\frac{\sin\alpha}{6.6} = \frac{\sin 37.2°}{5.471037}$$

$$\sin\alpha = \frac{6.6\sin 37.2°}{5.471037}$$

Thus $\alpha \approx 46.83°$ or $\alpha \approx 133.17°$. (Note that we used the unrounded result for c in computing α. In this calculation c is an intermediate result and should be unrounded to increase the accuracy of α.) However, α cannot be obtuse because it isn't opposite the longest side. So $\alpha \approx 46.83°$, and

$$\beta \approx 180° - 37.2° - 46.83° = 95.97°\,.$$

▲

In Example 5, after calculating c, we chose to determine the acute angle α instead of β, a possible obtuse angle. If we had selected β instead, we would have gotten

$$\frac{\sin \beta}{9} = \frac{\sin 37.2^\circ}{5.471037}$$

$$\sin \beta = \frac{9 \sin 37.2^\circ}{5.471037}.$$

So $\beta \approx 84.03^\circ$ or $\beta \approx 95.97^\circ$, and it appears that there are two possible triangles. See Figure 7.14.

Figure 7.14

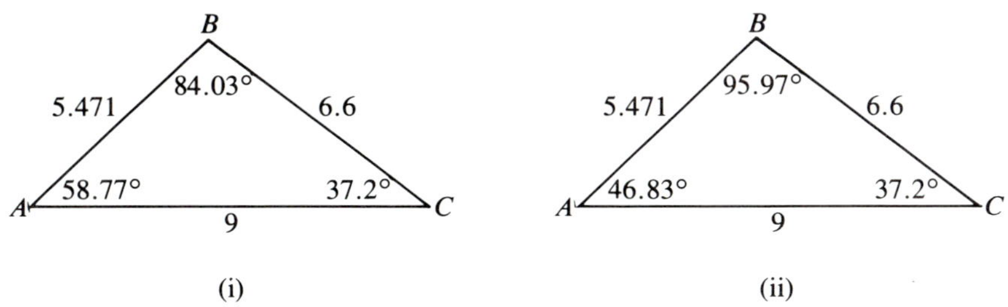

But given two sides and the included angle, only one triangle can be constructed. Which is correct? Neither can be eliminated by examining the relationships between angle measures and opposite sides. In each triangle the largest side is opposite the largest angle, and the smallest side is opposite the smallest angle. We have to note that, for the triangle in Figure 7.14(i), the law of sines fails:

$$\frac{\sin 58.77^\circ}{6.6} \neq \frac{\sin 37.2^\circ}{5.471}!$$

(Check it.)

By calculating α before β, we avoided such confusion. Thus, when an option exists, *determine the acute angle first*.

A similar situation arises when we try to solve a triangle, given three sides. We may avoid trouble by first calculating the two smallest angles (which must be acute).

▲ ***Exercise*** Solve triangle ABC if $a = 10$, $b = 15$, and $\gamma = 35^\circ$.

Answer $c \approx 8.902$, $\alpha \approx 40.11^\circ$, $\beta \approx 104.89^\circ$

Problem Set 7.2

1. In $\triangle ABC$, $a = 10$, $b = 30$, and $\gamma = 25^\circ$. Find c.
2. In $\triangle ABC$, $a = 12$, $c = 25$, and $\beta = 65^\circ$. Find b.
3. In $\triangle ABC$, $b = 20$, $c = 25$, and $\alpha = 28^\circ$. Find a.

4. In $\triangle ABC$, $a = 15$, $b = 28$, and $\gamma = 112°$. Find c.
5. In $\triangle ABC$, $a = 4$, $b = 5$, and $c = 7$. Find γ.
6. In $\triangle ABC$, $a = 12$, $b = 5$, and $c = 10$. Find α.
7. In $\triangle ABC$, $a = 17$, $b = 9$, and $c = 15$. Find β.
8. In $\triangle ABC$, $a = 3$, $b = 6$, and $c = 8$. Find γ.
9. In $\triangle ABC$, $a = 6$, $b = 12$, and $c = 5$. Find β.
10. In $\triangle ABC$, $a = 7$, $b = 8$, and $c = 16$. Find γ.

In Problems 11–14, find the area (to four significant digits) of triangle *ABC* for the given conditions.

11. $a = 8.5$, $b = 7$, $c = 14$
12. $a = 18$, $b = 15$, $c = 20$
13. $a = 6$, $b = 5$, $c = 7$
14. $a = 12.2$, $b = 7.9$, $c = 10$

In Problems 15–24, solve triangle *ABC*.

15. $a = 10$, $b = 15$, $c = 12$
16. $a = 5$, $b = 6$, $c = 2$
17. $b = 8$, $c = 10$, $\gamma = 55°$
18. $a = 5$, $c = 7$, $\gamma = 65°$
19. $a = 10$, $b = 15$, $\alpha = 38°$
20. $b = 6$, $c = 9$, $\beta = 20°$
21. $a = 5$, $b = 8$, $\gamma = 42°$
22. $a = 4$, $c = 7$, $\beta = 112°$
23. $a = 6$, $\alpha = 28°$, $\beta = 95°$
24. $b = 9$, $\alpha = 91°$, $\gamma = 12°$

25. The dimensions of a triangular metal plate are 8.55 inches, 10.2 inches, and 15.1 inches. What is its area (to three significant digits)?
26. A surveyor determines the dimensions of a triangular plot of land to be 175 feet, 120 feet, and 155 feet. What is its area (to three significant digits)?
27. From a fixed point the distance to the entrance of the Apennine Tunnel in Italy is 13 miles, and the distance to the exit is 10 miles. If the angle between the lines of sight is 55° and the tunnel is straight, how long (to two significant digits) is the tunnel?
28. In Cincinnati's Riverfront Stadium the distance from homeplate to the center field wall (straightaway, as measured through second base) is 404 feet. How far (to three significant digits) is it from this point to third base? (Consecutive bases are 90 feet apart.)
29. Two ships are sighted from a point on the coast. If one ship is 2.5 miles away, the other is 3.2 miles away, and the angle between the lines of sight is 115°, how far apart (to two significant digits) are the two ships?
30. Two cars simultaneously leave from the same point traveling along straight highways that intersect at an angle of 75°. If one car is traveling 55 miles per hour and the other 45 miles per hour, how far apart (to two significant digits) are the two cars after 15 minutes?

31. Prove the converse of the Pythagorean Theorem. That is, for triangle ABC, if $c^2 = a^2 + b^2$, then $\angle C$ is a right angle.

32. Prove Heron's formula by using the following steps:

i. Write the area as $\frac{1}{2}bc \sin \alpha$.

ii. Show that $\sin \alpha = \sqrt{(1 + \cos \alpha)(1 - \cos \alpha)}$ and substitute for $\sin \alpha$.

iii. Substitute for $\cos \alpha$ by using the law of cosines.

iv. Use algebra to simplify, and substitute s for $(a + b + c)/2$. *Hint*:

$$\frac{a+b-c}{2} = \frac{a+b+c-2c}{2} = \frac{a+b+c}{2} - c = s - c\,.$$

33. The straight-line distance, in miles, between two points P_1 and P_2 in North America is given by

$$d = 3960\sqrt{2 - 2 \cos \phi_1 \cos \phi_2 \cos \theta - 2 \sin \phi_1 \sin \phi_2}\,,$$

where ϕ_i is the latitude of P_i, θ_i is the longitude of P_i, and $\theta = \theta_2 - \theta_1$.

(a) Find the straight-line distance (to three significant digits) between New York City (48.8°N, 73.9°W) and Los Angeles (34.2°N, 118.2°W).

(b) On a sphere, the shortest distance between two points is along the arc that passes through the points and has the center of the sphere as its center. Use this, the law of cosines, and the result of part (a) to compute the surface distance (to three significant digits) from New York City to Los Angeles.

34. In $\triangle ABC$, $a = 8$, $b = 6$, and $\alpha = 75°$. Instead of finding β first, then γ, and then c, find c first by using $a^2 = b^2 + c^2 - 2bc \cos \alpha$.

7.3 VECTORS

Some quantities, like speed and weight, are one-dimensional. Other quantities, however, like velocity and force, are two-dimensional, needing a direction to describe them. A quantity that has both a magnitude and a direction is called a **vector**. Graphically a vector is represented by a directed line segment. In Figure 7.15 we see vectors $\overrightarrow{PQ}$ and $\overrightarrow{SR}$. The points P and S are called **initial points** (or **tails**), and points Q and R are called **terminal points** (or **heads**).

Figure 7.15

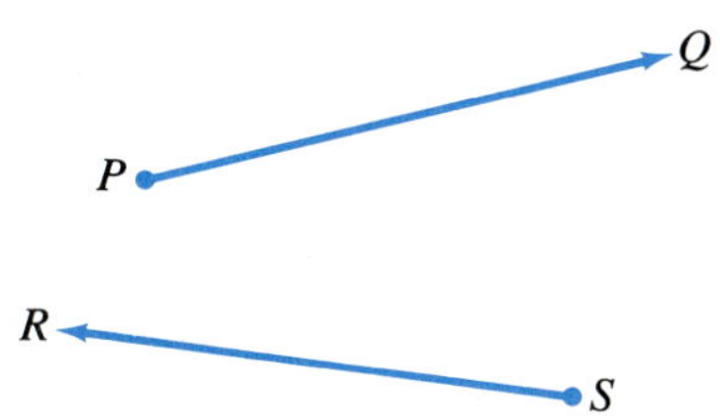

Note that the notation $\overrightarrow{PQ}$ immediately tells us which point is the tail and which is head. Sometimes, however, we will use a single letter to represent a vector. When doing so, we will use an arrow, like $\vec{V}$ for a vector named V. Some books use boldface, for example, **V**, to denote a vector.

The **magnitude** of a vector $\vec{V}$, denoted by $|\vec{V}|$, is the length of the directed line segment. We will study vectors in the plane and may use the distance formula to find magnitude.

EXAMPLE 1 Determine the magnitude of the vector in Figure 7.16.

Figure 7.16

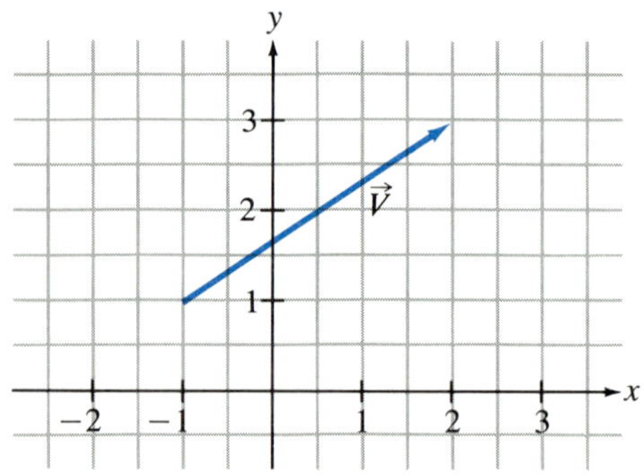

SOLUTION The vector $\vec{V}$ has its tail at $(-1, 1)$ and its head at $(2, 3)$. Hence, using the distance formula, the magnitude is

$$|\vec{V}| = \sqrt{(-1-2)^2 + (1-3)^2} = \sqrt{9+4} = \sqrt{13} \quad (\approx 3.606).$$

▲

▲ *Exercise* Determine the magnitude of the vector in Figure 7.17.

Figure 7.17

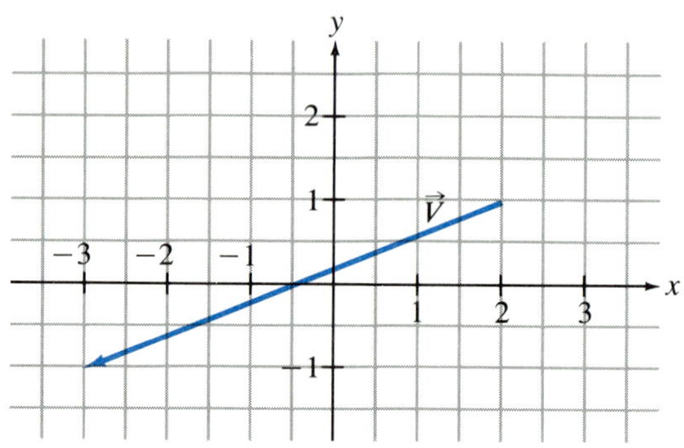

Answer $\sqrt{29}$ (≈ 5.385)

The **direction** of a vector is the positive angle that the vector makes with the positive part of the x-axis when the tail is placed at the origin. The directions of several vectors are shown in Figure 7.18.

Figure 7.18

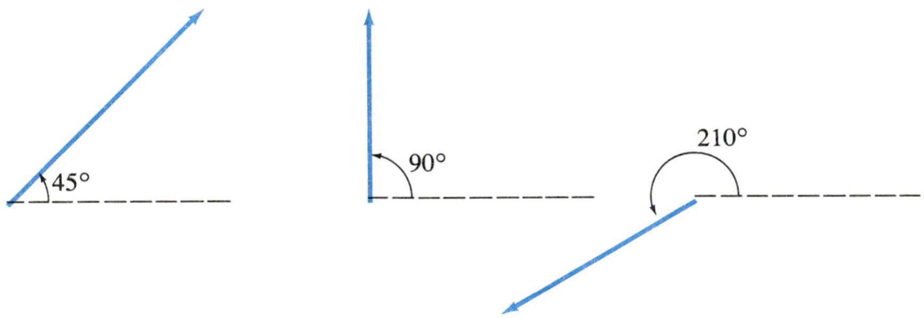

EXAMPLE 2 Determine the direction of the vector in Figure 7.19, to the nearest hundredth of a degree.

Figure 7.19

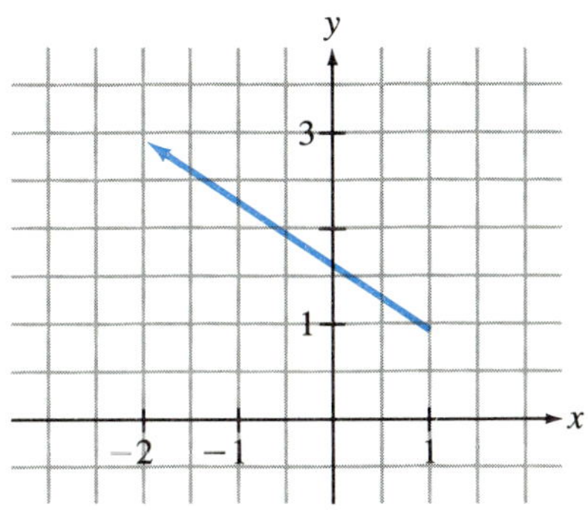

SOLUTION We don't have to move the vector's tail to the origin. The direction will be the angle θ noted in Figure 7.20.

Figure 7.20

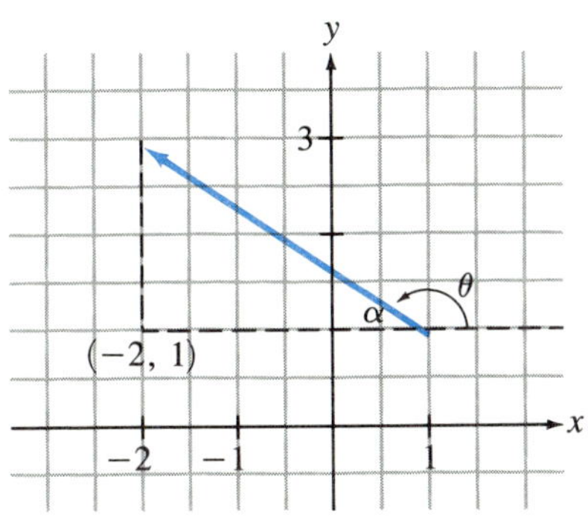

Note that we have constructed a right triangle. The length of the side opposite α is 2, and the side adjacent has length 3. Thus,

$$\tan \alpha = \frac{2}{3}.$$

So,

$$\alpha \approx 33.69°.$$

Thus, the direction is

$$\theta = 180° - \alpha \approx 146.31°.$$

▲

▲ *Exercise* Determine the direction of the vector in Figure 7.21, to the nearest hundredth of a degree.

Figure 7.21

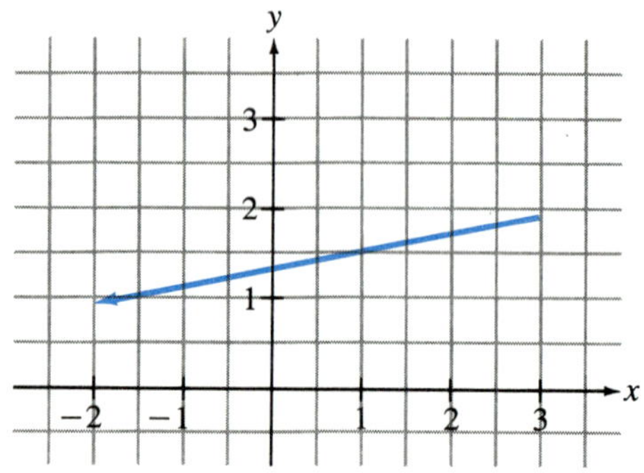

Answer 191.31°

A special vector that must be noted is the vector that has a magnitude of zero, the **zero vector**. Of course, such a vector cannot be represented by a directed line segment. It must be represented by a point, and it has any direction.

Two vectors are called **equal** provided that they have the same magnitude and the same direction. Vectors $\vec{U}$ and $\vec{V}$ in Figure 7.22 are equal. So two equal vectors have the same length and lie along the same or parallel lines. Therefore, a given vector may be moved to another position without consequence if the direction is maintained.

Figure 7.22

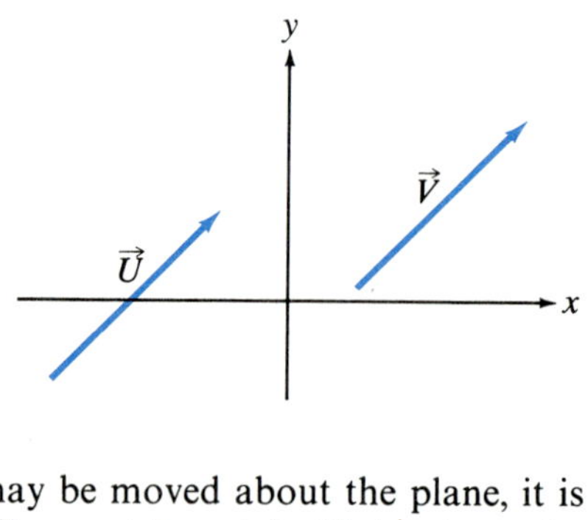

Since vectors may be moved about the plane, it is convenient to move them so that their tails are at the origin. If $A = (x_1, y_1)$ and $B = (x_2, y_2)$, then the vector $\overrightarrow{AB}$, when its tail is at the origin, will have a head of

$$(x_2 - x_1, y_2 - y_1).$$

When a vector has its tail at the origin and its head at (a, b), it is sometimes denoted $\langle a, b\rangle$, called the **component form**. The numbers a and b are called, respectively, the horizontal and vertical components of the vector. Geometrically the horizontal component is the projection of the head onto the x-axis, and the vertical component is the projection of the head onto the y-axis.

EXAMPLE 3 Determine the component form of $\overrightarrow{AB}$ if $A = (-5, 4)$ and $B = (-2, 5)$.

SOLUTION We take the coordinates of the head and subtract the coordinates of the tail. We get

$$\overrightarrow{AB} = \langle -2-(-5), 5-4 \rangle = \langle 3, 1 \rangle.$$

▲

▲ *Exercise* Determine the component form of $\overrightarrow{AB}$ if $A = (-3, 1)$ and $B = (2, 0)$.

Answer $\langle 5, -1 \rangle$

Note that for a vector $\vec{V}$ in component form $\langle a, b \rangle$, its magnitude r is

$$r = \sqrt{a^2 + b^2}.$$

Further, its direction θ is such that

$$\tan \theta = \frac{b}{a}.$$

The horizontal and vertical components may be determined from r and θ. From Figure 7.23 we see that

$$\cos \theta = \frac{a}{r} \quad \text{and} \quad \sin \theta = \frac{b}{r}.$$

Figure 7.23

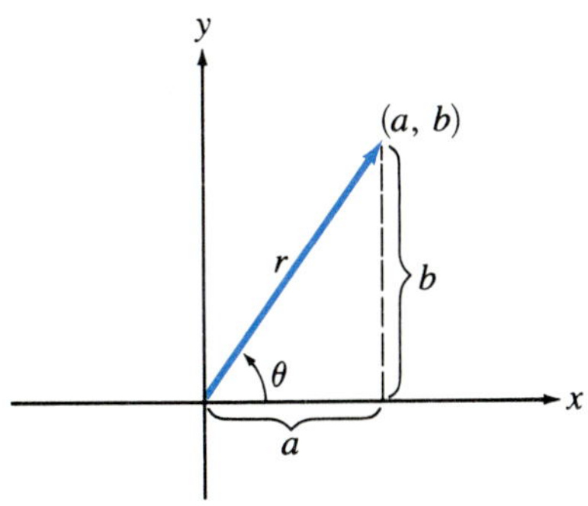

Therefore,

$$a = r \cos \theta \quad \text{and} \quad b = r \sin \theta,$$

where a and b are the horizontal and vertical components, respectively.

EXAMPLE 4 Determine the component form of a vector of magnitude 6 whose direction is 210°.

SOLUTION The horizontal component is

$$6 \cos 210° = 6\left(\frac{-\sqrt{3}}{2}\right) = -3\sqrt{3} \quad (\approx -5.196).$$

The vertical component is

$$6 \sin 210° = 6\left(\frac{-1}{2}\right) = -3.$$

The component form is

$$\langle -3\sqrt{3}, -3\rangle \approx \langle -5.196, -3\rangle.$$

▲

▲ *Exercise* Determine the component form of a vector of magnitude 4 whose direction is 120°.

Answer $\langle -2, 2\sqrt{3}\rangle \approx \langle -2, 3.464\rangle$

Let $\vec{U}$ and $\vec{V}$ be vectors with component forms $\langle a, b\rangle$ and $\langle c, d\rangle$, respectively. The **sum** of $\vec{U}$ and $\vec{V}$ is

$$\vec{U} + \vec{V} = \langle a + c, b + d\rangle.$$

Thus, we add two vectors by summing the horizontal and the vertical components. So if $\vec{U} = \langle -1, 2\rangle$ and $\vec{V} = \langle 3, 4\rangle$, then $\vec{U} + \vec{V} = \langle 2, 6\rangle$.

There is a geometric interpretation of the sum of two vectors. Consider the two vectors pictured in Figure 7.24.

Figure 7.24

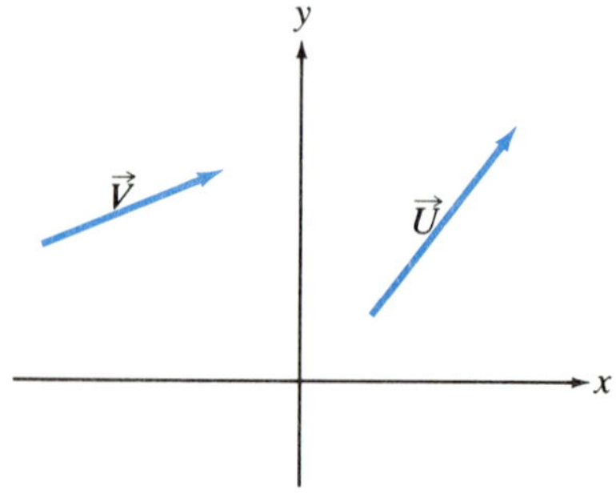

Let's move $\vec{V}$ so that the tail of $\vec{V}$ coincides with the head of $\vec{U}$. See Figure 7.25. Then $\vec{U} + \vec{V}$ is the vector that has its tail at the tail of $\vec{U}$ and its head at the head of $\vec{V}$. See Figure 7.26.

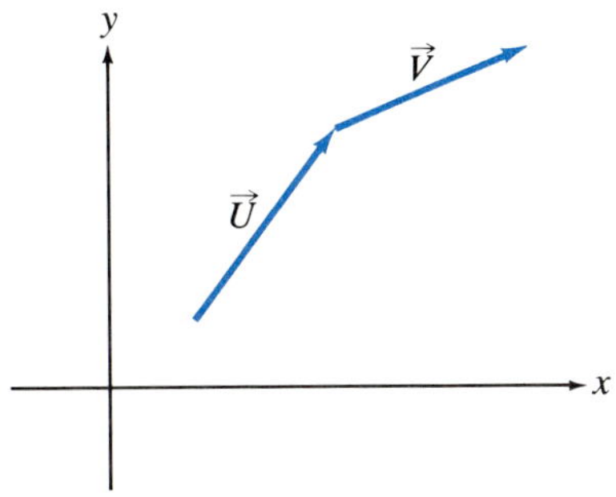

Figure 7.25

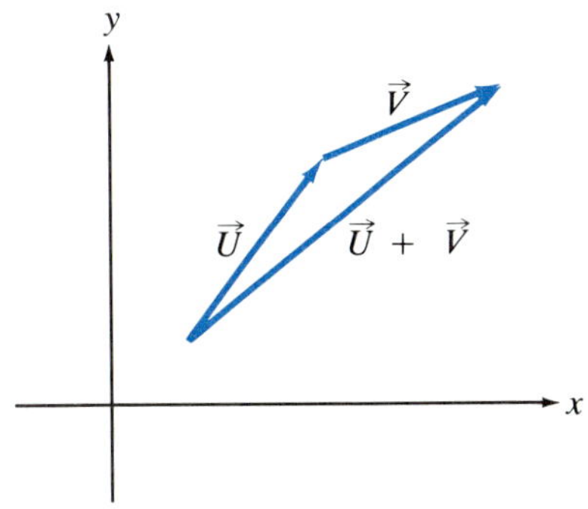

Figure 7.26

We may see that this interpretation is correct by looking at the components. Let $\langle a, b\rangle$ be the component form of $\vec{U}$ and let $\langle c, d\rangle$ be the component form of $\vec{V}$. If the tail of $\vec{U}$ is at $P(x, y)$, then the head is at $Q(x + a, y + b)$. When $\vec{V}$ is moved so that its tail is at $Q(x + a, y + b)$, its head will be at $R(x + a + c, y + b + d)$. See Figure 7.27. We see that the vector $\overrightarrow{PR}$ has a component form of $\langle a + c, b + d\rangle$, so it must equal $\vec{U} + \vec{V}$.

Figure 7.27

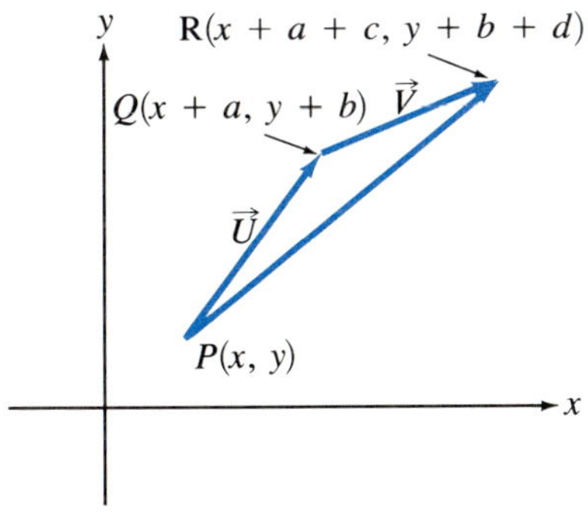

Another geometric interpretation of the sum of two vectors is also possible. Given the two vectors in Figure 7.24, let's move $\vec{V}$ so that its tail coincides with the tail of $\vec{U}$. See Figure 7.28.

Figure 7.28

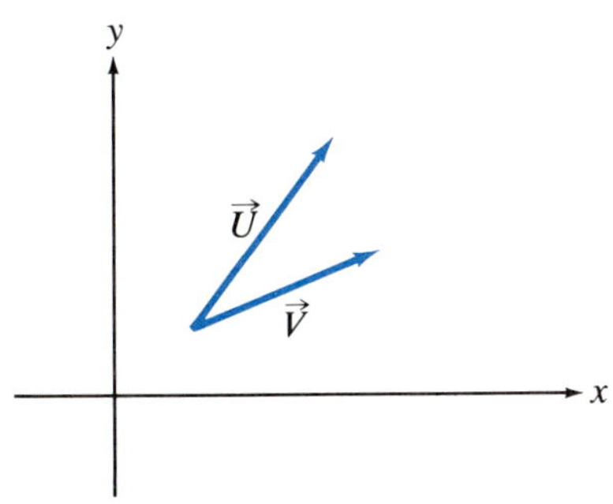

Then $\vec{U} + \vec{V}$ is the vector that forms the diagonal of the parallelogram generated by $\vec{U}$ and $\vec{V}$, as shown in Figure 7.29. This method of constructing the sum is called the **parallelogram rule**.

Figure 7.29

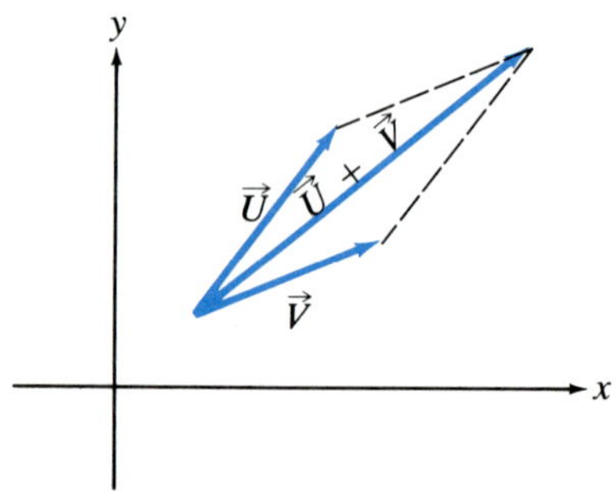

▲ *Exercise* Draw the vector $\vec{u} + \vec{v}$. (See Figure 7.30.)

Figure 7.30

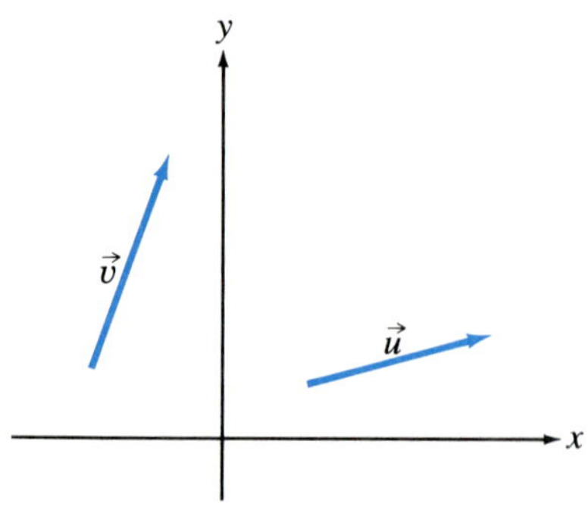

Answer See Figure 7.31.

Figure 7.31

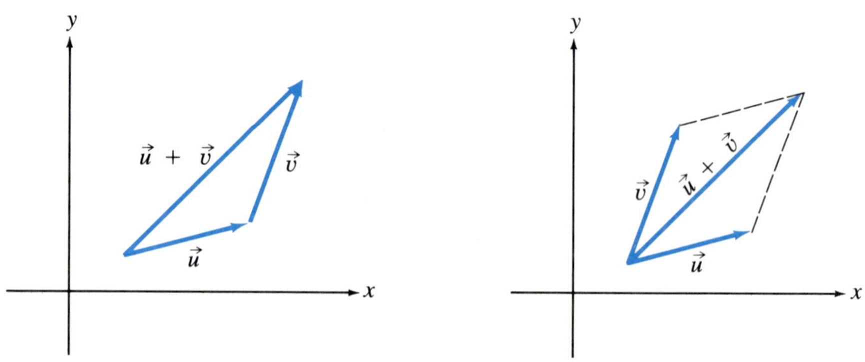

Problem Set 7.3

In Problems 1–4, determine the magnitude and direction of the vector.

1.

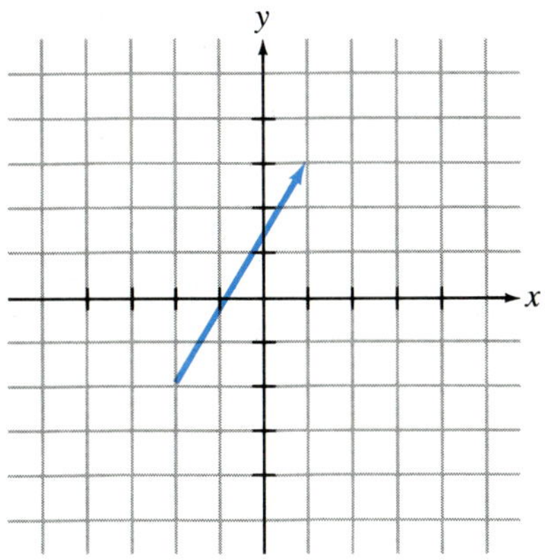

2.

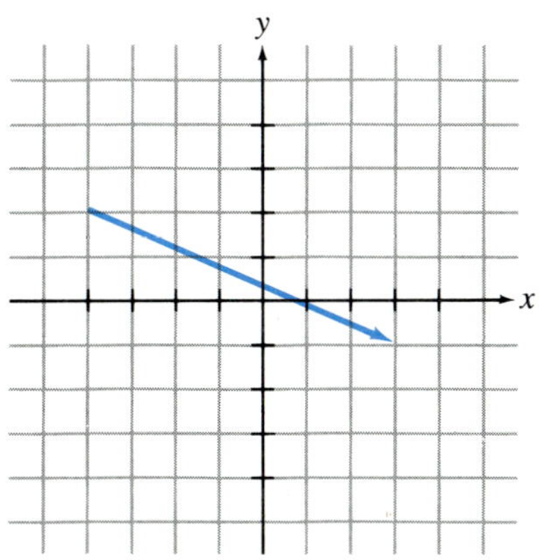

3.

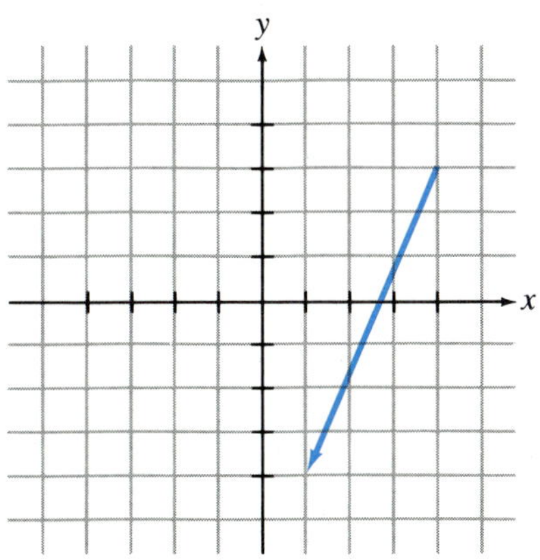

4.

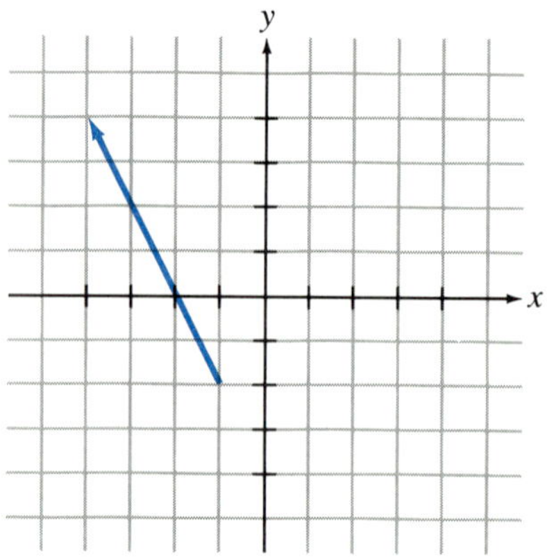

In Problems 5–8, determine the component form of $\overrightarrow{AB}$.

5. $A = (2.5, -1)$, $B = (-2, 3.4)$

6. $A = (-3, -1.5)$, $B = (-1.3, 2)$

7. $A = (3, 4)$, $B = (4.5, -6)$

8. $A = (-18, 20)$, $B = (-10, -12)$

In Problems 9–16, determine the component form of the vector with the specified magnitude and direction.

9. magnitude 5, direction 135°

10. magnitude 4, direction 225°

11. magnitude 6, direction 330°

12. magnitude 8, direction 120°

13. magnitude 8.2, direction 28.2°

14. magnitude 6.5, direction 118°

15. magnitude 2.7, direction 210.1°

16. magnitude 5.4, direction 310°

In Problems 17–20, determine $\vec{u} + \vec{v}$.

17. $\vec{u} = \langle 2, -5 \rangle$, $\vec{v} = \langle -1, 3 \rangle$

18. $\vec{u} = \langle -1, -4 \rangle$, $\vec{v} = \langle 5, 6 \rangle$

19. $\vec{u} = \langle -1.6, -2.3 \rangle$, $\vec{v} = \langle -1.8, 5.1 \rangle$

20. $\vec{u} = \langle 1.5, -1 \rangle$, $\vec{v} = \langle 3.5, -3 \rangle$

In Problems 21–24, draw the vector $u + v$.

21.

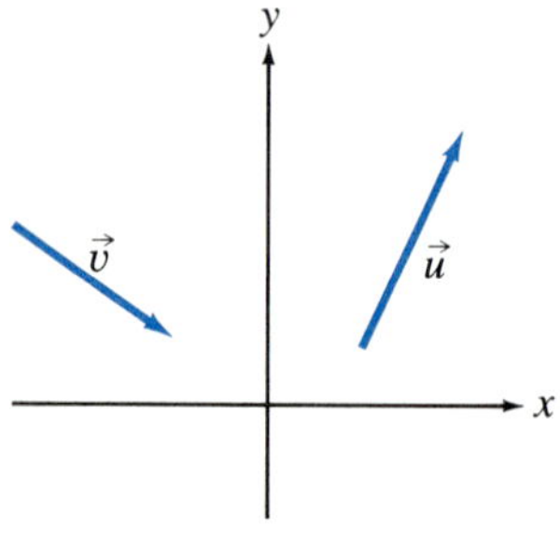

22.

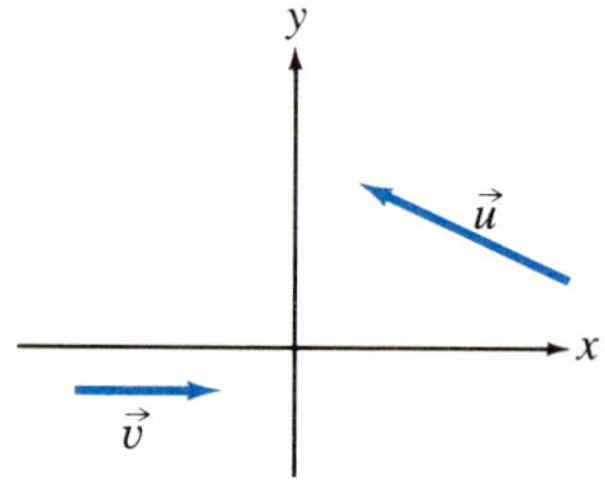

23.

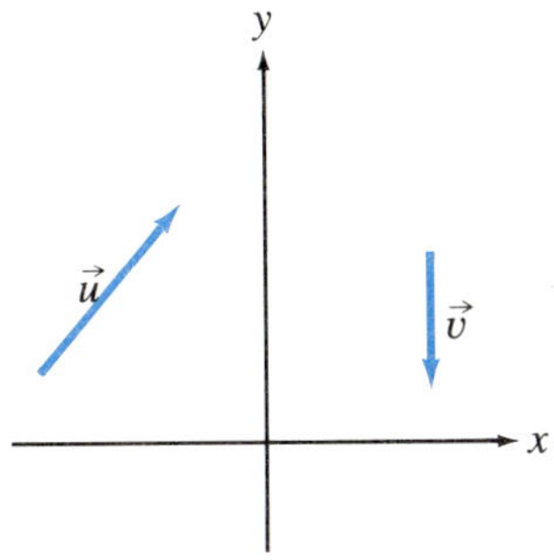

24.

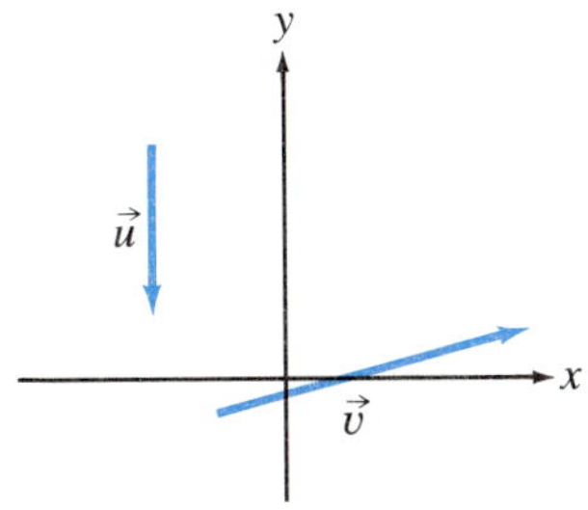

7.4 APPLICATIONS OF VECTORS

If two forces are acting on a body simultaneously, then the single force that is equivalent to the two is called the **resultant force**. The resultant force is the sum of the two forces. So the resultant force, a vector, is the sum of two vectors.

EXAMPLE 1 Two forces, one of 200 pounds and the other of 150 pounds, act on a body at an angle of 70.0°. Determine the magnitude of the resultant force (to three significant digits) and the angle it makes with the 200-pound force (to the nearest tenth of a degree).

SOLUTION Let $\vec{U}$ be the 200-pound force and $\vec{V}$ be the 150-pound force. We have the situation given in Figure 7.32.

Figure 7.32

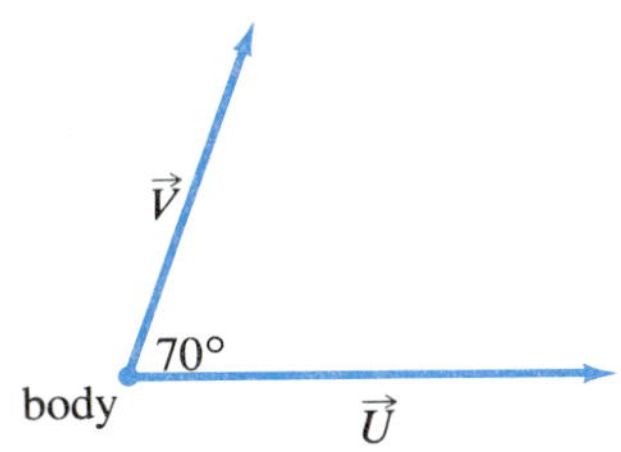

Placing the body at the origin, $\vec{U}$ along the x-axis, and the tail of $\vec{V}$ at the head of $\vec{U}$, we get Figure 7.33.

By the law of cosines,

$$\begin{aligned}|\vec{U} + \vec{V}|^2 &= |\vec{U}|^2 + |\vec{V}|^2 - 2|\vec{U}||\vec{V}| \cos 110^\circ \\ &= 200^2 + 150^2 - 2(200)(150) \cos 110^\circ \\ &= 83021.209 .\end{aligned}$$

Figure 7.33

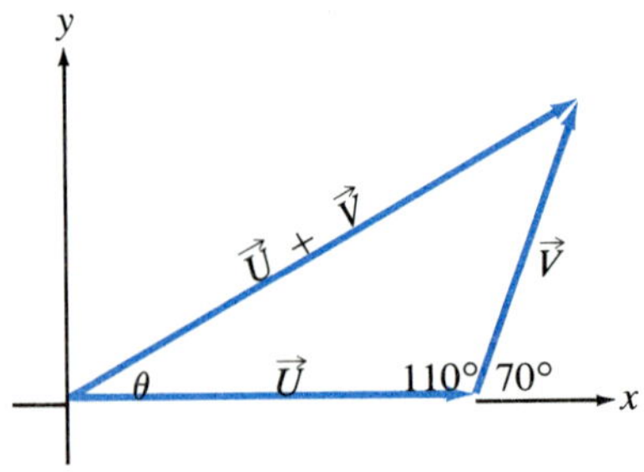

Thus, $$|\vec{U} + \vec{V}| = \sqrt{83021.209} = 288.13401 \approx 288.$$

To find θ we may use the law of sines. We get

$$\frac{\sin \theta}{150} = \frac{\sin 110°}{288.13401}.$$

$$\sin \theta = 0.4891956$$

$$\theta \approx 29.3°.$$

The resultant force is approximately 288 pounds and makes and angle of about 29.3° with the 200-pound force.

▲

We can work Example 1 by using component forms as well. Consider the situation in Figure 7.34.

Figure 7.34

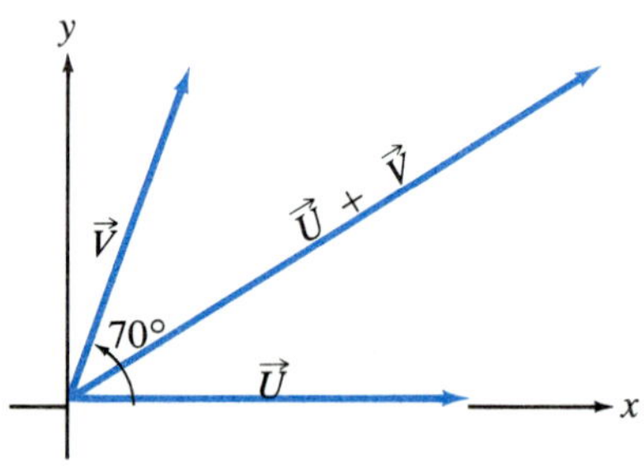

We have $$\vec{U} = \langle 200, 0 \rangle,$$

and

$$\vec{V} = \langle 150 \cos 70°, 150 \sin 70° \rangle = \langle 51.303022, 140.95389 \rangle.$$

Therefore,

$$\vec{U} + \vec{V} = \langle 251.303022, 140.95389 \rangle.$$

The magnitude of this vector is

$$|\vec{U} + \vec{V}| = \sqrt{251.303022^2 + 140.95389^2} \approx 288.$$

Its direction θ is such that

$$\tan\theta = \frac{140.95389}{251.303022}.$$

Hence $$\theta \approx 29.3^\circ.$$

The resultant force is approximately 288 pounds and makes an angle of about 29.3° with the 200-pound force.

▲ ***Exercise*** Two forces, one of 150 pounds and the other of 100 pounds, act on a body at an angle of 50.0°. Determine the magnitude (to three significant digits) of the resultant force and the angle it makes with the 150-pound force (to the nearest tenth of a degree).

Answer 228 lb; 19.7°

Before proceeding, some additional terms must be introduced. The heading of a plane is the direction in which it is pointed, and its airspeed is its speed with no wind. If there is a wind, then the actual course (direction the plane is moving) and the groundspeed (speed *with* wind) will differ from the heading and airspeed. In fact, the course vector (course and groundspeed) will be the sum of the heading vector (heading and airspeed) and wind vector (wind direction and speed). See Figure 7.35.

Figure 7.35

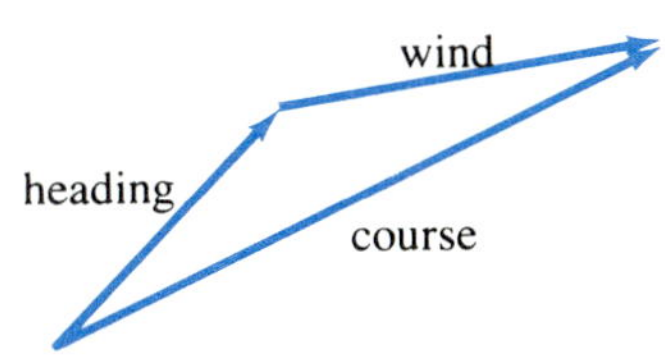

Further, the heading and course are frequently given in terms of **bearing**. The bearing identifies the number of degrees east or west of due north or due south. For example, N40°E means that the direction is 40° east of due north and S20°W means 20° west of due south. See Figure 7.36.

Figure 7.36

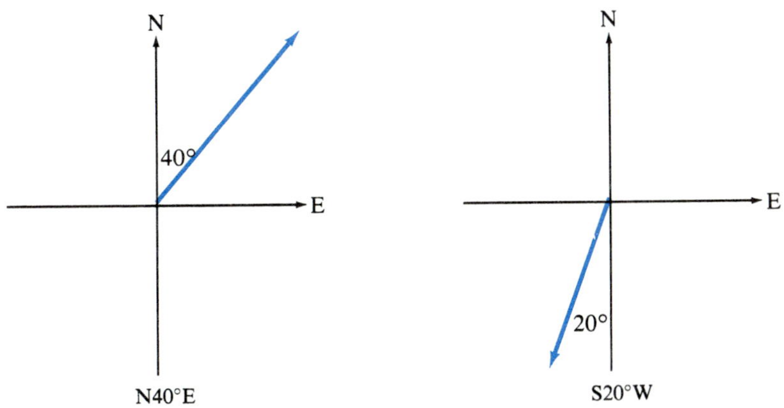

EXAMPLE 2 Suppose an airplane is heading northeast with an airspeed of 110 miles per hour. It is flying against a 30-miles-per-hour headwind from due north. What is its groundspeed (to two significant digits), and what is its course (to the nearest degree)?

SOLUTION Let $\vec{H}$ represent the plane's heading and airspeed, and let $\vec{W}$ represent the wind direction and speed. If the positive y direction is taken as due north, we have the situation given in Figure 7.37. The vector $\vec{C} = \vec{H} + \vec{W}$ is the course vector, representing course and groundspeed.

Figure 7.37

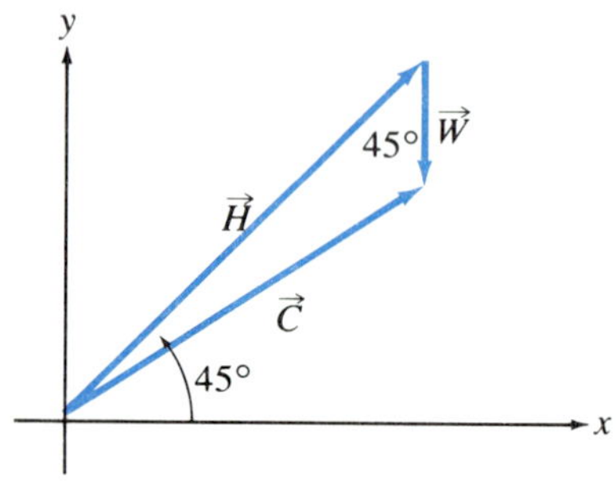

We use the law of cosines to determine the magnitude of $\vec{C}$. We get

$$\begin{aligned} |\vec{C}|^2 &= |\vec{H}|^2 + |\vec{W}|^2 - 2|\vec{H}||\vec{W}| \cos 45^\circ \\ &= 110^2 + 30^2 - 2(110)(30) \cos 45^\circ \\ &= 8333.0952. \end{aligned}$$

Thus,

$$|\vec{C}| = \sqrt{8333.0952} = 91.285789 \approx 91.$$

So the groundspeed is approximately 91 miles per hour. Now, letting α be the measure of the angle opposite W, we use the law of sines to get

$$\frac{\sin \alpha}{30} = \frac{\sin 45^\circ}{91.285789}$$

$$\sin \alpha = \frac{30 \sin 45^\circ}{91.285789} = 0.2323823.$$

So $\alpha \approx 13^\circ$. Subtracting this from 45° gives the direction of $\vec{C}$ as 32°. This gives a bearing of N58°E.

▲

▲ *Exercise* Suppose an airplane is heading due north with an airspeed of 150 miles per hour. It is flying with a 20-miles-per-hour tailwind from the southwest. What is its groundspeed (to two significant digits), and what is its course (to the nearest degree)?

Answer groundspeed is 160 mph (nearest ten mph); course is N5°E

EXAMPLE 3 Suppose an airplane is flying along the course N75°W with a groundspeed of 175 miles per hour. If the wind velocity is 20 miles per hour from N20°E, what is the plane's heading (to the nearest degree) and airspeed (to two significant digits)?

SOLUTION We have the situation given in Figure 7.38.

Figure 7.38

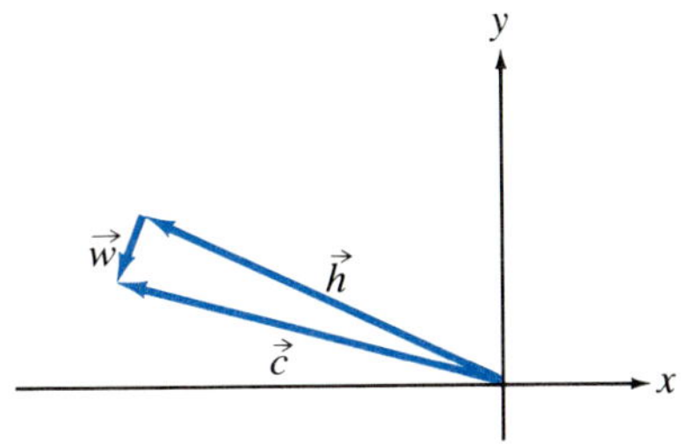

Instead of solving the triangle we will work with component forms. See Figure 7.39.

Figure 7.39

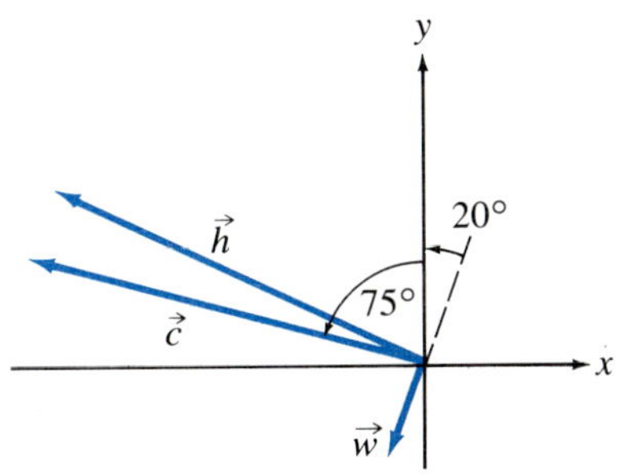

Let

$$\vec{h} = \langle x, y \rangle .$$

Since $\vec{w}$ has magnitude 20 and direction $270° - 20° = 250°$,

$$\vec{w} = \langle 20 \cos 250°, 20 \sin 250° \rangle .$$

Since $\vec{c}$ has magnitude 175 and direction $90° + 75° = 165°$,

$$\vec{c} = \langle 175 \cos 165°, 175 \sin 165° \rangle .$$

Since $\vec{h} + \vec{w} = \vec{c}$, we have

$$x + 20 \cos 250° = 175 \cos 165°$$

$$x = 175 \cos 165° - 20 \cos 250° = -162.19662,$$

and

$$y + 20 \sin 250° = 175 \sin 165°$$

$$y = 175 \sin 165° - 20 \sin 250° = 64.087185.$$

Thus,

$$\vec{h} = \langle -162.19662, 64.087185 \rangle .$$

Its magnitude is

$$|\vec{h}| = \sqrt{(-162.19662)^2 + (64.087185)^2} \approx 170 \quad \text{(nearest ten).}$$

Its direction θ is such that

$$\tan \theta = \frac{64.087185}{-162.19662} .$$

The reference angle θ' is 21.559984°, so

$$\theta = 180° - \theta' \approx 158° .$$

Hence, the heading is N68°W, and the airspeed is 170 miles per hour (nearest ten mph).

▲

In Example 3 we chose to use component forms instead of the triangle. This may be easier when we are being asked to determine one of the vectors that makes up the resultant.

▲ *Exercise* Suppose an airplane is flying along the course S10°W with a groundspeed of 195 miles per hour. If the wind velocity is 25 miles per hour from N15°W, what is the plane's heading (to the nearest degree) and airspeed (to two significant digits)?

Answer airspeed is 170 mph (nearest ten mph); heading is S14°W

Problem Set 7.4

1. Two forces, one of 125 pounds and the other of 110 pounds, act on a body at an angle of 115.0°. Determine the magnitude of the resultant force (to three significant digits) and the angle (to the nearest tenth of a degree) it makes with the 110-pound force.

2. Two forces, one of 60 pounds and the other of 75 pounds, act on a body at an angle of 125°. Determine the magnitude of the resultant force (to two significant digits) and the angle (to the nearest degree) it makes with the 75-pound force.

3. Suppose an airplane is heading southwest with an airspeed of 160 miles per hour. It is flying against a 20-miles-per-hour headwind from due west. What is its groundspeed (to two significant digits) and what is its course (to the nearest degree)?

4. Suppose a boat is heading southeast with a stillwater speed of 18 knots. It is sailing with a 10-knot current from due north. What is its actual speed (to two significant digits) and what is its course (to the nearest degree)?

5. Suppose an airplane is heading N10°E with an airspeed of 175 mph. If its course is N25°E with a groundspeed of 190 mph, what is the wind velocity (to two significant digits and to the nearest degree)?

6. Suppose an airplane is heading N30°W with an airspeed of 170 mph. If its course is N50°W with a groundspeed of 150 mph, what is the wind velocity (to two significant digits and to the nearest degree)?

7. Suppose the wind velocity is 15 mph from the northwest. If a pilot wishes to fly due south, what should be the heading (to the nearest degree) and what will be the groundspeed (to two significant digits) if the airplane has an airspeed of 155 miles per hour?

8. Suppose the wind velocity is 30 miles per hour from S20°E. If a pilot wishes to fly along a course S85°E, what should be the heading (to the nearest degree) and what will be the groundspeed (to two significant digits) if the airplane has an airspeed of 265 miles per hour?

Suppose a weight is supported by two cables attached to a horizontal beam. See Figure 7.40(i). Then there are three forces acting on point *P*: two in the directions of the cables (whose magnitudes are the tensions of the cables) and one straight down due to gravity (whose magnitude is the weight of the object). See Figure 7.40(ii). Because the weight is at rest, the resultant of the three forces acting upon it must be zero. Thus, the resultant of the two forces along the cables must be exactly opposite the force due to gravity. This information is needed in Problems 9–12.

Figure 7.40

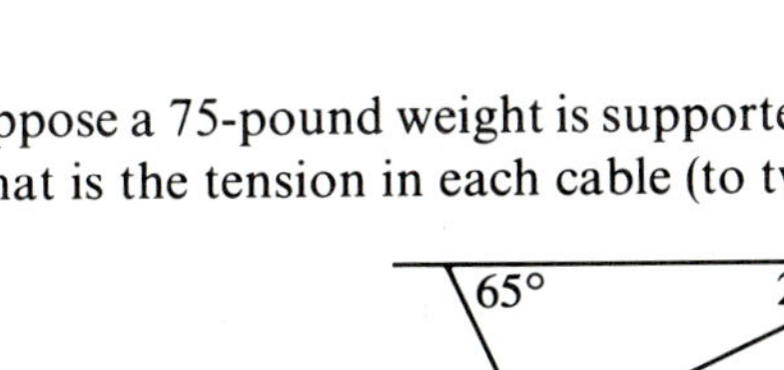

(i)

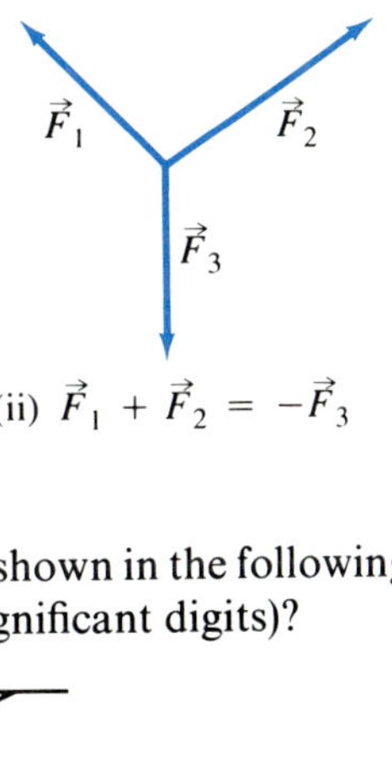

(ii) $\vec{F}_1 + \vec{F}_2 = -\vec{F}_3$

9. Suppose a 75-pound weight is supported as shown in the following figure. What is the tension in each cable (to two significant digits)?

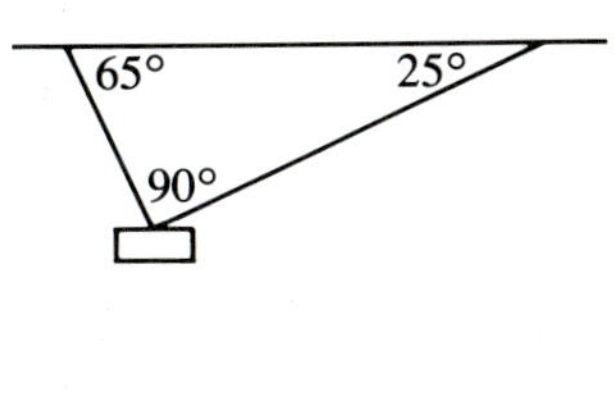

10. Suppose a 50-pound weight is supported as shown in the figure at the top of page 270. What is the tension in each cable (to two significant digits)?

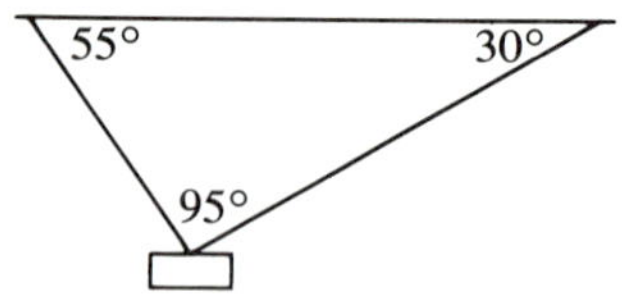

11. Suppose the ends of a wire are tied to a horizontal beam, and a 20-kilogram weight is suspended from the midpoint of the wire. If each end of the wire makes a 35° angle with the beam, what is the tension in the wire (to two significant digits)?

12. Suppose the ends of a rope 10 feet long are tied to a horizontal beam at points 2 feet apart. A 15-pound weight is suspended from the midpoint of the rope. What is the tension in the rope (to two significant digits)?

Suppose a body is at rest on an inclined plane. The force due to gravity is the weight of the object directed straight down. See Figure 7.41(i). The length of the projection of this vector onto the incline is called the *parallel component*. It represents the force exerted by the object down the incline. See Figure 7.41(ii). This information is needed in Problems 13 and 14.

Figure 7.41

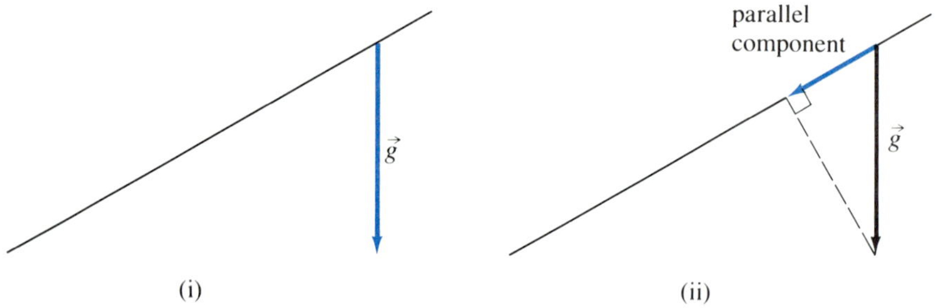

13. To two significant digits, what force, parallel to the roadway, is required to keep a car weighing 3000 pounds from rolling down a roadway inclined 10°?

14. To two significant digits, what force, parallel to the roadway, is required to keep a 2000-pound car from rolling down a roadway that rises 78 feet along a 300-foot section?

CHAPTER 7 REVIEW

Key Terms, Concepts, and Formulas

adding vectors, page 258

ambiguous case, page 241

area of a triangle, page 238

$K = \frac{1}{2}bc \sin \alpha = \frac{1}{2}ac \sin \beta = \frac{1}{2}ab \sin \gamma$

bearing, page 265

component form, page 256

course, heading, and wind vectors, page 265

direction, page 254

equal vectors, page 256

Heron's formula, page 249

$K = \sqrt{s(s-a)(s-b)(s-c)}$

initial point or head, page 253

law of cosines, page 245

$c^2 = a^2 + b^2 - 2ab \cos \gamma$

law of sines, page 240

$\frac{\sin \alpha}{a} = \frac{\sin \beta}{b} = \frac{\sin \gamma}{c}$

magnitude, page 254

parallelogram rule, page 260

resultant force, page 263

terminal point or tail, page 253

vector, page 253

Problem Set

1. Specify "true" or "false."
   (a) A triangle may be solved if we know any three of its parts.
   (b) In solving a triangle, the ambiguous case may arise when we have two sides and the included angle.
   (c) The area of a circle is an example of a vector quantity.
   (d) The component form of a vector is derived when the vector's tail is at the origin.
   (e) If two vectors are parallel and have the same length, then they must be equal.

2. In $\triangle ABC$, $\alpha = 25°$, $\beta = 68°$, and $a = 12$. Find b.

3. In $\triangle ABC$, $a = 7.5$, $c = 9.2$, and $\gamma = 47°$. Find α.

4. In $\triangle ABC$, $a = 7$, $c = 20$, and $\alpha = 35°$. Find γ.

5. In $\triangle ABC$, $a = 20$, $b = 25$, and $\gamma = 87°$. Find c.

6. In $\triangle ABC$, $a = 16$, $b = 18$, and $c = 15$. Find β.

In Problems 7–10, solve triangle *ABC*.

7. $a = 5, b = 20, c = 18$

8. $b = 5, c = 6, \beta = 55°$

9. $a = 8, b = 5, \gamma = 49°$

10. $b = 5, \alpha = 35°, \beta = 72°$

In Problems 11–14, find the area (to four significant digits) of triangle *ABC* for the given conditions.

11. $a = 35$, $b = 28$, and $\gamma = 65°$

12. $a = 25$, $c = 27$, and $\alpha = 65°$

13. $b = 16.5$, $\alpha = 28°$, and $\gamma = 54°$

14. $a = 26$, $b = 28$, and $c = 21$

15. A vertical pole is standing on a incline of 10°. If it casts a 25-foot shadow down the slope when the angle of elevation of the sun is 65°, how tall (to two significant digits) is the pole?

16. A tower 85 feet high sits on the edge of a cliff on the bank of a river. From the top of the tower the angle of depression to the opposite shore is 60°. From the bottom of the tower the angle of depression is 42°. How wide (to two significant digits) is the river?

17. From the bottom of a building the angle of elevation to the top of a tree is 67°. From the top of the building the angle of depression is 25°. How tall (to two significant digits) is the tree if the building is 55 feet high?

18. Two markers are placed on opposite sides, and at the edges, of a crater. An observer sights the two markers, noting that the angle between the lines of sight is 68°. If one marker is 155 meters from the observer and the other is 120 meters away, how wide (to two significant digits) is the crater?

19. A straight tunnel is to be built through a mountain. From a fixed point, the distance to the proposed entrance is 1.56 miles, and the distance to the proposed exit is 2.47 miles. If the angle between the lines of sight is 38.2°, how long (to three significant digits) will the tunnel be?

20. A vertical tower 150 feet tall stands on an incline of 12.5°. Two guy wires stretch from the top of the tower to points 65.5 feet directly above and directly below the base of the tower. How long (to three significant digits) are the two guy wires?

21. Determine the magnitude and the direction of the vector in the following figure.

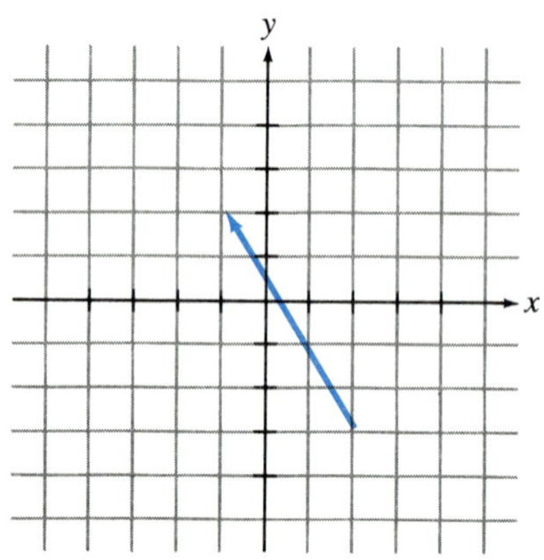

22. Determine the component form of $\overrightarrow{AB}$ if $A = (-3.5, -1)$ and $B = (4, 3)$.

In Problems 23 and 24, determine the component form of the vector with the specified magnitude and direction.

23. magnitude 6, direction 225°

24. magnitude 6.5, direction 112°

25. Determine $\vec{U} + \vec{V}$ if $\vec{U} = \langle -3, 5 \rangle$ and $\vec{V} = \langle 6, 1 \rangle$.

26. Draw the vector $\vec{u} + \vec{v}$ using the figure at the top of page 273.

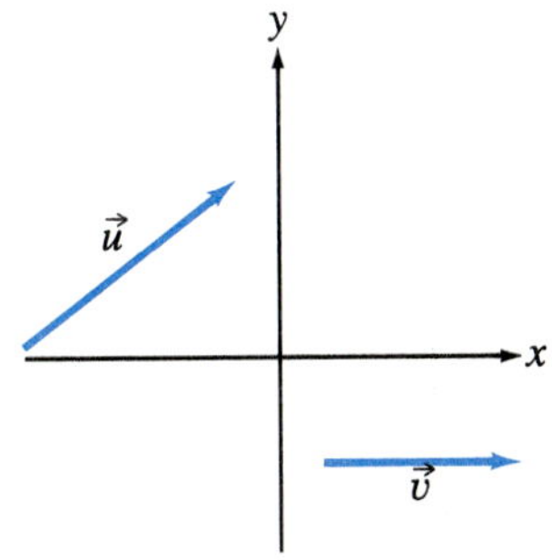

27. Two forces, one of 55 pounds and the other of 85 pounds, act on a body at an angle of 120°. Determine the magnitude of the resultant force (to two significant digits) and the angle (to the nearest degree) it makes with the 85-pound force.

28. Suppose an airplane is heading southeast with an airspeed of 120 miles per hour. It is flying with a 15-miles-per-hour tailwind from due west. What is its groundspeed (to two significant digits) and what is its course (to the nearest degree)?

29. Suppose an airplane is heading N35°E with an airspeed of 450 miles per hour. If its course is N33°E with a groundspeed of 460 miles per hour, what is the wind velocity (to two significant digits and to the nearest degree)?

30. Suppose the wind velocity is 20 miles per hour from S15°W. If a pilot wishes to fly along a course S80°W, what should be the heading (to the nearest degree) and what will be the groundspeed (to two significant digits) if the airplane has an airspeed of 165 miles per hour?

8
COMPLEX NUMBERS AND POLAR COORDINATES

CHAPTER OUTLINE

So far all variables we have considered have been *real* variables. That is, they have represented real numbers only. Hence there are no solutions of $x^2 = -7$ because the square of a real number is always nonnegative. Similarly, if we use the quadratic formula to solve a quadratic equation and the discriminant $b^2 - 4ac$ is negative, then there are no real solutions. For some applications we need a system of numbers in which such equations have solutions. The first two sections introduce such a system and show how trigonometry simplifies the arithmetic operations of this system. In the last two sections we study another coordinate system. As in the Cartesian coordinate system, which we have used extensively, points will be identified with ordered pairs. In the new system, however, one of the entries will be an angle.

8.1 COMPLEX NUMBERS

The complex number system needs a number whose square is negative. We thus define i to be a number such that

$$i^2 = -1.$$

(Of course, this number is not a real number since its square is negative.) In view of the above we will represent i as

$$i = \sqrt{-1}.$$

The set of all numbers of the form

$$a + bi,$$

where a and b are real numbers, is called the set of **complex numbers** and is denoted C. If $z = a + bi$, then a is called the **real part** of z, and b is called the **imaginary part** of z. A complex number such as $a + (-b)i$ will be written as $a - bi$. Also, we will agree that

$$0i = 0 \quad \text{and} \quad a + 0i = a.$$

Hence every real number is also a complex number.

Two complex numbers $a + bi$ and $c + di$ are said to be equal provided that $a = c$ and $b = d$. In other words, equal complex numbers must have equal real parts and equal imaginary parts.

For complex numbers the operations of addition and multiplication will be defined so that when the complex numbers are also real, we will get the sum and product expected. In this way, the set of complex numbers, with these operations, is just an extension of the real number system.

The sum of two complex numbers is defined as follows:

$$(a + bi) + (c + di) = (a + c) + (b + d)i.$$

Thus, to add complex numbers we merely add the real parts and add the imaginary parts. Note that this result may be obtained by viewing each complex number as a binomial and adding like terms.

The product of two complex numbers is defined as

$$(a + bi)(c + di) = (ac - bd) + (ad + bc)i.$$

This product may be achieved by treating each complex number as a binomial and using the fact that $i^2 = -1$:

$$\begin{aligned}(a + bi)(c + di) &= ac + adi + bci + bdi^2 \\ &= ac + adi + bci + bd(-1) \\ &= (ac - bd) + (ad + bc)i.\end{aligned}$$

EXAMPLE 1 Perform the indicated operation, expressing the result in the form $a + bi$.

(a) $(2 + 3i) + (5 + 4i)$ (b) $(4 + 2i) + (-2 - 3i)$
(c) $(5 + i)(3 + 2i)$ (d) $(3 - 2i)(4 + 3i)$

SOLUTION (a) $(2 + 3i) + (5 + 4i) = (2 + 5) + (3i + 4i) = 7 + 7i$
(b) $(4 + 2i) + (-2 - 3i) = (4 - 2) + (2i - 3i) = 2 - i$
(c) $(5 + i)(3 + 2i) = 15 + 10i + 3i + 2i^2 = 15 + 13i + 2(-1) = 13 + 13i$
(d) $(3 - 2i)(4 + 3i) = 12 + 9i - 8i - 6i^2 = 12 + i - 6(-1) = 18 + i$

▲

▲ *Exercise* Perform the indicated operation, expressing the result in the form $a + bi$.

(a) $(1 + 2i) + (5 + 3i)$ (b) $(3 + 2i) + (6 - 5i)$
(c) $(1 + 3i)(2 + 4i)$ (d) $(2 - 3i)(4 + 5i)$

Answer (a) $6 + 5i$ (b) $9 - 3i$ (c) $-10 + 10i$ (d) $23 - 2i$

Before discussing subtraction and division we need to discuss inverses. We begin by noting that there exist two complex numbers 0 and 1 ($0 + 0i$ and $1 + 0i$, respectively) such that

$$z + 0 = 0 + z = z \quad \text{and} \quad z \cdot 1 = 1 \cdot z = z$$

for all complex numbers z. This leads us to inverse elements.

For every complex number z, there exists a complex number denoted $-z$ such that

$$z + (-z) = (-z) + z = 0,$$

and for every $z \neq 0$, there exists a complex number denoted $1/z$ such that

$$z\left(\frac{1}{z}\right) = \left(\frac{1}{z}\right)z = 1.$$

We may see this by letting $z = a + bi$. Since

$$(a + bi) + (-a - bi) = 0 + 0i = 0,$$

$-z = -a - bi$. Also,

$$\begin{aligned}(a + bi)\left(\frac{a}{a^2 + b^2} - \frac{b}{a^2 + b^2}i\right) &= \frac{a^2}{a^2 + b^2} - \frac{ab}{a^2 + b^2}i + \frac{ab}{a^2 + b^2}i \\ &\quad - \frac{b^2}{a^2 + b^2}i^2 \\ &= \frac{a^2}{a^2 + b^2} + \frac{b^2}{a^2 + b^2} = 1.\end{aligned}$$

Thus, if $z \neq 0$,

$$\frac{1}{z} = \frac{a}{a^2 + b^2} - \frac{b}{a^2 + b^2}\,i.$$

The complex number $-z$ is called the *additive inverse* of z. And the complex number $1/z$ is called the *multiplicative inverse* of z.

The multiplicative inverse is perhaps easiest to describe with the use of conjugates. Let $z = a + bi$. Then the **conjugate** of z, denoted $\bar{z}$ (read "z bar"), is defined by

$$\bar{z} = a - bi.$$

For example, the conjugates of $2 + 3i$ and $5 - 4i$ are $2 - 3i$ and $5 + 4i$, respectively. (Of course, the conjugate of a real number is just itself.)

Note that

$$z\bar{z} = (a + bi)(a - bi) = a^2 - abi + abi - b^2i^2 = a^2 + b^2.$$

Thus the multiplicative inverse of z may be written as

$$\frac{1}{z} = \frac{a}{a^2 + b^2} - \frac{b}{a^2 + b^2}\,i = \frac{1}{a^2 + b^2}(a - bi) = \frac{1}{z\bar{z}} \cdot \bar{z}.$$

The operations of subtraction and division are defined in terms of addition and multiplication, respectively. If $z = a + bi$ and $w = c + di$, then

$$\begin{aligned} z - w &= z + (-w) \\ &= (a + bi) + (-c - di) \\ &= (a - c) + (b - d)i. \end{aligned}$$

Hence, to subtract complex numbers we subtract real parts and subtract imaginary parts. As with addition, we may view each complex number as a binomial and combine like terms.

To divide, note that

$$\begin{aligned} z \div w &= z \cdot \frac{1}{w} = z \cdot \frac{1}{w\bar{w}} \cdot \bar{w} \\ &= \frac{z}{w} \cdot \frac{\bar{w}}{\bar{w}}. \end{aligned}$$

Thus, when dividing we may view the quotient as a fraction and simply multiply both the numerator and denominator by the conjugate of the denominator.

EXAMPLE 2 Perform the indicated operation, expressing the result in the form $a + bi$.

(a) $(5 + 2i) - (3 + i)$ (b) $(6 - 3i) - (2 - i)$
(c) $(5 + i) \div (3 + 2i)$ (d) $(3 - 2i) \div (4 - 3i)$

SOLUTION For parts (a) and (b) we merely use the definition and combine like terms.

(a) $(5 + 2i) - (3 + i) = 5 - 3 + 2i - i = 2 + i$
(b) $(6 - 3i) - (2 - i) = 6 - 2 - 3i + i = 4 - 2i$
(c) We have

$$(5 + i) \div (3 + 2i) = \frac{5 + i}{3 + 2i}.$$

Multiplying both the numerator and the denominator by the conjugate of the denominator, we get

$$\frac{5+i}{3+2i} = \frac{5+i}{3+2i} \cdot \frac{3-2i}{3-2i} = \frac{15 - 7i - 2i^2}{9 - 4i^2}$$
$$= \frac{17 - 7i}{13} = \frac{17}{13} - \frac{7}{13}i.$$

(d) We have

$$(3 - 2i) \div (4 - 3i) = \frac{3-2i}{4-3i} \cdot \frac{4+3i}{4+3i} = \frac{12 + i - 6i^2}{16 - 9i^2}$$
$$\frac{18 + i}{25} = \frac{18}{25} + \frac{1}{25}i.$$

▲

▲ *Exercise* Perform the indicated operation, expressing the result in the form $a + bi$.

(a) $(7 + 2i) - (3 + 5i)$ (b) $(11 - 3i) - (7 - 2i)$
(c) $(3 + 2i) \div (4 + i)$ (d) $(5 - 4i) \div (2 - 5i)$

Answer (a) $4 - 3i$ (b) $4 - i$ (c) $\frac{14}{17} + \frac{5}{17}i$ (d) $\frac{30}{29} + \frac{17}{29}i$

Just as there are two real numbers whose square is 4, there are two complex numbers whose square is -4.

$$(2i)^2 = 2^2 i^2 = 4(-1) = -4,$$
$$(-2i)^2 = (-2)^2 i^2 = 4(-1) = -4.$$

The radical symbol refers to only one of these. For any positive number a,

$$\sqrt{-a} = \sqrt{a}\, i.$$

Hence

$$\sqrt{-4} = \sqrt{4}\, i = 2i.$$

EXAMPLE 3 Find the complex solutions.

(a) $x^2 = -9$ (b) $x^2 + x + 1 = 0$

SOLUTION (a) Using the method of extraction we have

$$x^2 = -9$$
$$x = \pm\sqrt{-9} = \pm 3i.$$

Therefore, the solutions are $3i$ and $-3i$.

(b) Using the quadratic formula we get

$$x = \frac{-1 \pm \sqrt{1-4}}{2} = \frac{-1 \pm \sqrt{-3}}{2} = \frac{-1 \pm \sqrt{3}\,i}{2}.$$

Hence the solutions are $\frac{-1}{2} + \frac{\sqrt{3}}{2}i$ and $\frac{-1}{2} - \frac{\sqrt{3}}{2}i$.

▲

Note that the equations of Example 3 have no real solutions, but with the allowance of complex solutions, there must be at least one. (Exactly one occurs when, using the quadratic formula, the discriminant $b^2 - 4ac = 0$.)

Let's also note that, in Example 3, the conjugate of each solution is the other solution. Will this always occur for quadratic equations? If all the coefficients are real, the answer is "yes." If z is a complex root of a quadratic equation with real coefficients, then $\bar{z}$ is also a complex root.

▲ *Exercise* Find the complex solutions of the equation.

(a) $x^2 = -36$ (b) $x^2 - x + 2 = 0$

Answer (a) $6i, -6i$ (b) $\frac{1}{2} + \frac{\sqrt{7}}{2}i, \frac{1}{2} - \frac{\sqrt{7}}{2}i$

Problem Set 8.1

In Problems 1–24, perform the indicated operation, expressing the result in the form $a + bi$.

1. $(6 + 4i) + (5 + 2i)$
2. $(12 + i) + (6 + 7i)$
3. $(4 - 7i) + (-3 + 2i)$
4. $(-5 - 6i) + (-2 - 5i)$
5. $(5 + 7i) - (3 + 2i)$
6. $(6 + 2i) - (7 + 4i)$
7. $(-3 + 2i) - (1 - i)$
8. $(11 - i) - (-5 + 3i)$
9. $(2 + 3i)(5 + 4i)$
10. $(5 + 2i)(7 + i)$
11. $(5 - 2i)(-6 + 3i)$
12. $(-4 - 3i)(2 + 3i)$
13. $(-4 + 5i) \div (3 + 5i)$
14. $(5 + 3i) \div (4 + 2i)$

15. $\dfrac{6 - i}{-3 + 2i}$

16. $\dfrac{-4 + 3i}{5 - 4i}$

17. $(2 + i)(2 - i)$

18. $5(2 - 3i)$

19. $(2 + 3i)^2$

20. $(1 - 2i)^2$

21. i^{36}

22. i^{50}

23. i^{27}

24. i^{17}

25. Show that addition and multiplication of complex numbers is commutative. That is, show that $z_1 + z_2 = z_2 + z_1$ and $z_1 z_2 = z_2 z_1$ for any complex numbers z_1 and z_2.

26. Show that $z_1(z_2 + z_3) = z_1 z_2 + z_1 z_3$ for any complex numbers z_1, z_2, and z_3.

27. Show that $i^n = 1, i, -1$, or $-i$ if the remainder upon dividing n by 4 is 0, 1, 2, or 3, respectively.

28. Show that for any complex number z, $\bar{\bar{z}} = z$.

29. Show that for any complex numbers z and w, $\overline{z \cdot w} = \bar{z} \cdot \bar{w}$.

30. Let f be a function with domain C (the set of complex numbers) defined by $f(x) = x^2 + 2x + 5$. Determine each of the following:

(a) $f(i)$ (b) $f(1 + i)$ (c) $f(2 - i)$

In Problems 31–40, find the complex solutions.

31. $x^2 = -16$

32. $x^2 = -25$

33. $(x + 3)^2 = -4$

34. $(x - 2)^2 = -9$

35. $x^2 + 3x + 5 = 0$

36. $x^2 - 5x + 7 = 0$

37. $2x^2 - 4x + 3 = 0$

38. $3x^2 + 2x + 2 = 0$

39. $2x + 1 + 3i = 6 - i$

40. $3x - 2 - 5i = 7 + 3i$

8.2 TRIGONOMETRIC FORM OF A COMPLEX NUMBER

We use a number line to represent the real numbers. To represent complex numbers, we use a Cartesian coordinate system. We associate the complex number $a + bi$ with the point (a, b), and vice versa. Some complex numbers are plotted in Figure 8.1.

When a Cartesian coordinate system is used to represent complex numbers, it is called a **complex plane**. The x-axis is called the **real axis**, and the

Figure 8.1

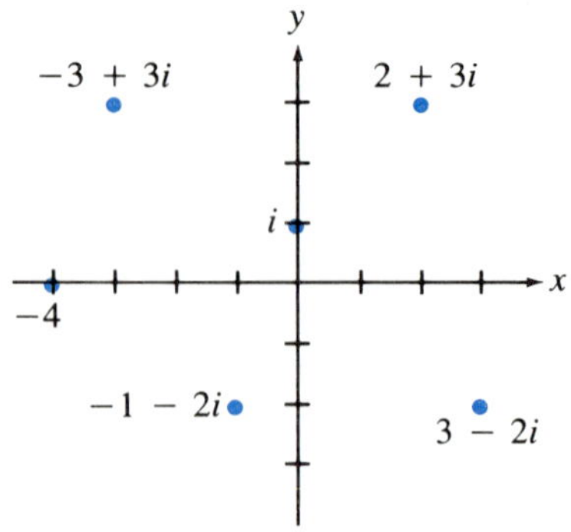

y-axis is called the **imaginary axis**. The distance from (a, b) to the origin is called the **absolute value** of $a + bi$ and is denoted by $|a + bi|$. By the distance formula we have

$$|a + bi| = \sqrt{a^2 + b^2}.$$

For example,

$$|2 - 3i| = \sqrt{2^2 + (-3)^2} = \sqrt{13}.$$

Let $a + bi$ be any nonzero complex number. Let $P = (a, b)$ be its representation in the complex plane, and let r be its absolute value. Now let θ be any angle in standard position whose terminal side passes through P. (θ is usually measured in radians.) See Figure 8.2.

Figure 8.2

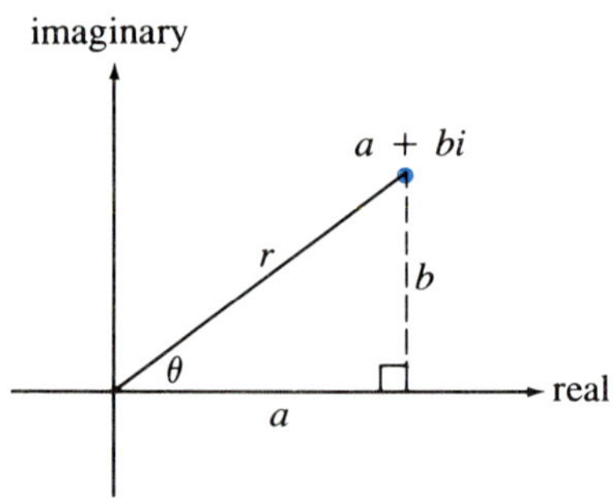

Thus θ is an angle whose tangent is b/a if $a \neq 0$. From the definitions of the trigonometric ratios,

$$\cos\theta = \frac{a}{r} \quad \text{and} \quad \sin\theta = \frac{b}{r}.$$

Hence

$$a = r\cos\theta \quad \text{and} \quad b = r\sin\theta,$$

and the complex number $a + bi$ may be written

$$a + bi = (r\cos\theta) + (r\sin\theta)i.$$

That is,

$$a + bi = r(\cos\theta + i\sin\theta). \tag{8.1}$$

The right side of equation (8.1) is called the **trigonometric form of a complex number**. Of course, since there are infinitely many choices for θ, the trigonometric form is not unique.

EXAMPLE 1 Write the complex number in trigonometric form, as specified.

(a) $3 - 3i$, exactly
(b) $3 + 5i$, approximately (with r to four significant digits and θ to the nearest thousandth)

SOLUTION (a) The absolute value of $3 - 3i$ is

$$r = \sqrt{3^2 + (-3)^2} = 3\sqrt{2}.$$

Angle θ must be one whose tangent is $\frac{-3}{3} = -1$. So the reference angle has a tangent of 1 and must be $\pi/4$. And because $3 - 3i$ is in the fourth quadrant, θ must be a fourth-quadrant angle. So let's choose $\theta = -\pi/4$. Thus

$$3 - 3i = 3\sqrt{2}\left(\cos\left(\frac{-\pi}{4}\right) + i\sin\left(\frac{-\pi}{4}\right)\right).$$

(b) The absolute value is

$$r = \sqrt{3^2 + 5^2} = \sqrt{34} \approx 5.831.$$

Angle θ must be a first-quadrant angle whose tangent is $\frac{5}{3}$. If we choose θ to be acute,

$$\theta = \operatorname{Arctan}\frac{5}{3} \approx 1.030.$$

Thus,

$$3 + 5i \approx 5.831(\cos 1.030 + i\sin 1.030).$$

▲

▲ ***Exercise*** Write the complex number in trigonometric form, as specified.

(a) $-1 + \sqrt{3}\,i$, exactly (b) $-2 - 3i$, approximately

Answer (a) $2\left(\cos\frac{2\pi}{3} + i\sin\frac{2\pi}{3}\right)$ (b) $3.606(\cos 4.124 + i\sin 4.124)$

EXAMPLE 2 Write the complex number in the form $a + bi$, as specified.

(a) $4\left(\cos\frac{\pi}{6} + i\sin\frac{\pi}{6}\right)$, exactly

(b) $2.7(\cos 1.51 + i\sin 1.51)$, approximately

SOLUTION (a) We simply have to substitute the proper trigonometric values. We get

$$4\left(\cos\frac{\pi}{6} + i\sin\frac{\pi}{6}\right) = 4\left(\frac{\sqrt{3}}{2} + i\frac{1}{2}\right) = 2\sqrt{3} + 2i.$$

(b) Approximating cos 1.51 and sin 1.51 and multiplying by 2.7 gives

$$2.7(\cos 1.51 + i\sin 1.51) \approx 0.1640 + 2.695i.$$

▲

▲ *Exercise* Write the complex number in the form $a + bi$, as specified.

(a) $2\left(\cos\frac{4\pi}{3} + i\sin\frac{4\pi}{3}\right)$, exactly

(b) $5.2(\cos 2.5 + i\sin 2.5)$, approximately

Answer (a) $-1 - \sqrt{3}i$ (b) $-4.166 + 3.112i$

Addition and subtraction of complex numbers is better accomplished when they are in the form $a + bi$. But the trigonometric form is quite suitable for multiplication and division. Letting

$$z_1 = r_1(\cos\theta_1 + i\sin\theta_1) \quad \text{and} \quad z_2 = r_2(\cos\theta_2 + i\sin\theta_2),$$

we have

$$\begin{aligned} z_1 z_2 &= r_1(\cos\theta_1 + i\sin\theta_1)\cdot r_2(\cos\theta_2 + i\sin\theta_2) \\ &= r_1 r_2(\cos\theta_1\cos\theta_2 + i\cos\theta_1\sin\theta_2 + i\sin\theta_1\cos\theta_2 \\ &\quad + i^2\sin\theta_1\sin\theta_2) \\ &= r_1 r_2((\cos\theta_1\cos\theta_2 - \sin\theta_1\sin\theta_2) + i(\sin\theta_1\cos\theta_2 \\ &\quad + \sin\theta_2\cos\theta_1)), \end{aligned}$$

but by the formulas for the cosine of a sum and the sine of a sum,

$$\cos\theta_1\cos\theta_2 - \sin\theta_1\sin\theta_2 = \cos(\theta_1 + \theta_2),$$

and
$$\sin\theta_1\cos\theta_2 + \sin\theta_2\cos\theta_1 = \sin(\theta_1 + \theta_2).$$

Therefore,

$$z_1 z_2 = r_1 r_2(\cos(\theta_1 + \theta_2) + i\sin(\theta_1 + \theta_2)).$$

Thus, to multiply two complex numbers in trigonometric form, we multiply the absolute values and add the angles. Similarly,

$$\frac{z_1}{z_2} = \frac{r_1}{r_2}(\cos(\theta_1 - \theta_2) + i \sin(\theta_1 - \theta_2)).$$

EXAMPLE 3 Let $z_1 = 5(\cos(\pi/4) + i \sin(\pi/4))$ and $z_2 = 4(\cos \pi + i \sin \pi)$. Perform the indicated operation.

(a) $z_1 z_2$ (b) $\dfrac{z_1}{z_2}$

SOLUTION We have

(a) $z_1 z_2 = 5 \cdot 4\left(\cos\left(\dfrac{\pi}{4} + \pi\right) + i \sin\left(\dfrac{\pi}{4} + \pi\right)\right) = 20 \cos\left(\dfrac{5\pi}{4} + i \sin \dfrac{5\pi}{4}\right)$

(b) $\dfrac{z_1}{z_2} = \dfrac{5}{4}\left(\cos\left(\dfrac{\pi}{4} - \pi\right) + i \sin\left(\dfrac{\pi}{4} - \pi\right)\right) = \dfrac{5}{4}\left(\cos\left(\dfrac{-3\pi}{4}\right) + i \sin\left(\dfrac{-3\pi}{4}\right)\right)$

▲

▲ ***Exercise*** Let $z_1 = 3(\cos(\pi/2) + i \sin(\pi/2))$ and $z_2 = 2(\cos(\pi/3) + i \sin(\pi/3))$. Perform the indicated operation.

(a) $z_1 z_2$ (b) $\dfrac{z_1}{z_2}$

Answer (a) $6\left(\cos \dfrac{5\pi}{6} + i \sin \dfrac{5\pi}{6}\right)$ (b) $\dfrac{3}{2}\left(\cos \dfrac{\pi}{6} + i \sin \dfrac{\pi}{6}\right)$

De Moivre's Theorem tells us how to raise a complex number in trigonometric form to a positive integral power. It is not a surprising result since such exponentiation is merely a repeated multiplication. It says that, for any positive integer n,

$$(r(\cos \theta + i \sin \theta))^n = r^n(\cos n\theta + i \sin n\theta).$$

EXAMPLE 4 Evaluate the expression as specified.

(a) $(\sqrt{3} - i)^5$, exactly (b) $(2 + i)^6$, approximately

SOLUTION (a) Let's begin by writing $\sqrt{3} - i$ in trigonometric form. Its absolute value is $r = 2$, and θ is a fourth-quadrant angle whose tangent is $-1/\sqrt{3}$. Let's choose $\theta = -\pi/6$. Hence

$$\sqrt{3} - i = 2\left(\cos\left(\frac{-\pi}{6}\right) + i \sin\left(\frac{-\pi}{6}\right)\right).$$

By De Moivre's Theorem,

$$\begin{aligned}(\sqrt{3} - i)^5 &= \left(2\left(\cos\left(\frac{-\pi}{6}\right) + i \sin\left(\frac{-\pi}{6}\right)\right)\right)^5 \\ &= 2^5\left(\cos\left(\frac{-5\pi}{6}\right) + i \sin\left(\frac{-5\pi}{6}\right)\right) \\ &= 32\left(\frac{-\sqrt{3}}{2} + i \cdot \frac{-1}{2}\right) = -16\sqrt{3} - 16i.\end{aligned}$$

(b) To write the trigonometric form of $2 + i$, we note that the absolute value is $r = \sqrt{5}$, and we may choose $\theta = \text{Arctan } \frac{1}{2}$. Thus

$$2 + i = \sqrt{5}(\cos\theta + i \sin\theta).$$

By De Moivre's Theorem,

$$\begin{aligned}(2 + i)^6 &= (\sqrt{5})^6(\cos 6\theta + i \sin 6\theta) \\ &= 125.0(-0.936 + i \cdot 0.352) \\ &\approx -117.0 + 44.00i\end{aligned}$$

▲

▲ ***Exercise*** Evaluate the expression as specified.

(a) $(1 + i)^7$, exactly (b) $(1 - 3i)^8$, approximately

Answer (a) $8 - 8i$ (b) $-8432 + 5376i$

If z is a complex number and n is a positive integer, then the complex number w is called an nth root of z provided that

$$w^n = z.$$

We may use De Moivre's Theorem to find these roots. Let

$$z = r(\cos\theta + i \sin\theta) \quad \text{and} \quad w = s(\cos\alpha + i \sin\alpha)$$

be the trigonometric forms. Since $w^n = z$, by De Moivre's Theorem,

$$s^n(\cos n\alpha + i \sin n\alpha) = r(\cos\theta + i \sin\theta).$$

These two equal numbers have equal absolute values, so

$$\begin{aligned}s^n &= r \\ s &= \sqrt[n]{r}.\end{aligned}$$

Furthermore,

$$\cos n\alpha + i \sin n\alpha = \cos \theta + i \sin \theta .$$

Since two equal complex numbers must have equal real parts and equal imaginary parts,

$$\cos n\alpha = \cos \theta \quad \text{and} \quad \sin n\alpha = \sin \theta .$$

Since both the cosine and sine functions have periods of 2π, the preceding two equations have solutions if and only if the angles differ by a multiple of 2π. Thus

$$n\alpha - \theta = 2\pi k$$

for some integer k. Solving for α gives

$$\alpha = \frac{\theta + 2\pi k}{n} .$$

This leads us to the following theorem.

For any positive integer n, the complex number $z = r(\cos \theta + i \sin \theta)$ has exactly n distinct nth roots. These n roots are given by

$$\sqrt[n]{r}\left(\cos\left(\frac{\theta + 2\pi k}{n}\right) + i \sin\left(\frac{\theta + 2\pi k}{n}\right)\right),$$

where $k = 0, 1, 2, \ldots, n - 1$.

There are several observations to be made. First, note that the absolute value of each root is $\sqrt[n]{r}$. Hence all nth roots will be equal in absolute value. Also, note that each value of $k = 0, 1, 2, \ldots, n - 1$ gives a distinct root because none of the angles

$$\frac{\theta + 2\pi k}{n} = \frac{\theta}{n} + \frac{2\pi}{n} k$$

will be coterminal. On the other hand, other values of k will simply repeat these results. For example, when $k = n$, the angle is

$$\frac{\theta}{n} + 2\pi ,$$

which is coterminal with

$$\frac{\theta}{n},$$

the angle we get when $k = 0$. Hence $k = n$ doesn't yield a different nth root.

EXAMPLE 5 Find the cube roots of 1, exactly.

SOLUTION We must first write 1 in trigonometric form. We have $r = 1$ and may choose $\theta = 0$, so

$$1 = 1(\cos 0 + i \sin 0).$$

By the theorem on nth roots we will get three roots:

$$\sqrt[3]{1}\left(\cos\left(\frac{0 + 2\pi k}{3}\right) + i \sin\left(\frac{0 + 2\pi k}{3}\right)\right)$$

for $k = 0$, 1, and 2. These values of k yield

$$1(\cos 0 + i \sin 0) = 1,$$
$$1\left(\cos\frac{2\pi}{3} + i \sin\frac{2\pi}{3}\right) = \frac{-1}{2} + \frac{\sqrt{3}}{2}i,$$
$$1\left(\cos\frac{4\pi}{3} + i \sin\frac{4\pi}{3}\right) = \frac{-1}{2} - \frac{\sqrt{3}}{2}i.$$

▲

The graph of the three cube roots determined in Example 5 is shown in Figure 8.3.

Figure 8.3

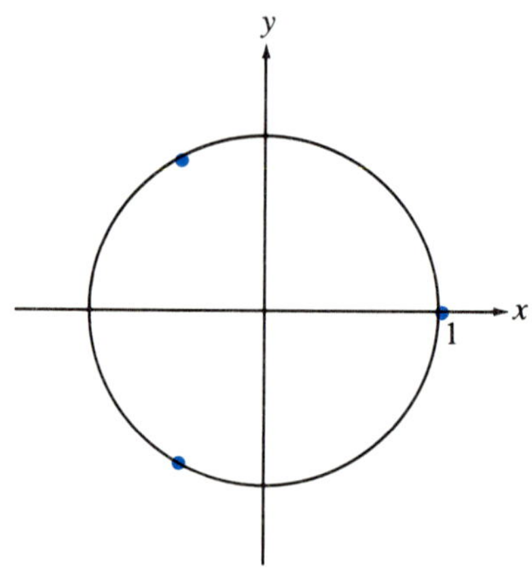

Note that the roots lie on a circle and are equally spaced along it. This will always be the case. The n distinct nth roots of $z = r(\cos\theta + i \sin\theta)$, given by

$$\sqrt[n]{r}\left(\cos\left(\frac{\theta + 2\pi k}{n}\right) + i \sin\left(\frac{\theta + 2\pi k}{n}\right)\right),$$

where $k = 0, 1, 2, \ldots, n - 1$, will be equally spaced (at intervals of $2\pi/n$) on a circle of radius $\sqrt[n]{r}$.

EXAMPLE 6 Find the fourth roots of $5 + i$, approximately.

SOLUTION We write $5 + i$ in trigonometric form. We have $r = \sqrt{26}$ and choose $\theta = \text{Arctan } \frac{1}{5}$. So

$$5 + i = \sqrt{26}(\cos \theta + i \sin \theta).$$

By the complex root theorem, the roots are

$$\sqrt[8]{26}\left(\cos\left(\frac{\theta + 2\pi k}{4}\right) + i \sin\left(\frac{\theta + 2\pi k}{4}\right)\right)$$

for $k = 0, 1, 2$, and 3. Upon substituting the various values of k we get these approximate roots:

$$1.501 + 0.07413i,$$

$$-0.07413 + 1.501i,$$

$$-1.501 - 0.07413i,$$

$$0.07413 - 1.501i.$$

▲

▲ *Exercise*

(a) Find the cube roots of -1, exactly.
(b) Find the fourth roots of $3 - i$, approximately.

Answers

(a) $\frac{1}{2} + \frac{\sqrt{3}}{2}i, -1, \frac{1}{2} - \frac{\sqrt{3}}{2}i$
(b) $1.329 - 0.1071i, 0.1071 + 1.329i, -1.329 + 0.1071i, -0.1071 - 1.329i$

Problem Set 8.2

In Problems 1–12, write the complex number in trigonometric form, exactly.

1. $1 + i$

2. $\sqrt{3} + i$

3. $2 - 2i$

4. $-3 + \sqrt{3}i$

5. $-4 + 4\sqrt{3}i$

6. $\sqrt{6} - \sqrt{2}i$

7. $-5\sqrt{2} - 5\sqrt{2}i$

8. $-\sqrt{10} - \sqrt{30}i$

9. $5i$

10. $-2i$

11. 3

12. -7

In Problems 13–20, write the complex number in trigonometric form, approximately.

13. $4 - 3i$

14. $6 - 5i$

15. $5 + 2i$

16. $3 + 6i$

17. $-7 - 2i$

18. $-3 - 5i$

19. $-5 + 12i$

20. $-2 + 8i$

In Problems 21–26, write the complex number in the form $a + bi$, exactly.

21. $2\left(\cos\frac{3\pi}{4} + i\sin\frac{3\pi}{4}\right)$

22. $2\left(\cos\frac{\pi}{6} + i\sin\frac{\pi}{6}\right)$

23. $5\left(\cos\left(\frac{-5\pi}{6}\right) + i\sin\left(\frac{-5\pi}{6}\right)\right)$

24. $4\left(\cos\left(\frac{-\pi}{3}\right) + i\sin\left(\frac{-\pi}{3}\right)\right)$

25. $6(\cos\pi + i\sin\pi)$

26. $5\left(\cos\frac{\pi}{2} + i\sin\frac{\pi}{2}\right)$

In Problems 27–30, write the complex number in the form $a + bi$, approximately.

27. $3(\cos 1.4 + i\sin 1.4)$

28. $4(\cos(-2.61) + i\sin(-2.61))$

29. $8(\cos(-1.2) + i\sin(-1.2))$

30. $6(\cos 4.2 + i\sin 4.2)$

In Problems 31–34, find $z_1 z_2$ and z_1/z_2, using the methods described in Section 8.2, and write the result in the form $a + bi$. To check, write z_1 and z_2 in the form $a + bi$, and use the methods described in Section 8.1.

31. $z_1 = 5\left(\cos\frac{\pi}{3} + i\sin\frac{\pi}{3}\right),\ z_2 = 2\left(\cos\left(\frac{-\pi}{6}\right) + i\sin\left(\frac{-\pi}{6}\right)\right)$

32. $z_1 = 3\left(\cos\frac{3\pi}{4} + i\sin\frac{3\pi}{4}\right),\ z_2 = 4(\cos\pi + i\sin\pi)$

33. $z_1 = 4\left(\cos\frac{2\pi}{3} + i\sin\frac{2\pi}{3}\right),\ z_2 = 2\left(\cos\frac{\pi}{2} + i\sin\frac{\pi}{2}\right)$

34. $z_1 = 3(\cos 1.21 + i\sin 1.21),\ z_2 = 2(\cos 3.5 + i\sin 3.5)$

In Problems 35–42, evaluate the expression by using De Moivre's Theorem, as specified.

35. $(5 - 5i)^4$, exactly

36. $(1 + \sqrt{3}\,i)^5$, exactly,

37. $(-\sqrt{3} + i)^7$, exactly

38. $(-2 + 2i)^6$, exactly

39. $(2 - 3i)^6$, approximately

40. $(5 + 4i)^3$, approximately

41. $(-1 + 5i)^5$, approximately

42. $(3 + 4i)^4$, approximately

In Problems 43–50, find the roots indicated, as specified.

43. cube roots of 8, exactly

44. cube roots of -27, exactly

45. square roots of $1 - \sqrt{3}\,i$, exactly

46. square roots of $1 + \sqrt{3}\,i$, exactly

47. square roots of $-4 - i$, approximately

48. square roots of $2 + i$, approximately

49. cube roots of $5 + i$, approximately

50. fourth roots of $2 - 5i$, approximately

8.3 POLAR COORDINATES

Let us begin with a plane containing the nonnegative part of a coordinate line, and let O denote the origin. Excluding the origin, each point P in the plane may be assigned coordinates (r, θ), where θ is the measure of an angle whose initial side is the coordinate line and whose terminal side is $\overrightarrow{OP}$, and where r is the length of the segment $\overrightarrow{OP}$. See Figure 8.4. The coordinates for the origin will be $(0, \theta)$, where θ is any angle.

Figure 8.4

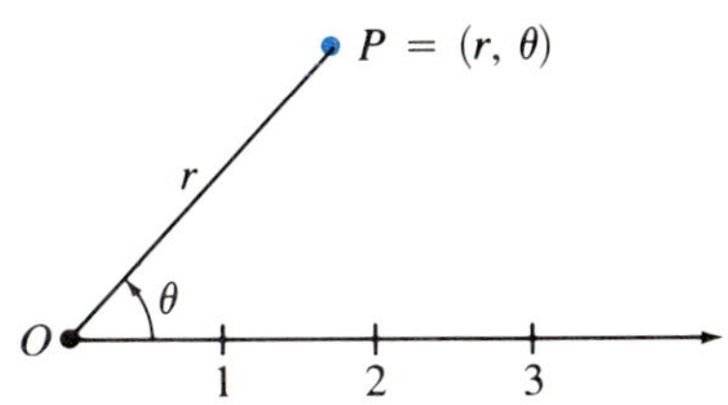

This system for identifying points in the plane is called the **polar coordinate system**. The origin is called the **pole**, and the coordinate line is called the **polar axis**. In Figure 8.5 (page 292) we have identified some points using this system.

Note that the measure may be greater than one rotation or may be negative. Consequently, with the polar coordinate system, the *representation of points is not unique*. For example, the point $(2, -3\pi/4)$ in Figure 8.5(ii) could have been represented as $(2, 5\pi/4)$, as well as in many other ways.

Additional representations of points may be made by allowing for negative values for the first coordinate. Each point P may be assigned coordinates (r, θ), with $r < 0$, where θ is the measure of an angle whose initial side is the polar axis and whose terminal side is the extension of $\overrightarrow{OP}$ through the pole, and $|r| = OP$. See Figure 8.6 (page 292).

The first entry of a coordinate pair, then, is a directed distance. In locating the point (r, θ) with $r < 0$, we measure the angle θ, but determine the distance on the opposite side. We have identified some points in Figure 8.7 (page 293).

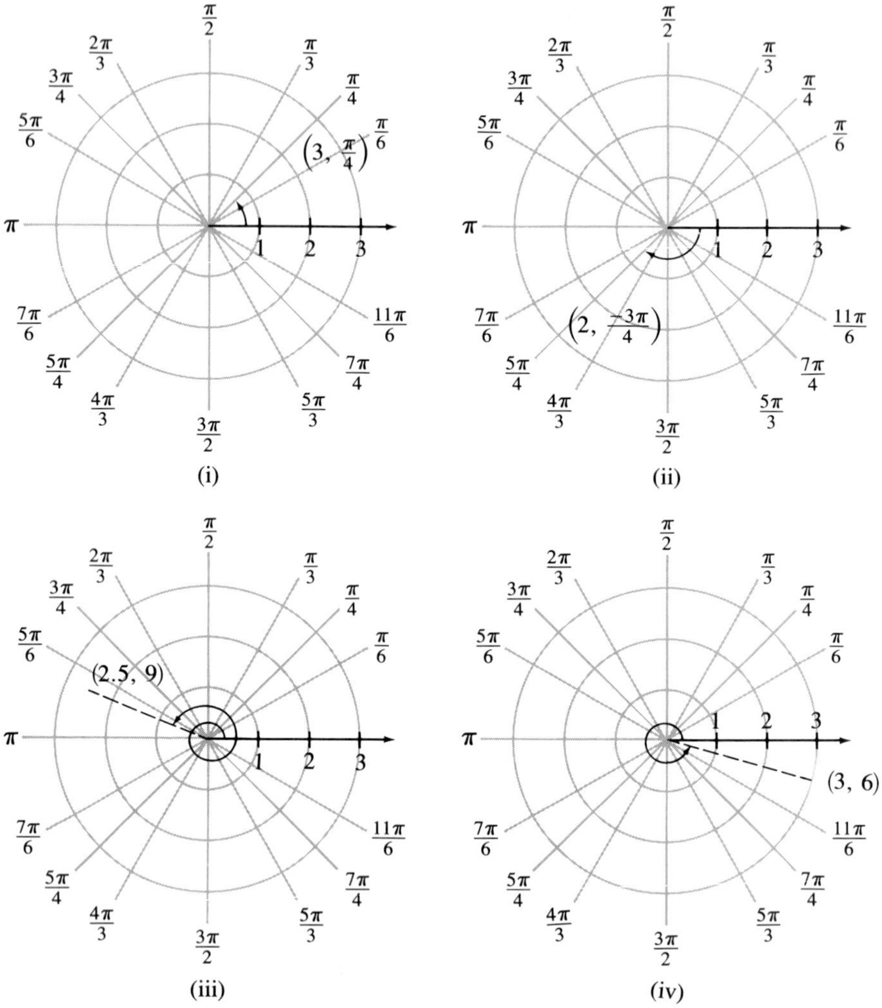

Figure 8.5

Figure 8.6
$r < 0$

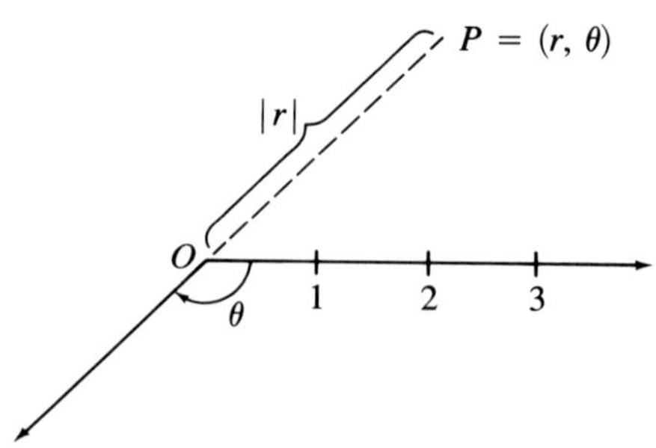

So we see that we may represent a point in many different ways, using multiple angles and negative first coordinates. The point $(-3, \pi/4)$ in Figure 8.7(i) could have been written as $(3, 5\pi/4)$ or as $(3, -3\pi/4)$ or as $(-3, -7\pi/4)$, etc.

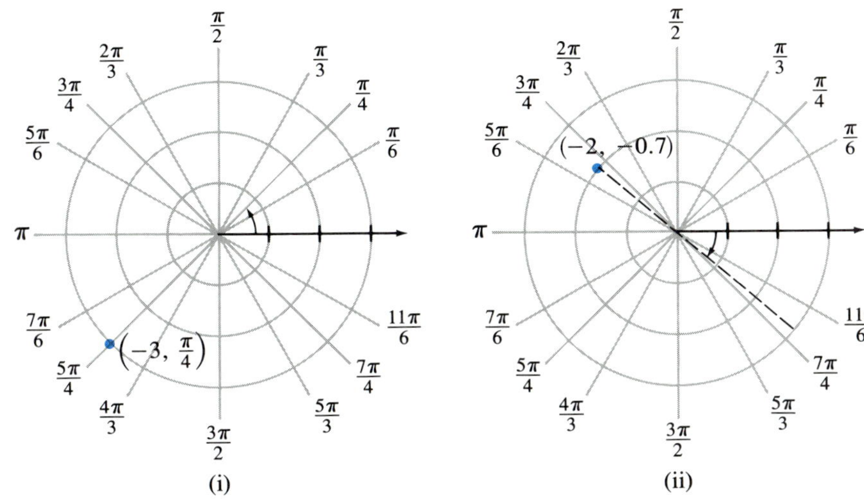

Figure 8.7

▲ *Exercise* Using a polar coordinate system, plot the following points: $(2, \pi/3)$, $(-2.5, 5\pi/6)$, $(3, -\pi/3)$, $(1, 7\pi/6)$, $(-2, -\pi/4)$.

Answer See Figure 8.8.

Figure 8.8

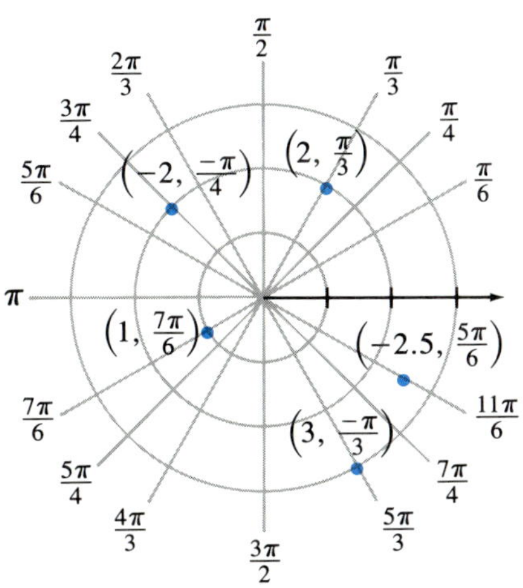

Suppose we superimpose a polar coordinate system on a rectangular (Cartesian) coordinate system by taking the polar axis to be the nonnegative part of the x-axis. See Figure 8.9 (page 294). Under such circumstances, each point in the plane has a rectangular representation (x, y) and a polar representation (r, θ). What is the connection between the two?

Let's first consider a point P whose polar representation is (r, θ). What is its rectangular representation (x, y)? See Figure 8.10 (page 294).

Figure 8.9

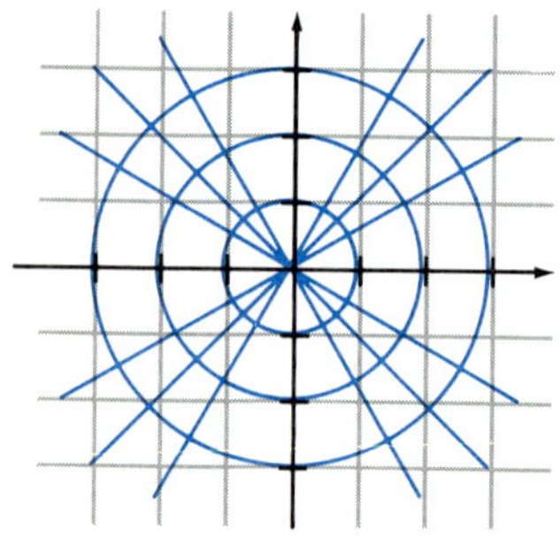

Figure 8.10

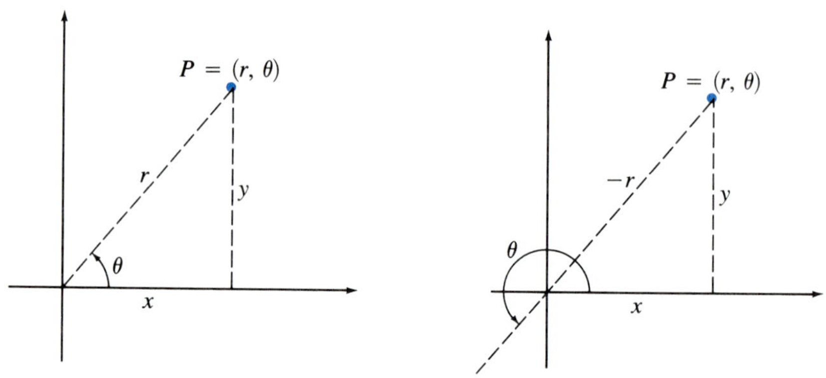

(i) $r > 0$ (ii) $r < 0$

In Figure 8.10(i) we see that

$$\cos\theta = \frac{x}{r} \quad \text{and} \quad \sin\theta = \frac{y}{r}.$$

Hence

$$x = r\cos\theta \quad \text{and} \quad y = r\sin\theta.$$

When $r < 0$, we have the situation in Figure 8.10(ii), where

$$\cos(\theta - \pi) = \frac{x}{-r} \quad \text{and} \quad \sin(\theta - \pi) = \frac{y}{-r}.$$

Thus

$$x = -r\cos(\theta - \pi) \quad \text{and} \quad y = -r\sin(\theta - \pi).$$

But

$$\begin{aligned}\cos(\theta - \pi) &= \cos\theta\cos\pi + \sin\theta\sin\pi \\ &= -\cos\theta,\end{aligned}$$

and

$$\begin{aligned}\sin(\theta - \pi) &= \sin\theta\cos\pi - \sin\pi\cos\theta \\ &= -\sin\theta.\end{aligned}$$

Consequently,

$$x = r \cos \theta \quad \text{and} \quad y = r \sin \theta, \qquad \textbf{(8.2)}$$

which is the same formula we obtained when $r > 0$. So the equations in (8.2) will be used to convert from polar to rectangular coordinates.

EXAMPLE 1 Convert to rectangular coordinates, as specified.

(a) $\left(3, \dfrac{\pi}{6}\right)$, exactly (b) $(-2, 2.1)$, approximately

SOLUTION (a) Using the formulas from statement (8.2) we get

$$x = 3 \cos \frac{\pi}{6} = 3 \cdot \frac{\sqrt{3}}{2} = \frac{3\sqrt{3}}{2},$$

$$y = 3 \sin \frac{\pi}{6} = 3 \cdot \frac{1}{2} = \frac{3}{2}.$$

Hence the rectangular representation is $(3\sqrt{3}/2, 3/2)$.

(b) We have

$$x = -2 \cos 2.1 \approx 1.010,$$

$$y = -2 \sin 2.1 \approx -1.726.$$

In rectangular coordinates, then, the point is approximately $(1.010, -1.726)$.

▲

▲ ***Exercise*** Convert to rectangular coordinates, as specified.

(a) $\left(4, \dfrac{\pi}{4}\right)$, exactly (b) $(-3, 2.4)$, approximately

Answer (a) $(2\sqrt{2}, 2\sqrt{2})$ (b) $\approx(2.212, -2.026)$

Let's now consider the problem of converting from rectangular to polar. Suppose that point P has a rectangular representation of (x, y). What is its polar representation (r, θ)? From the preceding Figure 8.10 we see that

$$r = \pm\sqrt{x^2 + y^2} \quad \text{and} \quad \tan \theta = \frac{y}{x}, \qquad x \neq 0.$$

Of course, the polar representation is not unique. Furthermore, since r may be positive or negative, we must be careful to choose θ in the proper quadrant. If r is to be positive, then θ must be in the same quadrant as (x, y). If r is to be negative, then θ must be in the opposite quadrant.

EXAMPLE 2 (a) Convert $(1, \sqrt{3})$ to polar coordinates, exactly, with $r > 0$ and $0 \le \theta < 2\pi$.
(b) Convert $(-3, 5)$ to polar coordinates, with $r < 0$ and $0 \le \theta < 2\pi$.

SOLUTION (a) First we note that

$$r = \sqrt{1^2 + \sqrt{3}^2} = 2.$$

Next, we must find θ such that $0 \le \theta < 2\pi$ and

$$\tan\theta = \sqrt{3}.$$

Since $(1, \sqrt{3})$ is in the first quadrant, we choose the first-quadrant angle $\theta = \pi/3$. So the polar representation is $(2, \pi/3)$.

(b) We get

$$r = -\sqrt{(-3)^2 + 5^2} = -\sqrt{34}.$$

Since r is negative and $(-3, 5)$ is in the second quadrant, we must find θ in the opposite (fourth) quadrant such that

$$\tan\theta = \frac{-5}{3}.$$

Arctan $\frac{5}{3}$ is the reference angle, so $\theta = 2\pi - \text{Arctan}\,\frac{5}{3}$. Hence, the exact polar representation is $(-\sqrt{34}, 2\pi - \text{Arctan}\,\frac{5}{3})$, which is approximately $(-5.831, 5.253)$.

▲

▲ *Exercise*

(a) Convert (2, 2) to polar coordinates, exactly, with $r > 0$ and $0 \le \theta < 2\pi$.
(b) Convert $(4, -5)$ to polar coordinates, with $r < 0$ and $0 \le \theta < 2\pi$.

Answer

(a) $\left(2\sqrt{2}, \frac{\pi}{4}\right)$

(b) $\left(-\sqrt{41}, \pi - \text{Arctan}\,\frac{5}{4}\right)$, which is approximately $(-6.403, 2.246)$

Problem Set 8.3

1. Identify each point (r, θ) on the polar coordinate system shown in the following figure, subject to the conditions that $r > 0$ and $0 \le \theta < 2\pi$.

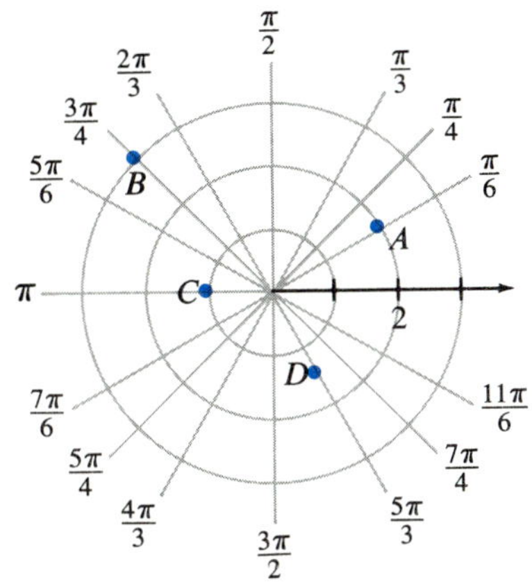

2. Using a polar coordinate system, plot the following points:

$$A\left(2, \frac{\pi}{6}\right), B\left(-3, \frac{2\pi}{3}\right), C\left(1, \frac{-3\pi}{4}\right), D\left(-2, \frac{-\pi}{4}\right)$$

In Problems 3–10, convert the polar representation to rectangular coordinates, as specified.

3. $\left(3, \frac{\pi}{3}\right)$, exactly

4. $\left(4, \frac{3\pi}{4}\right)$, exactly

5. $\left(-2, \frac{5\pi}{6}\right)$, exactly

6. $\left(-3, \frac{-2\pi}{3}\right)$, exactly

7. $(2.5, 1.2)$, approximately

8. $(4.8, 0.7)$, approximately

9. $(-5, -2.4)$, approximately

10. $(-6, 4.21)$, approximately

In Problems 11–18, convert the rectangular representation to polar coordinates, with $r > 0$ and $0 \le \theta < 2\pi$.

11. $(-2, 2\sqrt{3})$

12. $(-1, 1)$

13. $(-3, -6)$

14. $(-5, -4)$

15. $(6, -5)$

16. $(3, -2)$

17. $(2.5, 4)$

18. $(3, 5.2)$

In Problems 19–26, convert the rectangular representation to polar coordinates, with $r < 0$ and $0 \le \theta < 2\pi$.

19. $(-1, -1)$

20. $(-\sqrt{3}, -1)$

21. $(-3, 5)$

22. $(-5, 7)$

23. $(2, -3)$

24. $(8, -7)$

25. $(4, 5)$

26. $(5, 1)$

8.4 GRAPHS IN POLAR COORDINATES

An equation in the variables r and θ is called a **polar equation.** Its graph will be the set of all points in the plane whose polar representations (r, θ) are solutions of the equation. As a simple example, consider the polar equation

$$r = 1.$$

The solutions are all points with a first coordinate of 1. Since θ doesn't appear in the equation, the second coordinates take on all values. Hence the graph consists of all points located one unit from the pole. Of course, this is a circle of radius 1, centered at the pole. See Figure 8.11(i).

Figure 8.11

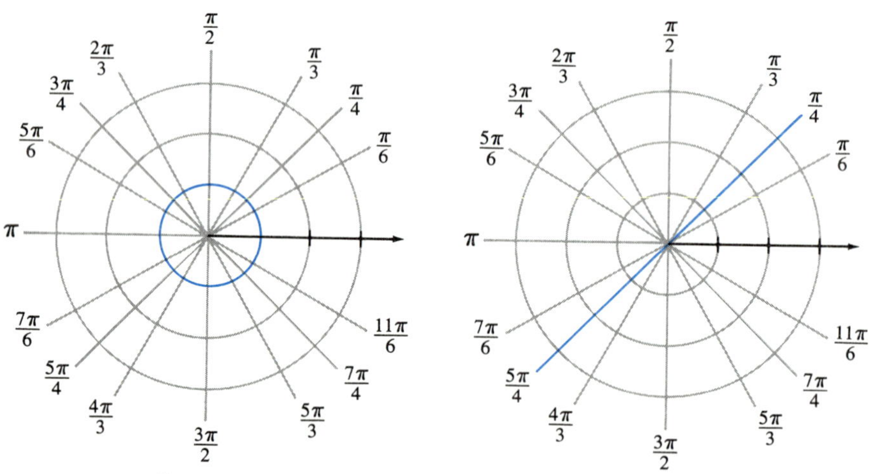

(i) $r = 1$ (ii) $\theta = \frac{\pi}{4}$

Figure 8.11(ii) shows the graph of

$$\theta = \frac{\pi}{4}.$$

Since r doesn't appear in the equation, the first coordinates take on all values. Hence the graph is a line that intersects the pole, making an angle of $\pi/4$ with the axis.

Most graphs won't be quite so easy. To sketch a graph, we will plot points and connect them with a smooth, unbroken curve.

EXAMPLE 1 Graph $r = 1 + \sin\theta$.

SOLUTION Let's choose values of θ from 0 to 2π in increments of $\pi/6$ and compute the corresponding value of r to the nearest tenth. We get the following table:

θ	0	$\frac{\pi}{6}$	$\frac{\pi}{3}$	$\frac{\pi}{2}$	$\frac{2\pi}{3}$	$\frac{5\pi}{6}$	π	$\frac{7\pi}{6}$	$\frac{4\pi}{3}$	$\frac{3\pi}{2}$	$\frac{5\pi}{3}$	$\frac{11\pi}{6}$	2π
r	1.0	1.5	1.9	2.0	1.9	1.5	1.0	0.5	0.1	0.0	0.1	0.5	1.0

These points are plotted in Figure 8.12.

Figure 8.12

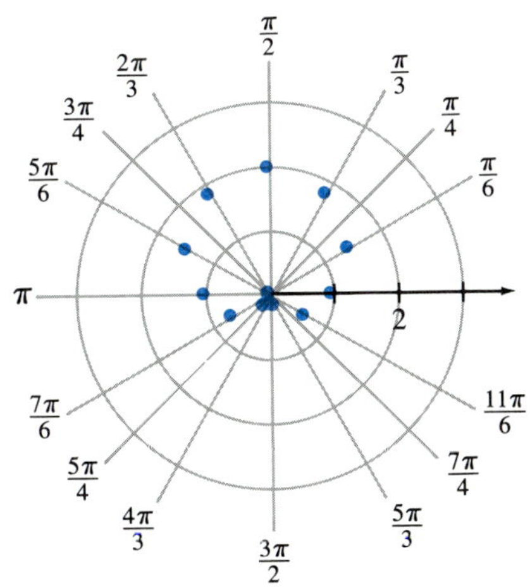

Connecting these points with a smooth curve gives the graph in Figure 8.13.

Figure 8.13
$r = 1 + \sin\theta$

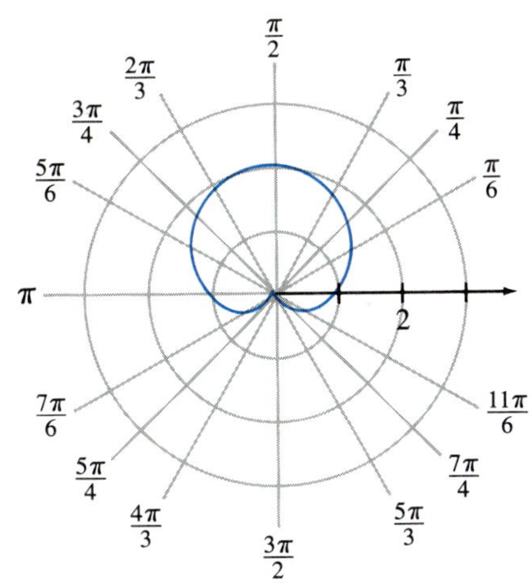

▲

Because of its heart-shaped appearance, the graph in Figure 8.13 is called a *cardioid*. Note that as θ assumes values in the interval $[2\pi, 4\pi]$ or in $[-2\pi, 0]$, the values of r will be the same as when θ assumes values in $[0, 2\pi]$. Consequently, the curve would be traced again.

▲ ***Exercise*** Graph $r = 1 + \cos\theta$.

Answer See Figure 8.14.

Figure 8.14

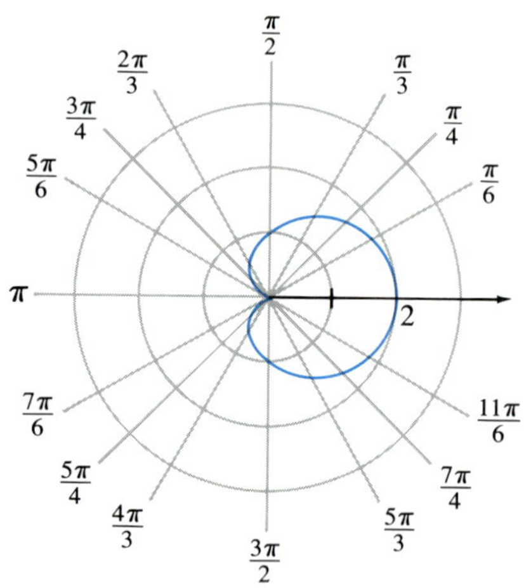

EXAMPLE 2 Graph $r = 1 - 2\cos\theta$.

SOLUTION Let's choose values of θ from 0 to 2π in increments of $\pi/6$ and compute the corresponding value of r. We get the following table, with r approximated to the nearest tenth.

θ	0	$\frac{\pi}{6}$	$\frac{\pi}{3}$	$\frac{\pi}{2}$	$\frac{2\pi}{3}$	$\frac{5\pi}{6}$	π	$\frac{7\pi}{6}$	$\frac{4\pi}{3}$	$\frac{3\pi}{2}$	$\frac{5\pi}{3}$	$\frac{11\pi}{6}$	2π
r	-1.0	-0.7	0.0	1.0	2.0	2.7	3.0	2.7	2.0	1.0	0.0	-0.7	-1.0

These points are plotted in Figure 8.15.

We want to connect the points in the order in which they appear in the table. This gives the graph of Figure 8.16.

▲

The graph in Figure 8.16 is called a *limaçon*. Note that the positive values of r trace the outer loop, whereas the inner loop is traced by negative values of r.

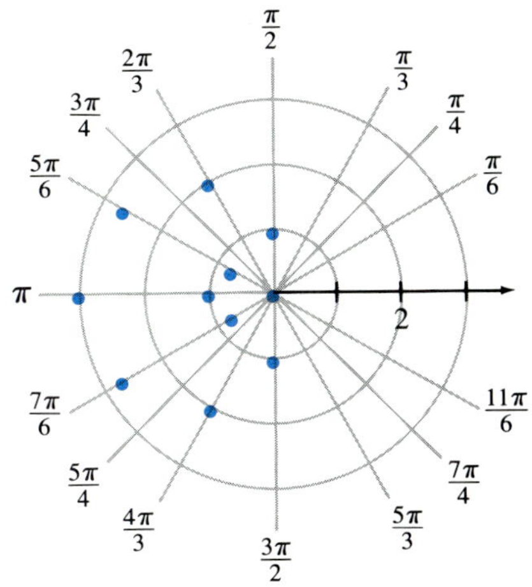

Figure 8.15

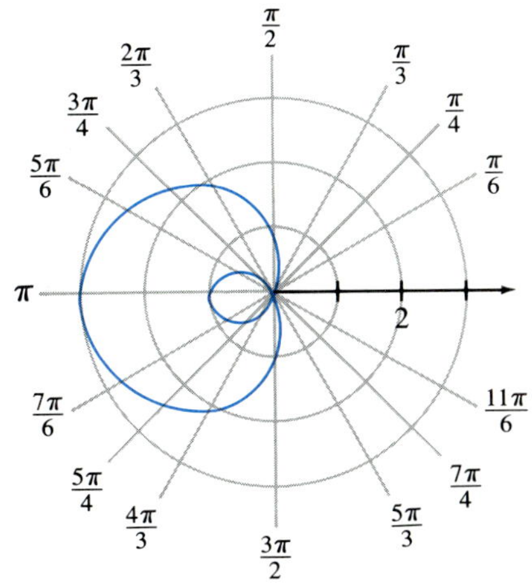

Figure 8.16
$r = 1 - 2\cos\theta$

▲ ***Exercise*** Graph $r = 1 - 2\sin\theta$.

Answer See Figure 8.17.

Figure 8.17

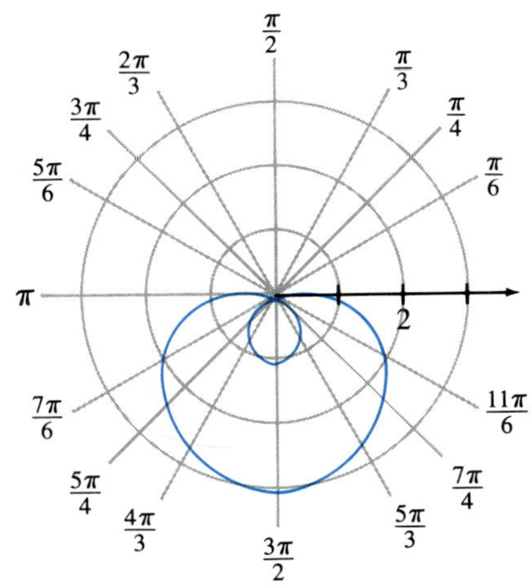

EXAMPLE 3 Graph $r = \cos\theta$.

SOLUTION Again we start with a table of values:

θ	0	$\frac{\pi}{6}$	$\frac{\pi}{3}$	$\frac{\pi}{2}$	$\frac{2\pi}{3}$	$\frac{5\pi}{6}$	π	$\frac{7\pi}{6}$	$\frac{4\pi}{3}$	$\frac{3\pi}{2}$	$\frac{5\pi}{3}$	$\frac{11\pi}{6}$	2π
$\approx r$	1.0	0.9	0.5	0.0	−0.5	−0.9	−1	−0.9	−0.5	0.0	0.5	0.9	1.0

These points are plotted in Figure 8.18 (page 302).

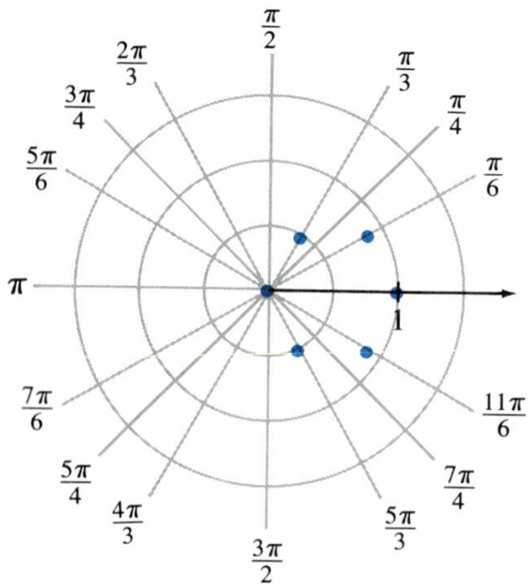

Figure 8.18

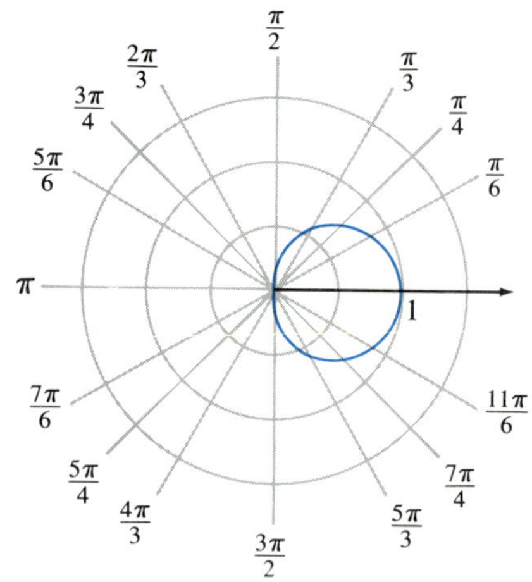

Figure 8.19
$r = \cos\theta$

Note that although our table has 13 solutions, our graph shows but 6. This is because each of the last 7 solutions in the table is just a different representation of an earlier solution. Connecting these points with a smooth curve gives the circle of Figure 8.19.

▲

▲ *Exercise* Graph $r = \sin\theta$.

Answer See Figure 8.20.

Figure 8.20

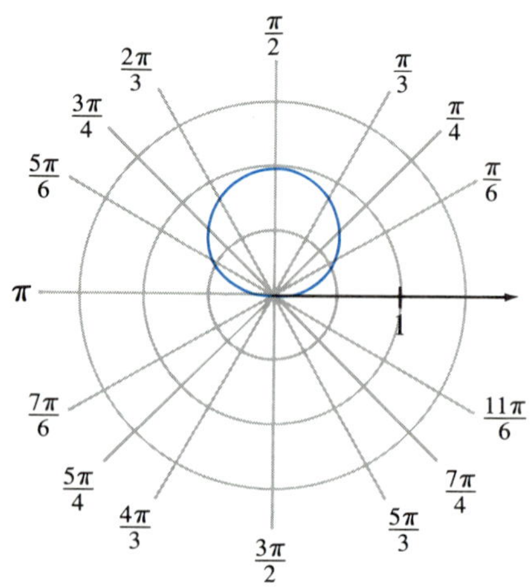

EXAMPLE 4 Graph $r = \sin 2\theta$.

SOLUTION We won't make a complete table of values. We'll try to examine several intervals to see what's happening, but let's begin with some first-quadrant angles. Since we are dealing with 2θ instead of θ, we'll choose increments of $\pi/12$. We get the following table:

θ	0	$\frac{\pi}{12}$	$\frac{\pi}{6}$	$\frac{\pi}{4}$	$\frac{\pi}{3}$	$\frac{5\pi}{12}$	$\frac{\pi}{2}$
$\approx r$	0.0	0.5	0.9	1.0	0.9	0.5	0.0

Plotting these points and connecting them with a smooth curve gives us the graph of Figure 8.21.

Figure 8.21

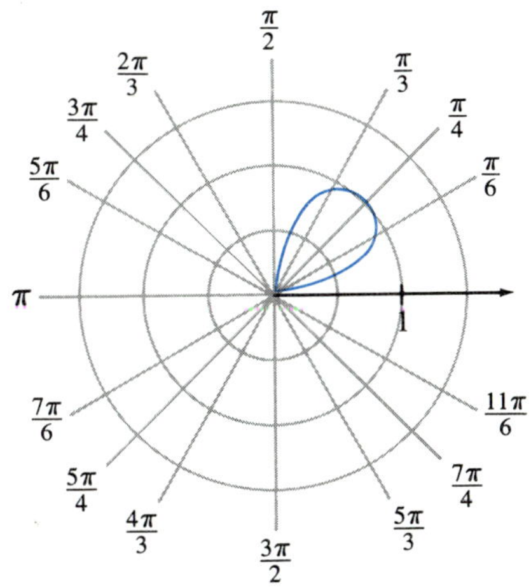

As θ assumes values in the interval $[\pi/2, \pi]$, $\sin 2\theta$, and hence r, will be negative. So we'll trace a curve in the fourth quadrant that has the same shape as that of the graph of Figure 8.21. When θ is in $[\pi, 3\pi/2]$, r is positive, and we trace this curve in the third quadrant. Finally, for θ in $[3\pi/2, 2\pi]$, r is negative, and we get this curve in the second quadrant. The complete graph is shown in Figure 8.22.

Figure 8.22
$r = \sin 2\theta$

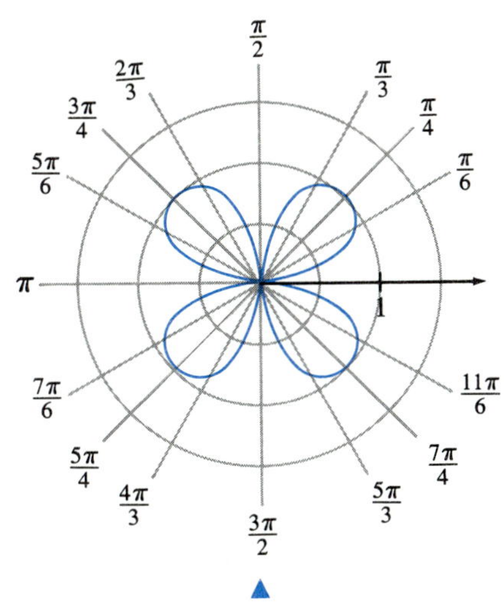

▲

The graph in Figure 8.22 is called a *four leaved rose.*

▲ ***Exercise*** Graph $r = \cos 2\theta$.

Answer See Figure 8.23.

Figure 8.23

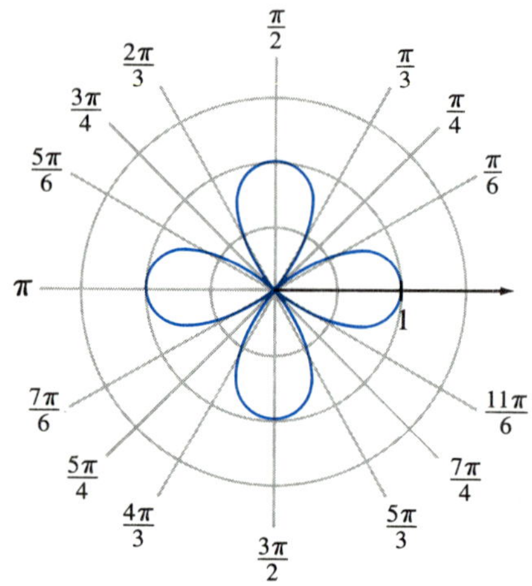

Problem Set 8.4

1. Describe the graph of $r = 4$.

2. Describe the graph of $\theta = \frac{-\pi}{4}$.

3. Find the polar equation of the circle of radius 2 with center at the origin.

4. Find the polar equation of the line passing through the rectangular point $(-1, \sqrt{3})$ and the origin.

In Problems 5–22, graph the equation.

5. $r = 2 + 2 \sin \theta$

6. $r = 2 + 2 \cos \theta$

7. $r = 1 + 2 \cos \theta$

8. $r = 1 + 2 \sin \theta$

9. $r = 2 + \sin \theta$

10. $r = 2 + \cos \theta$

11. $r = 1 - \cos \theta$

12. $r = 1 - \sin \theta$

13. $r = 2 \sin \theta$

14. $r = 2 \cos \theta$

15. $r = 2 \cos 2\theta$

16. $r = 2 \sin 2\theta$

17. $r = \sin 3\theta$

18. $r = \cos 3\theta$

19. $r = 2 \sec \theta$

20. $r = 2 \csc \theta$

21. $r = \theta$

22. $r = 2\theta$

23. Show that the graph of $r \sin \theta = k, k > 0$, is a straight line parallel to the polar axis.

24. Show that the graph of $r = k \sin \theta, k > 0$, is a circle centered at $\left(\frac{k}{2}, \frac{\pi}{2}\right)$.

CHAPTER 8 REVIEW

Key Terms and Concepts

absolute value, page 282

addition and multiplication of complex numbers, page 276

complex number, page 276

complex plane, page 281

conjugate, page 278

conversion, polar to rectangular, page 295

$x = r \cos \theta$ **and** $y = r \sin \theta$

conversion, rectangular to polar, page 295

$r = \pm\sqrt{x^2 + y^2}$ **and** $\tan \theta = \frac{y}{x}$

De Moivre's Theorem, page 285

polar coordinate system, page 291

polar equation, page 298

subtraction and division of complex numbers, page 278

trigonometric form, page 283

$a + bi = r(\cos \theta + i \sin \theta)$

Problem Set

1. Specify "true" or "false."

(a) The conjugate of $2 + 3i$ is $-2 - 3i$.

(b) Every quadratic equation has at least one complex solution.

(c) If $2 + 3i$ is a complex root of a quadratic equation with real coefficients, then $2 - 3i$ is also a complex root.

(d) The polar representation of a point is unique.

(e) The graph of $r = 5$ is a circle.

(f) The graph of $|\theta| = \frac{\pi}{3}$ is a line.

(g) For any complex numbers z and w, $\overline{z + w} = \bar{z} + \bar{w}$.

(h) If $\bar{z} = z$, then z is a real number.

(i) The absolute value of every complex number is a real number.

2. Let f be a function with domain C (the set of complex numbers) defined by $f(x) = x^2 - 2x - 4$. Determine each of the following:

(a) $f(i)$ (b) $f(i + 1)$ (c) $f(3 - i)$

In Problems 3–10, perform the indicated operation, expressing the result in the form $a + bi$.

3. $(5 + 2i) + (-3 + 5i)$
4. $(7 - i) - (2 + 6i)$
5. $(5 + i)(3 - 2i)$
6. $(-3 - 4i)(1 + i)$
7. $(4 + 2i) \div (3 - 2i)$
8. $\dfrac{6 + 5i}{5 + 2i}$
9. $(2 - 3i)^2$
10. i^{51}

In Problems 11–14, find the complex solutions.

11. $x^2 = -49$
12. $(2x - 1)^2 = -1$
13. $x^2 - 3x + 3 = 0$
14. $2x^2 - 6x + 7 = 0$

In Problems 15–18, write the complex number in trigonometric form, exactly.

15. $-4 + 4i$
16. $-2\sqrt{3} - 2i$
17. $3i$
18. $\sqrt{5} - \sqrt{15}i$

In Problems 19–22, write the complex number in trigonometric form, approximately.

19. $2 + 5i$
20. $6 - i$
21. $-7 - 6i$
22. $-3 + 5i$

23. Write the complex number $6\left(\cos\dfrac{5\pi}{6} + i\sin\dfrac{5\pi}{6}\right)$ in the form $a + bi$, exactly.

24. Write the complex number $4.5(\cos 2.75 + i \sin 2.75)$ in the form $a + bi$, approximately.

In Problems 25 and 26, find z_1z_2 and z_1/z_2, expressing the result in the form $a + bi$.

25. $z_1 = 4\left(\cos\dfrac{4\pi}{3} + i\sin\dfrac{4\pi}{3}\right)$ and $z_2 = 2\left(\cos\dfrac{\pi}{6} + i\sin\dfrac{\pi}{6}\right)$

26. $z_1 = 2.5(\cos 0.9 + i \sin 0.9)$ and $z_2 = 1.6(\cos 2.1 + i \sin 2.1)$

In Problems 27 and 28, evaluate the expression by using De Moivre's Theorem, as specified.

27. $(2 - 2\sqrt{3}\,i)^5$, exactly
28. $(1 + 5i)^6$, approximately

In Problems 29 and 30, find the roots indicated, as specified.

29. cube roots of i, exactly

30. cube roots of $2 - 3i$, approximately

In Problems 31–34, convert the polar representation to rectangular coordinates.

31. $\left(4, \frac{-\pi}{4}\right)$

32. $\left(-2, \frac{2\pi}{3}\right)$

33. $(4.1, 0.85)$

34. $(-3.5, -1.2)$

In Problems 35 and 36, convert the rectangular representation to polar coordinates, with $r > 0$ and $0 \leq \theta < 2\pi$.

35. $(-1, -1)$

36. $(6.5, 2)$

In Problems 37 and 38, convert the rectangular representation to polar coordinates, with $r < 0$ and $0 \leq \theta < 2\pi$.

37. $(-3, 3\sqrt{3})$

38. $(5, -3)$

In Problems 39–42, graph the equation.

39. $r = 2 + 3 \sin \theta$

40. $r = 2 + 3 \cos \theta$

41. $r = 2 - \cos \theta$

42. $r = 2 - \sin \theta$

APPENDIX

A.1 APPROXIMATIONS

When we use calculators, all irrational numbers and some rational numbers are displayed as approximations. For example, $\sqrt{2}$ might appear as 1.4142136. This has been rounded up to the nearest 10^{-7} (ten millionth) because the digit in the 10^{-8} position is greater than or equal to 5. (It's a 6; $\sqrt{2} \approx 1.4142135624$). And the number $\frac{1}{3}$ might appear as 0.3333333. This has been rounded down to the nearest 10^{-7} because the digit in the 10^{-8} position is less than 5. (It's a 3; $\frac{1}{3} \approx 0.3333333333$.)

Approximations also occur as a result of measurement. For example, we might say that the diameter of the top of a can is 4.75 centimeters. But because measuring devices and their use are somewhat inaccurate, this is only an approximation of the exact value. The diameter might really be 4.752 centimeters or 4.7491 centimeters, etc. To say that the diameter is 4.75 implies that it is correct to the nearest hundredth, lying between 4.745 and 4.755.

In discussing the accuracy of an approximation, we need to know about its **significant digits**—that is, those digits known to be correct. For example, if the *exact* diameter of our tabletop, whatever it is, when rounded to the nearest hundredth is 4.75, then the approximation 4.75 has three significant digits. All three digits are known to be correct (given rounding).

When we write an approximation, there will be several assumptions about the significant digits. First, all nonzero digits are assumed to be significant. Second, any zero between nonzero digits is also significant. Third, leading zeros are *not* significant. These concepts are illustrated as follows:

Approximation	*Number of significant digits*
53.251	5
200.5	4
0.0056	2

The only problems occur with *trailing* zeros.

Suppose that a distance is given as 5000 feet. Are the trailing zeros significant? If the true distance is between 4999.5 and 5000.5, then they are, but this may not be the case. This may be correct to the nearest thousands only. We may clarify the situation as follows. If all four digits are significant, then the number is correct to the nearest one, and we can write

$$5000 \text{ (nearest one)}.$$

On the other hand, if the zeros are not significant, and the only significant digit is the five, then the number is correct to the nearest thousand, and we may